Excursions in Classical Analysis

Pathways to Advanced Problem Solving and Undergraduate Research

Chapter 11 (Generating Functions for Powers of Fibonacci Numbers) is a reworked version of material from my same name paper in the *International Journal of Mathematical Education in Science and Technology*, Vol. 38:4 (2007) pp. 531–537.

Chapter 12 (Identities for the Fibonacci Powers) is a reworked version of material from my same name paper in the *International Journal of Mathematical Education in Science and Technology*, Vol. 39:4 (2008) pp. 534–541.

Chapter 13 (Bernoulli Numbers via Determinants)is a reworked version of material from my same name paper in the *International Journal of Mathematical Education in Science and Technology*, Vol. 34:2 (2003) pp. 291–297.

Chapter 19 (Parametric Differentiation and Integration) is a reworked version of material from my same name paper in the *International Journal of Mathematical Education in Science and Technology*, Vol. 40:4 (2009) pp. 559–570.

© *2010 by the Mathematical Association of America, Inc.*

Library of Congress Catalog Card Number 2010924991

ISBN 978-0-88385-768-7
e-ISBN 978-0-88385-935-3

Printed in the United States of America

Current Printing (last digit):
10 9 8 7 6 5 4 3 2 1

Excursions in Classical Analysis
Pathways to Advanced Problem Solving and Undergraduate Research

Hongwei Chen
Christopher Newport University

Published and Distributed by
The Mathematical Association of America

Committee on Books
Frank Farris, *Chair*

Classroom Resource Materials Editorial Board
Gerald M. Bryce, *Editor*

Michael Bardzell
William C. Bauldry
Diane L. Herrmann
Wayne Roberts
Susan G. Staples
Philip D. Straffin
Holly S. Zullo

CLASSROOM RESOURCE MATERIALS

Classroom Resource Materials is intended to provide supplementary classroom material for students—laboratory exercises, projects, historical information, textbooks with unusual approaches for presenting mathematical ideas, career information, etc.

101 Careers in Mathematics, 2nd edition edited by Andrew Sterrett

Archimedes: What Did He Do Besides Cry Eureka?, Sherman Stein

The Calculus Collection: A Resource for AP and Beyond, edited by Caren L. Diefenderfer and Roger B. Nelsen

Calculus Mysteries and Thrillers, R. Grant Woods

Conjecture and Proof, Miklós Laczkovich

Counterexamples in Calculus, Sergiy Klymchuk

Creative Mathematics, H. S. Wall

Environmental Mathematics in the Classroom, edited by B. A. Fusaro and P. C. Kenschaft

Excursions in Classical Analysis: Pathways to Advanced Problem Solving and Undergraduate Research, by Hongwei Chen

Exploratory Examples for Real Analysis, Joanne E. Snow and Kirk E. Weller

Geometry From Africa: Mathematical and Educational Explorations, Paulus Gerdes

Historical Modules for the Teaching and Learning of Mathematics (CD), edited by Victor Katz and Karen Dee Michalowicz

Identification Numbers and Check Digit Schemes, Joseph Kirtland

Interdisciplinary Lively Application Projects, edited by Chris Arney

Inverse Problems: Activities for Undergraduates, Charles W. Groetsch

Laboratory Experiences in Group Theory, Ellen Maycock Parker

Learn from the Masters, Frank Swetz, John Fauvel, Otto Bekken, Bengt Johansson, and Victor Katz

Math Made Visual: Creating Images for Understanding Mathematics, Claudi Alsina and Roger B. Nelsen

Ordinary Differential Equations: A Brief Eclectic Tour, David A. Sánchez

Oval Track and Other Permutation Puzzles, John O. Kiltinen

A Primer of Abstract Mathematics, Robert B. Ash

Proofs Without Words, Roger B. Nelsen

Proofs Without Words II, Roger B. Nelsen

She Does Math!, edited by Marla Parker

Solve This: Math Activities for Students and Clubs, James S. Tanton

Student Manual for Mathematics for Business Decisions Part 1: Probability and Simulation, David Williamson, Marilou Mendel, Julie Tarr, and Deborah Yoklic

Student Manual for Mathematics for Business Decisions Part 2: Calculus and Optimization, David Williamson, Marilou Mendel, Julie Tarr, and Deborah Yoklic

Teaching Statistics Using Baseball, Jim Albert

Visual Group Theory, Nathan C. Carter

Writing Projects for Mathematics Courses: Crushed Clowns, Cars, and Coffee to Go, Annalisa Crannell, Gavin LaRose, Thomas Ratliff, Elyn Rykken

MAA Service Center
P.O. Box 91112
Washington, DC 20090-1112
1-800-331-1MAA   FAX: 1-301-206-9789

Contents

Preface		**xi**
1	**Two Classical Inequalities**	**1**
	1.1 AM-GM Inequality	1
	1.2 Cauchy-Schwarz Inequality	8
	Exercises	12
	References	15
2	**A New Approach for Proving Inequalities**	**17**
	Exercises	22
	References	23
3	**Means Generated by an Integral**	**25**
	Exercises	30
	References	32
4	**The L'Hôpital Monotone Rule**	**33**
	Exercises	37
	References	38
5	**Trigonometric Identities via Complex Numbers**	**39**
	5.1 A Primer of complex numbers	39
	5.2 Finite Product Identities	41
	5.3 Finite Summation Identities	43
	5.4 Euler's Infinite Product	45
	5.5 Sums of inverse tangents	48
	5.6 Two Applications	49
	Exercises	51
	References	53
6	**Special Numbers**	**55**
	6.1 Generating Functions	55
	6.2 Fibonacci Numbers	56
	6.3 Harmonic numbers	58
	6.4 Bernoulli Numbers	61

	Exercises	69
	References	72
7	**On a Sum of Cosecants**	**73**
	7.1 A well-known sum and its generalization	73
	7.2 Rough estimates	74
	7.3 Tying up the loose bounds	76
	7.4 Final Remarks	79
	Exercises	79
	References	81
8	**The Gamma Products in Simple Closed Forms**	**83**
	Exercises	89
	References	91
9	**On the Telescoping Sums**	**93**
	9.1 The sum of products of arithmetic sequences	94
	9.2 The sum of products of reciprocals of arithmetic sequences	95
	9.3 Trigonometric sums	97
	9.4 Some more telescoping sums	101
	Exercises	106
	References	108
10	**Summation of Subseries in Closed Form**	**109**
	Exercises	117
	References	119
11	**Generating Functions for Powers of Fibonacci Numbers**	**121**
	Exercises	128
	References	130
12	**Identities for the Fibonacci Powers**	**131**
	Exercises	140
	References	142
13	**Bernoulli Numbers via Determinants**	**143**
	Exercises	149
	References	152
14	**On Some Finite Trigonometric Power Sums**	**153**
	14.1 Sums involving $\sec^{2p}(k\pi/n)$	154
	14.2 Sums involving $\csc^{2p}(k\pi/n)$	156
	14.3 Sums involving $\tan^{2p}(k\pi/n)$	157
	14.4 Sums involving $\cot^{2p}(k\pi/n)$	159
	Exercises	162
	References	164
15	**Power Series of $(\arcsin x)^2$**	**165**
	15.1 First Proof of the Series (15.1)	165
	15.2 Second Proof of the Series (15.1)	167
	Exercises	171

	References	173
16	**Six Ways to Sum $\zeta(2)$**	**175**
	16.1 Euler's Proof	176
	16.2 Proof by Double Integrals	177
	16.3 Proof by Trigonometric Identities	181
	16.4 Proof by Power Series	182
	16.5 Proof by Fourier Series	183
	16.6 Proof by Complex Variables	183
	Exercises	184
	References	187
17	**Evaluations of Some Variant Euler Sums**	**189**
	Exercises	197
	References	199
18	**Interesting Series Involving Binomial Coefficients**	**201**
	18.1 An integral representation and its applications	202
	18.2 Some Extensions	208
	18.3 Searching for new formulas for π	210
	Exercises	212
	References	215
19	**Parametric Differentiation and Integration**	**217**
	Example 1	217
	Example 2	218
	Example 3	219
	Example 4	220
	Example 5	220
	Example 6	222
	Example 7	223
	Example 8	224
	Example 9	225
	Example 10	226
	Exercises	228
	References	230
20	**Four Ways to Evaluate the Poisson Integral**	**231**
	20.1 Using Riemann Sums	232
	20.2 Using A Functional Equation	233
	20.3 Using Parametric Differentiation	234
	20.4 Using Infinite Series	235
	Exercises	235
	References	239
21	**Some Irresistible Integrals**	**241**
	21.1 Monthly Problem 10611	241
	21.2 Monthly Problem 11206	243
	21.3 Monthly Problem 11275	244

21.4 Monthly Problem 11277 245
21.5 Monthly Problem 11322 246
21.6 Monthly Problem 11329 247
21.7 Monthly Problem 11331 249
21.8 Monthly Problem 11418 250
    Exercises 252
    References 254

Solutions to Selected Problems   **255**

Index   **297**

About the Author   **301**

Preface

This book grew out of my work in the last decade teaching, researching, solving problems, and guiding undergraduate research. I hope it will benefit readers who are interested in advanced problem solving and undergraduate research. Math students often have trouble the first time they confront advanced problems like those that appear in Putnam competitions and in the journals of the Mathematical Association of America. And while many math students would like to do research in mathematics, they don't know how to get started. They aren't aware of what problems are open, and when presented with a problem outside of the realm of their courses, they find it difficult to apply what they know to this new situation. Historically, deep and beautiful mathematical ideas have often emerged from simple cases of challenging problems. Examining simple cases allows students to experience working with abstract ideas at a nontrivial level — something they do little of in standard mathematics courses. Keeping this in mind, the book illustrates creative problem solving techniques via case studies. Each case in the book has a central theme and contains kernels of sophisticated ideas connected to important current research. The book seeks to spell out the principles underlying these ideas and to immerse readers in the processes of problem solving and research.

The book aims to introduce students to advanced problem solving and undergraduate research in two ways. The first is to provide a colorful tour of classical analysis, showcasing a wide variety of problems and placing them in historical contexts. The second is to help students gain mastery in mathematical discovery and proof. Although one proof is enough to establish a proposition, students should be aware that there are (possibly widely) various ways to approach a problem. Accordingly, this book often presents a variety of solutions for a particular problem. Some reach back to the work of outstanding mathematicians like Euler; some connect to other beautiful parts of mathematics. Readers will frequently see problems solved by using an idea that might at first have seemed inapplicable, or by employing a specific technique that is used to solve many different kinds of problems. The book emphasizes the rich and elegant interplay between continuous and discrete mathematics by applying induction, recursion, and combinatorics to traditional problems in classical analysis.

Advanced problem solving and research involve not only deduction but also experimentation, guessing, and arguments from analogy. In the last two decades, computer algebra systems have become a useful tool in mathematical research. They are often used to experiment with various examples, to decide if a potential result leans in the desired direction, or to formulate credible conjectures. With the continuing increases of computing power and accessibility, experimental mathematics has not only come of age but is quickly maturing. Appealing to this trend, I try to provide in the book a variety of accessible problems in which computing plays a significant role. These experimentally discovered results are indeed based on rigorous mathematics. Two interesting examples are the WZ-method for telescoping in Chapter 9 and the Hermite-Ostrogradski Formula for searching for a new formula for π in Chapter 18.

In his classic "How to Solve It", George Pólya provides a list of guidelines for solving mathematical problems: learn to understand the problem; devise a plan to solve the problem; carry out that plan; and look back and check what the results indicate. This list has served as a standard rubric for several generations of mathematicians, and has guided this book as well. Accordingly, I have begun each topic by categorizing and identifying the problem at hand, then indicated which technique I will use and why, and ended by making a worthwhile discovery or proving a memorable result. I often take the reader through a method which presents rough estimates before I derive finer ones, and I demonstrate how the more easily solved special cases often lead to insights that drive improvements of existing results. Readers will clearly see how mathematical proofs evolve — from the specific to the general and from the simplified scenario to the theoretical framework.

A carefully selected assortment of problems presented at the end of the chapters includes 22 Putnam problems, 50 MAA Monthly problems, and 14 open problems. These problems are related not solely to the chapter topics, but they also connect naturally to other problems and even serve as introductions to other areas of mathematics. At the end of the book appear approximately 80 selected problem solutions. To help readers assimilate the results, most of the problem solutions contain directions to additional problems solvable by similar methods, references for further reading, and to alternate solutions that possibly involve more advanced concepts. Readers are invited to consider open problems and to consult publications in their quest to prove their own beautiful results.

This book serves as a rich resource for advanced problem solving and undergraduate mathematics research. I hope it will prove useful in students' preparations for mathematics competitions, in undergraduate reading courses and seminars, and as a supplement in analysis courses. Mathematicians and students interested in problem solving will find the collection of topics appealing. This book is also ideal for self study. Since the chapters are independent of one another, they may be read in any order. It is my earnest hope that some readers, including working mathematicians, will come under the spell of these interesting topics.

This book is accessible to anyone who knows calculus well and who cares about problem solving. However, it is not expected that the book will be easy reading for many math students straight out of first-year calculus. In order to proceed comfortably, readers will need to have some results of classical analysis at their fingertips, and to have had exposure to special functions and the rudiments of complex analysis. Some degree of mathematical maturity is presumed, and upon occasion one is required to do some careful thinking.

Acknowledgments

I wish to thank Gerald Bryce and Christopher Kennedy. Both read through the entire manuscript with meticulous and skeptical eyes, caught many of my errors, and offered much valuable advice for improvement. This book is much improved because of their dedication and I am deeply grateful to them. Naturally all possible errors are my own responsibility. I also wish to thank Don Albers at the Mathematical Association of America and the members of the Classroom Resource Materials Editorial Board for their kindness, thoughtful and constructive comments and suggestions. I especially wish to thank Susan Staples for carefully reviewing the book and suggesting its current extended title. Elaine Pedreira and Beverly Ruedi have helped me through many aspects of production and have shepherded this book toward a speedy publication. It is an honor and a privilege for me to learn from all of them. My colleagues Martin Bartelt, Brian Bradie and Ron Persky have read the first draft of several chapters and lent further moral support. Much of the work was accomplished during a sabbatical leave in the fall of 2007. My thanks go to Christopher Newport University for providing this sabbatical opportunity. Finally, I wish to thank my wife Ying and children Alex and Abigail for their patience, support and encouragement. I especially wish to thank Alex, who drew all of the figures in the book and commented on most of chapters from the perspective of an undergraduate.

Comments, corrections, and suggestions from readers are always welcome. I would be glad to receive these at hchen@cnu.edu. Thank you in advance.

1
Two Classical Inequalities

> *The heart of mathematics is its problems.* — P. R. Halmos
>
> *It appears to me that if one wants to make progress in mathematics, one should study the masters and not the pupils.* — N. H. Abel

Mathematical progress depends on a stream of new problems, but most new problems emerge simply from well-established principles. For students who want to master existing principles and build their own knowledge of mathematics, the moral is simple: "Learn from the masters!" Often, historical developments of a particular topic provide the best way to learn that topic. In this chapter, by examining various proofs of the arithmetic mean and geometric mean (AM-GM) inequality and the Cauchy-Schwarz inequality, we see the motivation necessary to work through all the steps of a rigorous proof. Some of these proofs contain brilliant ideas and clever insights from famous mathematicians. Personally, I still remember how exciting and stirring these proofs were when I first encountered them.

We begin with the AM-GM inequality.

1.1 AM-GM Inequality

Let $a_1, a_2, \ldots, a_n$ be positive real numbers. Then

$$\sqrt[n]{a_1 a_2 \cdots a_n} \leq \frac{a_1 + a_2 + \cdots + a_n}{n}, \tag{1.1}$$

with equality if and only if all a_k ($1 \leq k \leq n$) are equal.

First Proof. The following proof is based on induction. Let $H(n)$ stand for the hypothesis that inequality (1.1) is valid for n. For $n = 2$, we have

$$\sqrt{a_1 a_2} \leq \frac{a_1 + a_2}{2}. \tag{1.2}$$

This is equivalent to $a_1 + a_2 - 2\sqrt{a_1 a_2} \geq 0$ or $(\sqrt{a_1} - \sqrt{a_2})^2 \geq 0$, which is always true. Moreover, equality in (1.2) holds if and only if $a_1 = a_2$. Next, we show that $H(2)$ and $H(n)$ together imply $H(n+1)$. Let

$$A_{n+1} = \frac{a_1 + a_2 + \cdots + a_n + a_{n+1}}{n+1}.$$

Then
$$A_{n+1} = \frac{1}{2n}[(n+1)A_{n+1} + (n-1)A_{n+1}]$$
$$= \frac{1}{2n}[(a_1 + \cdots + a_n) + (a_{n+1} + \underbrace{A_{n+1} + \cdots + A_{n+1}}_{n-1 \text{ terms}})]$$
$$\geq \frac{1}{2n}\left(n\sqrt[n]{a_1 a_2 \cdots a_n} + n\sqrt[n]{a_{n+1} A_{n+1}^{n-1}}\right)$$
$$\geq \sqrt[2n]{a_1 a_2 \cdots a_n a_{n+1} A_{n+1}^{n-1}},$$

where the first inequality follows from $H(n)$, while the second inequality follows from $H(2)$ in the form
$$\sqrt{ab} \leq \frac{a+b}{2}$$
with
$$a = \sqrt[n]{a_1 a_2 \cdots a_n}, \qquad b = \sqrt[n]{a_{n+1} A_{n+1}^{n-1}}.$$

Now, $H(n+1)$ follows from
$$A_{n+1}^{2n} \geq a_1 a_2 \cdots a_{n+1} A_{n+1}^{n-1}.$$

As before, equality holds if and only if $a_1 = a_2 = \cdots = a_n$ and $a_{n+1} = A_{n+1}$, or equivalently, if and only if $a_1 = a_2 = \cdots = a_{n+1}$.

Following the above induction proof is not difficult. The challenges lie in how to regroup the terms appropriately and then where to apply the induction hypothesis. It is natural to ask whether we can find methods that will help us meet these challenges and help us in investigating more complicated questions later in the book and in the future. To this end, we present five more proofs of the AM-GM inequality, chosen specifically to emphasize different and creative proof techniques. By relating the proofs to simple mathematical facts, we will see that the basic ingredients of each proof are few and simple.

We start with a non-standard induction proof that is attributed to Cauchy. The proof consists of two applications of induction; one, a "leap-forward" induction, leads to the desired result for all powers of 2 while the other, a "fall-back" induction, from a positive integer to the preceding one, together with the leap-forward induction enables us to establish the result for all positive integers.

Second Proof. Again let $H(n)$ stand for the hypothesis that inequality (1.1) is valid for n. $H(2)$ follows directly from $(\sqrt{a_1} - \sqrt{a_2})^2 \geq 0$ as noted in the first proof. Moreover, $H(2)$ is *self-generalizing*: for example, $H(2)$ can be applied twice to get
$$(a_1 a_2 a_3 a_4)^{1/4} \leq \frac{(a_1 a_2)^{1/2} + (a_3 a_4)^{1/2}}{2} \leq \frac{a_1 + a_2 + a_3 + a_4}{4}.$$

This confirms $H(4)$, and then $H(4)$ can be used again with $H(2)$ to establish
$$(a_1 a_2 \cdots a_8)^{1/8} \leq \frac{(a_1 \cdots a_4)^{1/4} + (a_5 \cdots a_8)^{1/4}}{2} \leq \frac{a_1 + a_2 + \cdots + a_8}{8}.$$

1. Two Classical Inequalities

This is precisely $H(8)$. Continuing in this way k times yields $H(2^k)$ for all $k \geq 1$. Now, to prove $H(n)$ for all n between the powers of 2, let $n < 2^k$. The plan is to extend $(a_1, a_2, \ldots, a_n)$ into a longer sequence $(b_1, b_2, \ldots, b_{2^k})$ to which we can apply $H(2^k)$. To see this, take the case $n = 3$ first. Set $b_i = a_i$ $(i = 1, 2, 3)$ and let b_4 be the value that yields the equality

$$\frac{b_1 + b_2 + b_3 + b_4}{4} = \frac{a_1 + a_2 + a_3}{3}.$$

This leads to
$$b_4 = \frac{a_1 + a_2 + a_3}{3}.$$

Applying $H(4)$ gives

$$\sqrt[4]{a_1 a_2 a_3 \left(\frac{a_1 + a_2 + a_3}{3}\right)} \leq \frac{a_1 + a_2 + a_3 + (a_1 + a_2 + a_3)/3}{4} = \frac{a_1 + a_2 + a_3}{3}$$

which then simplifies to the desired result $H(3)$:

$$\sqrt[3]{a_1 a_2 a_3} \leq \frac{a_1 + a_2 + a_3}{3}.$$

In general, for $n < 2^k$, similar to the case $n = 3$, set

$$b_i = \begin{cases} a_i, & i = 1, 2, \ldots, n, \\ \frac{a_1 + \cdots + a_n}{n} = A, & i = n+1, \ldots, 2^k. \end{cases}$$

Applying $H(2^k)$ to $(b_1, b_2, \ldots, b_{2^k})$, we obtain

$$\left(a_1 \cdots a_n A^{2^k - n}\right)^{1/2^k} \leq \frac{a_1 + \cdots + a_n + (2^k - n)A}{2^k} = \frac{2^k A}{2^k} = A,$$

which simplifies to, after eliminating A on the left-hand side and then raising both sides to the power of $2^k/n$,

$$(a_1 a_2 \cdots a_n)^{1/n} \leq \frac{a_1 + a_2 + \cdots + a_n}{n}.$$

This proves $H(n)$ as desired. The condition for equality is derived just as easily.

Cauchy's proof is longer than the first proof, but because of the self-generalizing nature of $H(2)$, almost without help, we see how to regroup the terms and where to apply the induction hypothesis. Moreover, we find that the above leap-forward fall-back induction approach is actually equivalent to verifying the following two steps:

(I) $H(n) \Longrightarrow H(n-1)$,

(II) $H(n)$ and $H(2) \Longrightarrow H(2n)$.

This leads to a short instructive proof of (1.1) as follows:
To prove (I), set
$$A = \frac{a_1 + a_2 + \cdots + a_{n-1}}{n-1}.$$

Then $H(n)$ yields

$$\left(\prod_{k=1}^{n-1} a_k\right) A \leq \left(\frac{\sum_{k=1}^{n-1} a_k + A}{n}\right)^n = \left(\frac{(n-1)A + A}{n}\right)^n = A^n$$

and hence

$$\prod_{k=1}^{n-1} a_k \leq A^{n-1} = \left(\frac{a_1 + a_2 + \cdots + a_{n-1}}{n-1}\right)^{n-1}.$$

For (II), we have

$$\prod_{k=1}^{2n} a_k = \left(\prod_{k=1}^{n} a_k\right)\left(\prod_{k=n+1}^{2n} a_k\right)$$

$$\leq \left(\sum_{k=1}^{n} \frac{a_k}{n}\right)^n \left(\sum_{k=n+1}^{2n} \frac{a_k}{n}\right)^n$$

$$\leq \left(\sum_{k=1}^{2n} \frac{a_k}{2n}\right)^{2n},$$

where in the first inequality we use the induction hypothesis $H(n)$, and in the second inequality we use $H(2)$.

Let us now turn to an alternate approach to prove (1.1). This new approach easily avoids the hurdle of how to regroup the terms and where to apply the induction hypothesis.

Introduce new variables

$$A_k = \frac{a_k}{\sqrt[n]{a_1 a_2 \cdots a_n}}, \quad (1 \leq k \leq n)$$

which are normalized in the sense that

$$\prod_{k=1}^{n} A_k = 1.$$

Now, the AM-GM inequality says that if $A_1 A_2 \cdots A_n = 1$ then

$$A_1 + A_2 + \cdots + A_n \geq n, \tag{1.3}$$

with equality if and only if $A_1 = A_2 = \cdots = A_n = 1$. This observation yields the third induction proof.

Third Proof. Let $H(n)$ stand for the hypothesis that, subject to $A_1 A_2 \cdots A_n = 1$, inequality (1.3) is valid for n. For $n = 2$, if $A_1 A_2 = 1$, we have

$$A_1 + A_2 \geq 2 \iff \left(\sqrt{A_1} - \sqrt{A_2}\right)^2 \geq 0,$$

which is always true. Assume that $A_1 A_2 \cdots A_{n+1} = 1$ and set

$$B_k = A_k \ (1 \leq k \leq n-1), \quad B_n = A_n A_{n+1}.$$

1. Two Classical Inequalities

Since $B_1 B_2 \cdots B_n = 1$, $H(n)$ implies that

$$B_1 + B_2 + \cdots + B_n = A_1 + \cdots + A_{n-1} + A_n A_{n+1} \geq n.$$

We have to show that $A_1 + \cdots + A_n + A_{n+1} \geq n+1$. To see this, it suffices to prove that

$$A_n + A_{n+1} \geq A_n A_{n+1} + 1,$$

which is equivalent to $(A_n - 1)(A_{n+1} - 1) \leq 0$. While this does not generally solely follow from the assumption that $A_1 A_2 \cdots A_{n+1} = 1$, due to the symmetry of the A_k's, we may order the numbers $A_1, A_2, \ldots, A_{n+1}$ in advance such that $A_n \leq A_k \leq A_{n+1}$ for all $1 \leq k \leq n-1$. Thus, we have

$$A_n^{n+1} \leq A_1 A_2 \cdots A_{n+1} \leq A_{n+1}^{n+1}$$

and so $A_n \leq 1 \leq A_{n+1}$, which implies that $(A_n - 1)(A_{n+1} - 1) \leq 0$. This establishes the desired result $H(n+1)$. The equality case means that

$$A_1 = A_2 = \cdots = A_n A_{n+1} = 1 \quad \text{and} \quad (A_n - 1)(A_{n+1} - 1) = 0,$$

which implies $A_1 = \cdots = A_n = A_{n+1} = 1$.

The additive inequality (1.3) yields a quite different proof of (1.1). Indeed, it also connects to a larger theme, where normalization and orderings give us a systematic way to pass from an additive inequality to a multiplicative inequality. As another example of this process, we present Hardy's elegant proof of (1.1), which is based on a special ordering designed by a *smoothing transformation*. Let $f(a_1, a_2, \ldots, a_n)$ be a symmetric function, i.e., for all $\sigma(a_1, a_2, \ldots, a_n)$,

$$f(a_1, a_2, \ldots, a_n) = f(\sigma(a_1, a_2, \ldots, a_n)),$$

where $\sigma(a_1, a_2, \ldots, a_n)$ is a permutation of $(a_1, a_2, \ldots, a_n)$. The smoothing transformation states that if the symmetric function $f(a_1, a_2, \ldots, a_n)$ becomes larger as two of the variables are made closer in value while the sum is fixed, then $f(a_1, a_2, \ldots, a_n)$ is maximized when all variables are equal. Hardy's following proof provides one of the best applications of this principle.

Fourth Proof. Let

$$G = \sqrt[n]{a_1 a_2 \cdots a_n}, \quad A = \frac{a_1 + a_2 + \cdots + a_n}{n}. \tag{1.4}$$

If $a_1 = a_2 = \cdots = a_n$, then $A = G = a_1$. Hence (1.1) holds with equality. Suppose that $a_1, a_2, \ldots, a_n$ are not all equal. Without loss of generality, we may assume that

$$a_1 = \min\{a_1, a_2, \ldots, a_n\}, \quad a_2 = \max\{a_1, a_2, \ldots, a_n\}.$$

Then it is clear that

$$a_1 < G < a_2, \quad a_1 < A < a_2$$

and $a_1 + a_2 - A > a_1 > 0$. Now, consider a new set of n positive numbers

$$\{A, a_1 + a_2 - A, a_3, \ldots, a_n\} \tag{1.5}$$

and let G_1 and A_1 be the corresponding geometric and arithmetic means. Since

$$A(a_1 + a_2 - A) - a_1 a_2 = (A - a_1)(a_2 - A) > 0,$$

we have

$$A(a_1 + a_2 - A)a_3 \cdots a_n > a_1 a_2 \cdots a_n,$$

from which it follows that $G_1 > G$. On the other hand,

$$nA_1 = A + (a_1 + a_2 - A) + a_3 + \cdots + a_n = nA,$$

i.e., $A_1 = A$. Thus, we have shown that the transformation defined by (1.5) does not change the arithmetic mean while the geometric mean is increased. In particular, if all numbers in (1.5) are the same then we have

$$A = A_1 = G_1 > G,$$

which proves (1.1). Otherwise we continue to apply the transformation to the latest numbers. Notice that under the transformation, at least one more element in the underlying set of n positive numbers is replaced with the arithmetic mean. Thus, after applying the transformation up to k times, where $k \leq n - 1$, we arrive at a vector with n equal coordinates. Now, we have $G_k = A_k$ and

$$A = A_1 = \cdots = A_k = G_k > G_{k-1} > \cdots > G_1 > G.$$

This finally proves (1.1) as desired.

As an example of applying the smoothing transformation above, consider

$$(a_1, a_2, a_3, a_4) = (1, 7, 5, 3).$$

The transformation successively yields the sets of positive numbers

$$(1, 7, 5, 3) \implies (4, 4, 5, 3) \implies (4, 4, 4, 4).$$

In Hardy's proof, observe that the key ingredient is that the double inequalities $a_1 < A < a_2$ are put together to build the simple quadratic inequality $(A - a_1)(a_2 - A) > 0$. In fact, to prove symmetric inequalities, without loss of generality, we can always assume that $a_1 \leq a_2 \leq \cdots \leq a_n$. Taking this monotonicity as an additional bonus, Alzer obtained the following stunning proof of (1.1).

Fifth Proof. Let $a_1 \leq a_2 \leq \cdots \leq a_n$, G and A be given by (1.4). Then $a_1 \leq G \leq a_n$ and so there exists one k with $a_k \leq G \leq a_{k+1}$. It follows that

$$\sum_{i=1}^{k} \int_{a_i}^{G} \left(\frac{1}{t} - \frac{1}{G}\right) dt + \sum_{i=k+1}^{n} \int_{G}^{a_i} \left(\frac{1}{G} - \frac{1}{t}\right) dt \geq 0 \tag{1.6}$$

1. Two Classical Inequalities

or equivalently,

$$\sum_{i=1}^{n} \int_{G}^{a_i} \frac{1}{G}\, dt \geq \sum_{i=1}^{n} \int_{G}^{a_i} \frac{1}{t}\, dt.$$

Here the left-hand side gives

$$\sum_{i=1}^{n} \frac{a_i - G}{G} = n\left(\frac{A}{G} - 1\right),$$

while the right-hand side becomes

$$\sum_{i=1}^{n} (\ln a_i - \ln G) = \ln\left(\prod_{i=1}^{n} a_i\right) - n \ln G = 0.$$

Thus, $A/G - 1 \geq 0$, hence $A \geq G$, which gives our required inequality. In the case of equality, all integrals in (1.6) must be zero, which implies $a_1 = a_2 = \cdots = a_n = G$.

The final proof is due to George Pólya, who reported that the proof came to him in a dream. As a matter of fact, his proof yields the following stronger inequality:

Weighted Arithmetic Mean and Geometric Mean Inequality Let $a_1, a_2, \ldots, a_n$ and $p_1, p_2, \ldots, p_n$ be positive real numbers such that $p_1 + p_2 + \cdots + p_n = 1$. Then

$$a_1^{p_1} a_2^{p_2} \cdots a_n^{p_n} \leq p_1 a_1 + p_2 a_2 + \cdots + p_n a_n, \tag{1.7}$$

with equality if and only if all a_k ($1 \leq k \leq n$) are equal.

Pólya's Proof. It is well known that

$$x \leq \exp(x - 1) \quad \text{for all } x \in \mathbb{R}.$$

For $1 \leq k \leq n$, using this inequality repeatedly yields

$$a_k \leq \exp(a_k - 1) \quad \text{and} \quad a_k^{p_k} \leq \exp(p_k a_k - p_k)$$

and so

$$a_1^{p_1} a_2^{p_2} \cdots a_n^{p_n} \leq \exp\left(\sum_{k=1}^{n} p_k a_k - 1\right). \tag{1.8}$$

To obtain the desired arithmetic mean bound, the normalization process is used again. Let

$$b_k = \frac{a_k}{A} \quad \text{with} \quad A = p_1 a_1 + p_2 a_2 + \cdots + p_n a_n.$$

Replacing a_k in (1.8) by b_k yields

$$\left(\frac{a_1}{A}\right)^{p_1} \left(\frac{a_1}{A}\right)^{p_2} \cdots \left(\frac{a_n}{A}\right)^{p_n} \leq \exp\left(\sum_{k=1}^{n} p_k \frac{a_k}{A} - 1\right) = 1.$$

This is equivalent to (1.7). In the case of equality, noticing that

$$\frac{a_k}{A} < \exp(\frac{a_k}{A} - 1) \quad \text{unless} \quad \frac{a_k}{A} = 1,$$

we see that (1.8) is strict unless $a_k = A$ for all $k = 1, 2, \ldots, n$. Thus, one has equality in (1.7) if and only if all a_k's are equal.

Similar to the AM-GM inequality, the Cauchy-Schwarz inequality is another one of the most fundamental mathematical inequalities. It also turns out to have a remarkable number of proofs, some just as charming and informative as the proofs of the AM-GM inequality. To highlight this, we present four proofs of the Cauchy-Schwarz inequality.

1.2 Cauchy-Schwarz Inequality

Let a_k, b_k $(1 \leq k \leq n)$ be real numbers. Then

$$\left(\sum_{k=1}^{n} a_k b_k\right)^2 \leq \left(\sum_{k=1}^{n} a_k^2\right)\left(\sum_{k=1}^{n} b_k^2\right), \tag{1.9}$$

with equality if and only if there is a $t \in \mathbb{R}$ such that $a_k = t b_k$ $(1 \leq k \leq n)$.

First, we demonstrate that the Cauchy-Schwarz inequality is a consequence of the AM-GM inequality.

First Proof. Let $H(n)$ represent the hypothesis that inequality (1.9) is valid for n. For $n = 2$, by the AM-GM inequality, we have

$$\begin{aligned}(a_1 b_1 + a_2 b_2)^2 &= a_1^2 b_1^2 + 2(a_1 b_2)(a_2 b_1) + a_2^2 b_2^2 \\ &\leq a_1^2 b_1^2 + a_1^2 b_2^2 + a_2^2 b_1^2 + a_2^2 b_2^2 \\ &= (a_1^2 + a_2^2)(b_1^2 + b_2^2),\end{aligned}$$

which confirms $H(2)$. For the case of equality, appealing to the equality condition in the AM-GM inequality, we have $a_1 b_2 = a_2 b_1$, or equivalently, $a_i = t b_i$ $(i = 1, 2)$ with $t = a_1/a_2 = b_1/b_2$. Now, we prove that $H(2)$ and $H(n)$ together imply $H(n + 1)$. Indeed, using the AM-GM inequality and $H(n)$ yields

$$2\left(\sum_{k=1}^{n} a_k b_k\right) a_{n+1} b_{n+1} \leq 2 \sqrt{a_{n+1}^2 \left(\sum_{k=1}^{n} b_k^2\right)} \sqrt{b_{n+1}^2 \left(\sum_{k=1}^{n} a_k^2\right)}$$

$$\leq a_{n+1}^2 \left(\sum_{k=1}^{n} b_k^2\right) + b_{n+1}^2 \left(\sum_{k=1}^{n} a_k^2\right)$$

and hence

$$\left(\sum_{k=1}^{n+1} a_k b_k\right)^2 = \left(\sum_{k=1}^{n} a_k b_k\right)^2 + 2\left(\sum_{k=1}^{n} a_k b_k\right) a_{n+1} b_{n+1} + a_{n+1}^2 b_{n+1}^2$$

$$\leq \left(\sum_{k=1}^{n} a_k b_k\right)^2 + a_{n+1}^2 \left(\sum_{k=1}^{n} b_k^2\right) + b_{n+1}^2 \left(\sum_{k=1}^{n} a_k^2\right) + a_{n+1}^2 b_{n+1}^2$$

1. Two Classical Inequalities

$$\leq \left(\sum_{k=1}^{n} a_k^2\right)\left(\sum_{k=1}^{n} b_k^2\right) + a_{n+1}^2 \left(\sum_{k=1}^{n} b_k^2\right) + b_{n+1}^2 \left(\sum_{k=1}^{n} a_k^2\right) + a_{n+1}^2 b_{n+1}^2$$

$$= \left(\sum_{k=1}^{n+1} a_k^2\right)\left(\sum_{k=1}^{n+1} b_k^2\right).$$

For the case of equality, again, the equality condition in the AM-GM inequality implies

$$a_k b_{n+1} = b_k a_{n+1} \quad \text{and} \quad a_k = t b_k, \quad \text{for all } k = 1, 2, \ldots, n.$$

Therefore, $a_k = t b_k$ for all $k = 1, 2, \ldots, n, n+1$.

The next proof is based on normalization.

Second Proof. If all a_k or b_k are zero, we have equality in (1.9). Suppose that $a_i \neq 0$ and $b_j \neq 0$ for some $1 \leq i, j \leq n$. Set

$$A_k = \frac{a_k}{\left(\sum_{k=1}^{n} a_k^2\right)^{1/2}} \quad \text{and} \quad B_k = \frac{b_k}{\left(\sum_{k=1}^{n} b_k^2\right)^{1/2}},$$

which are normalized in the sense that

$$\sum_{k=1}^{n} A_k^2 = 1 \quad \text{and} \quad \sum_{k=1}^{n} B_k^2 = 1. \tag{1.10}$$

With these assignments, the inequality (1.9) is equivalent to

$$A_1 B_1 + A_2 B_2 + \cdots + A_n B_n \leq 1. \tag{1.11}$$

To see this, adding the well-known elementary inequalities

$$A_k B_k \leq \frac{1}{2}(A_k^2 + B_k^2) \quad (\Longleftrightarrow (A_k - B_k)^2 \geq 0)$$

for k from 1 to n, in view of (1.10), we have

$$A_1 B_1 + A_2 B_2 + \cdots + A_n B_n \leq \frac{1}{2}\left(\sum_{k=1}^{n} A_k^2 + \sum_{k=1}^{n} B_k^2\right) = 1.$$

Moreover, subject to (1.10), it is easy to see that (1.11) is equivalent to

$$(A_1 - B_1)^2 + (A_2 - B_2)^2 + \cdots + (A_n - B_n)^2 \geq 0.$$

Thus, the equality in (1.9) occurs if and only if $A_k = B_k$ for all $1 \leq k \leq n$; that is, if and only if

$$a_k = t b_k \quad \text{for all } 1 \leq k \leq n \text{ with } t = \left(\sum_{k=1}^{n} a_k^2 \Big/ \sum_{k=1}^{n} b_k^2\right)^{1/2}.$$

Once again, in terms of normalization, we see how to pass from a simple additive inequality to a multiplicative inequality. In the following, we show how to pass from a simple identity to the famous *Lagrange's identity*, which provides a "one-line proof" of the Cauchy-Schwarz inequality.

Third Proof. For $n = 2$, the Cauchy-Schwarz inequality just says
$$(a_1b_1 + a_2b_2)^2 \leq (a_1^2 + a_2^2)(b_1^2 + b_2^2).$$
This is a simple assertion of the elementary identity
$$(a_1^2 + a_2^2)(b_1^2 + b_2^2) - (a_1b_1 + a_2b_2)^2 = (a_1b_2 - a_2b_1)^2.$$
In general, let
$$D_n = \left(\sum_{k=1}^n a_k^2\right)\left(\sum_{k=1}^n b_k^2\right) - \left(\sum_{k=1}^n a_k b_k\right)^2.$$
We have already seen that D_2 is a perfect square. In view of the symmetry, we can rewrite D_n as
$$D_n = \frac{1}{2} \sum_{i=1}^n \sum_{j=1}^n (a_i^2 b_j^2 + a_j^2 b_i^2) - \sum_{i=1}^n \sum_{j=1}^n a_i b_i a_j b_j.$$
This quickly yields the well-known Lagrange's identity
$$D_n = \frac{1}{2} \sum_{i=1}^n \sum_{j=1}^n (a_i b_j - a_j b_i)^2.$$
Now the Cauchy-Schwarz inequality follows from Lagrange's identity immediately.

Finally, we present Schwarz's proof, which is based on one simple fact about quadratic functions. Let
$$\mathbf{x} = (a_1, a_2, \ldots, a_n), \quad \mathbf{y} = (b_1, b_2, \ldots, b_n).$$
Recall the inner product on $\mathbb{R}^n$ is defined by
$$(\mathbf{x}, \mathbf{y}) = a_1 b_1 + a_2 b_2 + \cdots + a_n b_n$$
and the norm is given by $\|\mathbf{x}\| = \sqrt{(\mathbf{x}, \mathbf{x})}$. In vector form, the Cauchy-Schwarz inequality becomes
$$(\mathbf{x}, \mathbf{y})^2 \leq \|\mathbf{x}\|^2 \|\mathbf{y}\|^2.$$
Schwarz observed that the quadratic function
$$f(t) = \|\mathbf{x}t + \mathbf{y}\|^2 = \|\mathbf{x}\|^2 t^2 + 2(\mathbf{x}, \mathbf{y})t + \|\mathbf{y}\|^2$$
is nonnegative for all $t \in \mathbb{R}$ and hence the discriminant
$$(\mathbf{x}, \mathbf{y})^2 - \|\mathbf{x}\|^2 \|\mathbf{y}\|^2 \leq 0.$$
This provides another short proof of the Cauchy-Schwarz inequality.

For readers interested in learning more about the Cauchy-Schwarz inequality, please refer to Michael Steele's fascinating book *The Cauchy-Schwarz Master Class* [5]. With the Cauchy-Schwarz inequality as the initial guide, this lively, problem-oriented book will coach one towards mastery of many more classical inequalities, including those of Hölder, Hilbert, and Hardy.

Analysis abounds with applications of these two fundamental inequalities. Let us single out two applications.

1. Two Classical Inequalities

Application 1. Prove that sequence $\left\{\left(1+\frac{1}{n}\right)^n\right\}$ is monotonic increasing and bounded.

Proof. By the AM-GM inequality, we have

$$\sqrt[n+1]{\left(1+\frac{1}{n}\right)^n \cdot 1} < \frac{1}{n+1}\left[n\left(1+\frac{1}{n}\right)+1\right] = 1 + \frac{1}{n+1}.$$

This is equivalent to

$$\left(1+\frac{1}{n}\right)^n < \left(1+\frac{1}{n+1}\right)^{n+1},$$

which shows that $\left(1+\frac{1}{n}\right)^n$ is monotonic increasing as required. To find an upper bound of the sequence, for $n > 5$, using the AM-GM inequality once more, we have

$$\sqrt[n+1]{\left(\frac{5}{6}\right)^6 \cdot 1^{n-5}} < \frac{1}{n+1}\left[6 \times \frac{5}{6} + (n-5) \cdot 1\right] = \frac{n}{n+1},$$

and so $(5/6)^6 < (n/(n+1))^{n+1}$, or equivalently, $(1+1/n)^{n+1} < (6/5)^6$. Therefore,

$$\left(1+\frac{1}{n}\right)^n < \left(1+\frac{1}{n}\right)^{n+1} < \left(\frac{6}{5}\right)^6 < 3, \qquad \text{(for } n > 5\text{)}.$$

In view of the monotonicity of the sequence, for all $n \geq 1$, we obtain $\left(1+\frac{1}{n}\right)^n < 3$. It is well known that the sequence converges to the natural base e.

Application 2. (Monthly Problem 11430, 2009). For real $x_1, \ldots, x_n$, show that

$$\frac{x_1}{1+x_1^2} + \frac{x_2}{1+x_1^2+x_2^2} + \cdots + \frac{x_n}{1+x_1^2+\cdots+x_n^2} < \sqrt{n}.$$

Solution Notice that

$$\frac{x_1}{1+x_1^2} + \cdots + \frac{x_n}{1+x_1^2+\cdots+x_n^2} \leq \frac{|x_1|}{1+x_1^2} + \cdots + \frac{|x_n|}{1+x_1^2+\cdots+x_n^2}.$$

Without lose of generality, we only consider the case when $x_1, \ldots, x_n$ are all nonnegative real numbers. Let S denote the left-hand side of the required inequality. Applying the Cauchy-Schwarz inequality yields

$$S^2 = \left(1 \cdot \frac{x_1}{1+x_1^2} + 1 \cdot \frac{x_2}{1+x_1^2+x_2^2} + \cdots + 1 \cdot \frac{x_n}{1+x_1^2+\cdots+x_n^2}\right)^2$$

$$\leq n \cdot \left[\left(\frac{x_1}{1+x_1^2}\right)^2 + \left(\frac{x_2}{1+x_1^2+x_2^2}\right)^2 + \cdots + \left(\frac{x_n}{1+x_1^2+\cdots+x_n^2}\right)^2\right].$$

For $k = 1$, we have

$$\left(\frac{x_1}{1+x_1^2}\right)^2 \leq \frac{x_1^2}{1+x_1^2} = 1 - \frac{1}{1+x_1^2},$$

and
$$\left(\frac{x_k}{1+x_1^2+\cdots+x_k^2}\right)^2 \leq \frac{x_k^2}{(1+x_1^2+\cdots+x_{k-1}^2)(1+x_1^2+\cdots+x_k^2)}$$
$$= \frac{1}{1+x_1^2+\cdots+x_{k-1}^2} - \frac{1}{1+x_1^2+\cdots+x_k^2},$$

for $2 \leq k \leq n$. Combining these inequalities results in

$$\left(\frac{x_1}{1+x_1^2}\right)^2 + \cdots + \left(\frac{x_n}{1+x_1^2+\cdots+x_n^2}\right)^2 \leq 1 - \frac{1}{1+x_1^2+\cdots+x_n^2} < 1.$$

Thus, we have $S^2 < n$, which implies the desired inequality.

We now conclude this chapter by introducing the *Gauss arithmetic-geometric mean* based on the AM-GM inequality. Let a and b be two distinct positive numbers. Define the successive arithmetic and geometric means by

$$a_0 = a, \quad a_{n+1} = \frac{a_n + b_n}{2}; \quad b_0 = b, \quad b_{n+1} = \sqrt{a_n b_n}, \quad (n = 1, 2, \ldots).$$

These sequences were mentioned by Lagrange in 1785 and six years later came to the attention of the 14-year old Gauss. By inequality (1.1) we see that $a_n \geq b_n$ for $n = 1, 2, \ldots$. It follows that

$$\sqrt{ab} \leq \cdots \leq b_n \leq b_{n+1} \leq a_{n+1} \leq a_n \leq \cdots \leq \frac{a+b}{2}$$

and
$$a_{n+1} - b_{n+1} \leq \frac{1}{2}(a_n - b_n).$$

Hence the sequences $\{a_n\}$ and $\{b_n\}$ both converge to the same limit, which is called the Gauss arithmetic-geometric mean and denoted by $\mathrm{agm}(a, b)$. In 1799 Gauss successfully found $\mathrm{agm}(1, \sqrt{2})$ to great precision, but it wasn't until 1818 that he finally established

$$\mathrm{agm}(a, b) = \left(\frac{2}{\pi} \int_0^{\pi/2} \frac{dx}{\sqrt{a^2 \sin^2 x + b^2 \cos^2 x}}\right)^{-1},$$

which later was used to found the branch of mathematics called the "theory of elliptic functions."

Exercises

1. Prove the AM-GM inequality by constructing a smoothing transformation in which the geometric mean is invariant while the arithmetic mean is decreased.

2. For positive real numbers a, b, x and y, show that
$$\frac{(a+b)^2}{x+y} \leq \frac{a^2}{x} + \frac{b^2}{y}.$$
Then use this *self-generalizing* inequality to prove the Cauchy-Schwarz inequality.

1. Two Classical Inequalities

3. Let $a_k > 0$ for all $1 \leq k \leq n$ and $a_1 a_2 \cdots a_n = 1$. Prove the following inequalities.

   (a) (1) $(1 + a_1)(1 + a_2) \cdots (1 + a_n) \geq 2^n$.

   (b) (2) $a_1^m + a_2^m + \cdots + a_n^m \geq \frac{1}{a_1} + \frac{1}{a_2} + \cdots + \frac{1}{a_n}$ for any positive integer $m \geq n-1$.

4. Let $a_k > 0$ for all $1 \leq k \leq n$ and $\sum_{k=1}^n a_k = 1$. Prove the following inequalities.

   (a) $\dfrac{a_1^2}{a_1 + a_2} + \dfrac{a_2^2}{a_2 + a_3} + \cdots + \dfrac{a_n^2}{a_n + a_1} \geq \dfrac{1}{2}$.

   (b) $\displaystyle\sum_{k=1}^n \dfrac{a_k}{2 - a_k} \geq \dfrac{n}{2n - 1}$.

   (c) $\displaystyle\sum_{k=1}^n \dfrac{a_k^m}{1 - a_k} \geq \dfrac{1}{(n-1)n^{m-2}}$, where $m > 1$.

   (d) $\displaystyle\sum_{k=1}^n \dfrac{a_k}{\sqrt{1 - a_k}} \geq \dfrac{1}{\sqrt{n-1}} \sum_{k=1}^n \sqrt{a_k}$.

   (e) $1 \leq \displaystyle\sum_{k=1}^n \dfrac{a_k}{\sqrt{1 + a_0 + \cdots + a_{k-1}} \sqrt{a_k + \cdots + a_n}} \leq \dfrac{\pi}{2}$, where $a_0 = 0$.

5. For all nonnegative sequences $a_1, a_2, \ldots$, and positive integers n and k, use leap-forward fall-back induction to prove

$$\left(\frac{a_1 + a_2 + \cdots + a_k}{k}\right)^n \leq \frac{a_1^n + a_2^n + \cdots + a_k^n}{k}.$$

This is a special case of the power mean inequality, which we will explore more in Chapter 3.

6. Let a_k ($k = 1, 2, \ldots, n$) be positive real numbers. Let A and G be the arithmetic and geometric means of a_k's respectively, and H be the *harmonic mean* of a_k's defined by

$$H = \frac{n}{1/a_1 + 1/a_2 + \cdots + 1/a_n}.$$

Prove that

$$nA + H \geq (n+1)G.$$

In addition, if $AH \geq G^n, n \geq 3$, prove that

$$\sum_{k=1}^n a_k^n \geq n\, G^{n(n-1)}.$$

7. Let $a_1 < a_2 < \cdots < a_n$ and $\sum_{k=1}^n 1/a_k \leq 1$. Prove that for any real number x and $n \geq 2$,

$$\left(\sum_{k=1}^n \frac{1}{a_k^2 + x^2}\right)^2 \leq \frac{1}{2(a_1(a_1 - 1) + x^2)}.$$

8. Let $\{a_n\}$ be a positive real sequence with $a_{i+j} \leq a_i + a_j$ for all $i, j = 1, 2, \ldots$. For each $n \in \mathbb{N}$, show that
$$a_1 + \frac{a_2}{2} + \cdots + \frac{a_n}{n} \geq a_n.$$

9. Let $a_1 \geq a_2 \geq \cdots \geq a_n > 0$. Let $\sum_{i=1}^{k} a_i \leq \sum_{i=1}^{k} b_i$ for all $1 \leq k \leq n$. Show that
   (a) $\sum_{i=1}^{n} a_i^2 \leq \sum_{i=1}^{n} b_i^2$,
   (b) $\sum_{i=1}^{n} a_i^3 \leq \sum_{i=1}^{n} a_i b_i^2$.

10. *The Schwab-Schoenberg Mean.* Let a and b be positive numbers with $a < b$. Define $a_0 = a, b_0 = b$, and for $n \geq 0$,
$$a_{n+1} = \frac{a_n + b_n}{2}, \quad b_{n+1} = \sqrt{a_{n+1} b_n}.$$

Show that $\{a_n\}$ and $\{b_n\}$ approach to a common limit and determine the value of the limit.

11. **Putnam Problem 1964-A5.** Prove that there is a positive constant K such that the following inequality holds for any sequence of positive numbers $a_1, a_2, a_3, \ldots$:
$$\sum_{n=1}^{\infty} \frac{n}{a_1 + a_2 + \cdots + a_n} \leq K \sum_{n=1}^{\infty} \frac{1}{a_n}.$$

12. **Monthly Problem 11145** [2005, 366; 2006, 766]. Find the least c such that if $n \geq 1$ and $a_1, \ldots, a_n > 0$ then
$$\sum_{k=1}^{n} \frac{k}{\sum_{j=1}^{k} 1/a_j} \leq c \sum_{k=1}^{n} a_k.$$

Remark. The inequality above can be generalized as: for any positive p,
$$\sum_{k=1}^{\infty} \left(\frac{k}{\sum_{j=1}^{k} 1/a_j} \right)^p \leq \left(\frac{p+1}{p} \right)^p \sum_{k=1}^{\infty} a_k^p.$$

13. **Putnam Problem 1978-A5.** Let $0 < x_k < \pi$ for $k = 1, 2, \ldots, n$ and set
$$x = \frac{x_1 + x_2 + \cdots + x_n}{n}.$$

Prove that
$$\prod_{k=1}^{n} \frac{\sin x_k}{x_k} \leq \left(\frac{\sin x}{x} \right)^n.$$

14. **Putnam Problem 2003-A2.** Let $a_1, a_2, \ldots, a_n$ and $b_1, b_2, \ldots, b_n$ be nonnegative real numbers. Show that
$$(a_1 a_2 \cdots a_n)^{1/n} + (b_1 b_2 \cdots b_n)^{1/n} \leq [(a_1 + b_1)(a_2 + b_2) \cdots (a_n + b_n)]^{1/n}.$$

15. **Carleman's Inequality.** Let $a_1, a_2, \ldots$, be a positive sequence with $0 < \sum_{n=1}^{\infty} a_n < \infty$. Prove that
$$\sum_{n=1}^{\infty} (a_1 a_2 \cdots a_n)^{1/n} \le e \sum_{n=1}^{\infty} a_n.$$

Remark. Pólya gave an elegant proof of this inequality that depends on little more than the AM-GM inequality. Several refinements of Carleman's inequality were obtained recently. See "On an infinity series for $(1 + 1/x)^x$ and its application" (*Int. J. Math. Math. Anal.*, **29**(2002) 675–680) for details.

References

[1] H. Alzer, A proof of the arithmetic mean-geometric mean inequality, *Amer. Math. Monthly*, **103**(1996) 585.

[2] E. Beckenbach and R. Bellman, *An Introduction to Inequalities*, The Mathematical Association of America, 1961.

[3] A. L. Cauchy, *Analyse Algebrique* (Note 2; Oeuves Completes), Ser. 2, Vol. 3, Paris, (1897) 375–377.

[4] G. H. Hardy, J. E. Littlewood and G. Pólya, *Inequalities*, Cambridge University Press, Cambridge, 1952.

[5] J. M. Steele, *The Cauchy-Schwarz Master Class*, Cambridge University Press, Cambridge, 2004.

2

A New Approach for Proving Inequalities

A beautiful approach for a specific problem frequently leads to wide applicability, thereby solving a host of new, related problems. We have seen in the preceding chapter some memorable approaches from a great variety of sources for proving the AM-GM inequality and the Cauchy-Schwarz inequality. There are many more different proofs recorded in the monograph [2] by Bullen et.al. In this regard, one may wonder whether there are any new approaches to re-prove these two inequalities? In this chapter, we introduce a *unified elementary approach* for proving inequalities [3, 4]. We will see that this new approach not only recovers these two classical inequalities, but also is strong enough to capture the famous Hölder, Triangle and Minkowski inequalities, as well as Ky Fan's inequality.

The unified elementary approach is rarely seen and depends on little more than the following well-known fact in analysis:

$$\inf_{x \in E} f(x) + \inf_{x \in E} g(x) \leq \inf_{x \in E} (f(x) + g(x)),$$

where E is a subset of $\mathbb{R}$. In general, this suggests that

$$\sum_{k=1}^{n} \inf_{x \in E} f_k(x) \leq \inf_{x \in E} \sum_{k=1}^{n} f_k(x). \tag{2.1}$$

where $f_k : E \subset \mathbb{R} \to \mathbb{R}$ $(k = 1, 2 \ldots, n)$. Moreover, if all $f_k(x)$ are convex in E, the convexity allows us to conclude:

Lemma 1. *Let all $f_k(x)$ be convex in E. Then the equality holds in (2.1) if and only if all infimums of the $f_k(x)$ and the infumum of $\sum_{k=1}^{n} f_k(x)$ are achieved at the same point in E.*

We leave the proof of Lemma 1 to the reader (See Exercise 1).

We first prove the AM-GM inequality in terms of (2.1).

Theorem 2.1. *Let $a_1, a_2, \ldots, a_n$ be positive real numbers. Then*

$$\sqrt[n]{a_1 a_2 \cdots a_n} \leq \frac{a_1 + a_2 + \cdots + a_n}{n}, \tag{2.2}$$

with equality if and only if all a_k's are equal.

Proof. Clearly, the key ingredient here is how to properly define the $f_k(x)$, which will then enable us to employ (2.1). In order to accomplish this, consider the logarithm of both sides of (2.2), which becomes

$$\sum_{k=1}^{n} \ln a_k \leq n \ln\left(\frac{\sum_{k=1}^{n} a_k}{n}\right).$$

Adding n to both sides yields

$$\sum_{k=1}^{n} (1 + \ln a_k) \leq n + n \ln\left(\frac{\sum_{k=1}^{n} a_k}{n}\right). \tag{2.3}$$

So to prove that (2.2) is true, it is enough to show (2.3) holds. With (2.1) in mind, we simply need $f_k(x)$ that has a minimum value of $1 + \ln a_k$; for example,

$$f_k(x) = a_k x - \ln x \quad \text{or} \quad f_k(x) = \frac{x}{a_k} + \ln x.$$

Once we determine the form of the $f_k(x)$, the rest becomes a routine calculation. Indeed, let $f_k(x) = a_k x - \ln x$, $E = (0, +\infty)$. Since

$$f_k'(x) = a_k - \frac{1}{x}, \quad f_k''(x) = \frac{1}{x^2},$$

f_k has a unique critical point at $x_{kc} = 1/a_k$, where it has minimum value

$$f_k(x_{kc}) = 1 + \ln a_k,$$

as desired. Similarly, the function

$$f(x) = \sum_{k=1}^{n} f_k(x) = \left(\sum_{k=1}^{n} a_k\right) x - n \ln x$$

has a unique critical point $x_c = n/\sum_{k=1}^{n} a_k$ and its minimum value is

$$f(x_c) = n + n \ln\left(\frac{\sum_{k=1}^{n} a_k}{n}\right).$$

Finally, applying (2.1), we establish (2.3) and thus confirm the desired inequality (2.2). Furthermore, since all the f_k are convex, by Lemma 1, we conclude that the equality in (2.2) holds if and only if all x_{kc}'s are equal to x_c; that is, all a_k's are equal.

Before proceeding, let us take another look at the proof. It seems that the key insight in the foregoing proof is how to construct the functions $f_k(x)$. Once $f_k(x)$ have been defined, the remaining steps are a straightforward exercise in calculating infimums. Next, we turn our attention to proving the Cauchy-Schwarz inequality in terms of (2.1).

Theorem 2.2. *Let a_k, b_k ($1 \leq k \leq n$) be real numbers. Then*

$$\left(\sum_{k=1}^{n} a_k b_k\right)^2 \leq \left(\sum_{k=1}^{n} a_k^2\right)\left(\sum_{k=1}^{n} b_k^2\right), \tag{2.4}$$

with equality if and only if there is a $t \in \mathbb{R}$ such that $a_k = tb_k$ ($1 \leq k \leq n$).

2. A New Approach for Proving Inequalities

Proof. Since $a_k b_k \leq |a_k||b_k|$, without loss of generality, we may assume that $a_k, b_k > 0$. In view of the algebraic identity

$$a_k^2 x + \frac{b_k^2}{x} = \left(a_k \sqrt{x} - \frac{b_k}{\sqrt{x}}\right)^2 + 2a_k b_k,$$

we define

$$f_k(x) = a_k^2 x + \frac{b_k^2}{x}, \quad E = (0, +\infty).$$

Now, $f_k(x)$ has minimum value of $2a_k b_k$, which occurs at $x_{kc} = b_k/a_k$. Furthermore, we find that

$$f(x) = \sum_{k=1}^{n} f_k(x) = \left(\sum_{k=1}^{n} a_k^2\right) x + \frac{\sum_{k=1}^{n} b_k^2}{x}$$

has a unique critical number at $x_c = (\sum_{k=1}^{n} b_k^2 / \sum_{k=1}^{n} a_k^2)^{1/2}$, and $f(x)$ has minimum value

$$f(x_c) = 2\left(\sum_{k=1}^{n} a_k^2 \sum_{k=1}^{n} b_k^2\right)^{1/2}.$$

Applying (2.1) yields

$$\sum_{k=1}^{n} a_k b_k \leq \left(\sum_{k=1}^{n} a_k^2 \sum_{k=1}^{n} b_k^2\right)^{1/2},$$

which is equivalent to (2.4). By Lemma 1, we have that equality holds in (2.4) if and only if all $x_{kc} = x_c$. Setting $t = x_c$ yields $a_k = t b_k$ for all $1 \leq k \leq n$.

We further show how to use (2.1) to prove the Triangle inequality, which is usually deduced via the Cauchy-Schwarz inequality.

Theorem 2.3. *Let a_k, b_k ($1 \leq k \leq n$) be real numbers. Then*

$$\left(\sum_{k=1}^{n} (a_k + b_k)^2\right)^{1/2} \leq \left(\sum_{k=1}^{n} a_k^2\right)^{1/2} + \left(\sum_{k=1}^{n} b_k^2\right)^{1/2}, \quad (2.5)$$

with equality if and only if there is a $t \in \mathbb{R}$ such that $a_k = t b_k$ ($1 \leq k \leq n$).

Proof. Since $a_k + b_k \leq |a_k| + |b_k|$, without loss of generality, we may assume that $a_k, b_k > 0$. Modifying the functions used in the proof of (2.4), we define

$$f_k(x) = (a_k + b_k)^2 x + \frac{a_k^2}{x}, \quad E = (0, \infty).$$

By completing the square, we find that $\min_E\{f_k(x)\} = f(x_{kc}) = 2a_k(a_k + b_k)$, where $x_{kc} = a_k/(a_k + b_k)$. Similarly, we obtain that

$$f(x) = \sum_{k=1}^{n} f_k(x) = \left(\sum_{k=1}^{n} (a_k + b_k)^2\right) x + \frac{\sum_{k=1}^{n} a_k^2}{x}$$

has a minimum value at $x_c = (\sum_{k=1}^n a_k^2)^{1/2}/(\sum_{k=1}^n (a_k + b_k)^2)^{1/2}$. In particular,

$$f(x_c) = 2 \left(\sum_{k=1}^n a_k^2\right)^{1/2} \left(\sum_{k=1}^n (a_k + b_k)^2\right)^{1/2}.$$

Inequality (2.1) thus ensures

$$2\sum_{k=1}^n a_k(a_k + b_k) \leq 2 \left(\sum_{k=1}^n a_k^2\right)^{1/2} \left(\sum_{k=1}^n (a_k + b_k)^2\right)^{1/2}. \tag{2.6}$$

Since

$$2\sum_{k=1}^n a_k(a_k + b_k) = \sum_{k=1}^n a_k^2 + \sum_{k=1}^n (a_k + b_k)^2 - \sum_{k=1}^n b_k^2,$$

(2.6) becomes

$$\sum_{k=1}^n a_k^2 + \sum_{k=1}^n (a_k + b_k)^2 - 2\left(\sum_{k=1}^n a_k^2\right)^{1/2} \left(\sum_{k=1}^n (a_k + b_k)^2\right)^{1/2} \leq \sum_{k=1}^n b_k^2.$$

Completing the square on the left-hand side, we have

$$\left\{\left(\sum_{k=1}^n (a_k + b_k)^2\right)^{1/2} - \left(\sum_{k=1}^n a_k^2\right)^{1/2}\right\}^2 \leq \sum_{k=1}^n b_k^2.$$

Taking the square root on both sides, we finally obtain

$$\left(\sum_{k=1}^n (a_k + b_k)^2\right)^{1/2} - \left(\sum_{k=1}^n a_k^2\right)^{1/2} \leq \left(\sum_{k=1}^n b_k^2\right)^{1/2},$$

which is equivalent to (2.5). By using Lemma 1 one more time, we see the equality holds in (2.5) if and only if all $x_{kc} = x_c$. Set $t = x_c$. Then all $a_k = tb_k$.

Further exploration along these lines suggests that (2.1) can be adapted to establish some more general classical inequalities. For example, replacing $f_k(x)$ by $p_k(a_k - \ln x)$ in the proof of Theorem 2.1 gives the following weighted arithmetic-geometric mean inequality

$$\prod_{k=1}^n a_k^{p_k/\sum_{i=1}^n p_i} \leq \frac{\sum_{k=1}^n p_k a_k}{\sum_{k=1}^n p_k}.$$

Moreover, for $p, q > 1, 1/p + 1/q = 1$, replacing $f_k(x)$ by $a_k x^p/p + b_k x^{-q}/q$ in the proof of Theorem 2.2 leads to Hölder's inequality

$$\sum_{k=1}^n a_k b_k \leq \left(\sum_{k=1}^n a_k^p\right)^{1/p} \left(\sum_{k=1}^n b_k^q\right)^{1/q}.$$

Observe that inequality (2.1) can be extended to

$$\sum_{k=1}^n \inf_{x \in E} f_k(x) \leq \inf_{x \in E} \sum_{k=1}^n f_k(x) \leq \sum_{k=1}^n f_k(y) \tag{2.7}$$

2. A New Approach for Proving Inequalities

for all $y \in E$. The relaxed right-hand side bound becomes more convenient for both proving and discovering inequalities. To demonstrate this, we present a new elementary proof of the Ky Fan inequality.

Theorem 2.4. *Let* $a_k \in (0, 1/2]$, $p_k > 0$ $(1 \le k \le n)$ *with* $\sum_{k=1}^{n} p_k = 1$. *Define*

$$A = \sum_{k=1}^{n} p_k a_k, \quad G = \prod_{k=1}^{n} a_k^{p_k}, \quad H = \left(\sum_{k=1}^{n} p_k/a_k\right)^{-1}$$

and

$$A' = \sum_{k=1}^{n} p_k(1 - a_k), \quad G' = \prod_{k=1}^{n} (1 - a_k)^{p_k}, \quad H' = \left(\sum_{k=1}^{n} p_k/(1 - a_k)\right)^{-1}.$$

Then

$$\frac{H}{H'} \le \frac{G}{G'} \le \frac{A}{A'}. \tag{2.8}$$

with equality if and only if all a_k's are equal.

Proof. We prove the first inequality and leave the proof of the second as an exercise. Motivated by the proof of Theorem 2.3 in [5], take

$$f_k(x) = p_k\left(\frac{x}{a_k} - \frac{1-x}{1-a_k} + \ln\frac{1-x}{x}\right), \quad E = (0, 1/2].$$

Since

$$f_k'(x) = p_k\left(\frac{1}{a_k(1-a_k)} - \frac{1}{x(1-x)}\right), \quad f_k''(x) = \frac{p_k(1-2x)}{x^2(1-x)^2} > 0,$$

$f_k(x)$ is strictly convex and has a unique critical point at $x = a_k \in E$. In light of (2.7), we have

$$f_k(a_k) = \ln\left(\frac{1-a_k}{a_k}\right)^{p_k} \le f_k(y) = p_k\left(\frac{y}{a_k} - \frac{1-y}{1-a_k} + \ln\frac{1-y}{y}\right)$$

and

$$\ln \prod_{k=1}^{n}\left(\frac{1-a_k}{a_k}\right)^{p_k} \le \left(\sum_{k=1}^{n}\frac{p_k}{a_k}\right)y - \left(\sum_{k=1}^{n}\frac{p_k}{1-a_k}\right)(1-y) + \ln\frac{1-y}{y}.$$

Thus,

$$\ln \prod_{k=1}^{n}\left(\frac{1-a_k}{a_k}\right)^{p_k} \le \frac{y}{H} - \frac{1-y}{H'} + \ln\frac{1-y}{y}.$$

Taking $y = H/(H + H') \in E$ leads to

$$\ln\frac{G'}{G} \le \ln\frac{H'}{H},$$

which is equivalent to the proposed first inequality. Since all f_k are strictly convex on E, by Lemma 1, we conclude that equality in (2.8) holds if and only if all a_k's are equal. This completes the proof.

Bellman observed that the fundamental results of mathematics are often inequalities rather than equalities. If he is right, then the ability to solve an inequality, especially those that differ from any previously encountered as exercises, is a valuable skill for anyone who does mathematics. Those whose appetites have been whetted by the approach presented here may want to examine other inequalities. To check recent advances in inequalities, the free online journal *Journal of Inequalities in Pure and Applied Mathematics* (jipam.vu.edu.au) is always a good resource.

Exercises

1. Let $E \subset \mathbb{R}$. Prove that
$$\inf_{x \in E} f(x) + \inf_{x \in E} g(x) \leq \inf_{x \in E} (f(x) + g(x)).$$
Moreover, if f and g are convex in E, show that equality holds if and only if f, g and $f + g$ achieve the infimum at same point.

2. **Minkowski's Inequality.** Let $a_k, b_k \geq 0$ for all $1 \leq k \leq n$ and $p > 1$. Use (2.1) to prove that
$$\left(\sum_{k=1}^{n} (a_k + b_k)^p \right)^{1/p} \leq \left(\sum_{k=1}^{n} a_k^p \right)^{1/p} + \left(\sum_{k=1}^{n} b_k^p \right)^{1/p}.$$

3. Let $a_k > 0$ for all $1 \leq k \leq n$. Use (2.1) to prove that
$$\left(\frac{\sum_{k=1}^{n} a_k}{n} \right)^{\sum_{k=1}^{n} a_k} \leq \prod_{k=1}^{n} a_k^{a_k}.$$

4. Let $a_k > 0$ and $b_k \geq 0$ for all $1 \leq k \leq n$. Use (2.1) to prove that
$$\prod_{k=1}^{n} \left(\frac{b_k}{a_k} \right)^{b_k} \geq \left(\frac{\sum_{k=1}^{n} b_k}{\sum_{k=1}^{n} a_k} \right)^{\sum_{k=1}^{n} b_k}.$$

5. **Monthly Problem 11189** [2005, 929; 2007, 645]. Let $a_1, a_2, \ldots, a_n$ be positive. Let $a_{n+1} = a_1$ and $p > 1$. Prove that
$$\sum_{k=1}^{n} \frac{a_k^p}{a_k + a_{k+1}} \geq \frac{p}{2} \sum_{k=1}^{n} a_k - \frac{p-1}{2^{p/(p-1)}} \sum_{k=1}^{n} (a_k + a_{k+1})^{1/(p-1)}.$$

6. Under the conditions stated in Theorem 2.4, prove that
$$\frac{G}{G'} \leq \frac{A}{A'}.$$

7. **An Interpolation Inequality.** Under the conditions stated in Theorem 2.4, prove that
$$\frac{A'}{G'} \leq \frac{1}{G + G'} \leq \frac{A}{G}.$$

8. Let $a_k \in (0, 1/2]$ $(1 \leq k \leq n)$, which do not all coincide. Define

$$f(x) = \prod_{k=1}^{n}\left(\frac{1}{a_k} + x\sum_{i=1}^{n}\left(\frac{1}{a_i} - \frac{1}{a_k}\right) - 1\right)^{-1/n}, \quad x \in (0, 1/n).$$

Show that $f(x)$ is continuous, strictly decreasing, and therefore

$$\frac{H}{1-H} = f(1/n) \leq f(x) \leq f(0) = \frac{G}{G'}, \quad x \in (0, 1/n).$$

References

[1] E. Beckenbach and R. Bellman, *An Introduction to Inequalities*, The Mathematical Association of America, 1961.

[2] P. S. Bullen, D. S. Mitrinovics and P. M. Vasic, *Means and their Inequalities*, Reidel, Dordrecht, 1988.

[3] H. Chen, A unified elementary approach to some classical inequalities, *Internat. J. Math. Ed. Sci. Tech.*, **31** (2000) 289–292.

[4] J. Sandor and V. Szabo, On an inequality for the sum infimums of functions, *J. Math. Anal. Appl.*, **204** (1996) 646–654.

[5] W. L. Wang, Some inequality involving means and their converses, *J. Math. Anal. Appl.*, **238** (1999) 567–579.

3

Means Generated by an Integral

For a pair of distinct positive numbers, a and b, a number of different quantities $M(a,b)$ are known as *means*:

1. the arithmetic mean: $A(a,b) = (a+b)/2$
2. the geometric mean: $G(a,b) = \sqrt{ab}$
3. the harmonic mean: $H(a,b) = 2ab/(a+b)$
4. the logarithmic mean: $L(a,b) = (b-a)/(\ln b - \ln a)$
5. the Heronian mean: $N(a,b) = (a + \sqrt{ab} + b)/3$
6. the centroidal mean: $T(a,b) = 2(a^2 + ab + b^2)/3(a+b)$

These are all *positively homogeneous*, in the sense that

$$M(\lambda a, \lambda b) = \lambda M(a,b) \quad \text{for all } \lambda > 0,$$

and *symmetric*, in the sense that $M(a,b) = M(b,a)$. Moreover, all of the named means satisfy a third property, called *intermediacy*, as well:

$$\min(a,b) \leq M(a,b) \leq \max(a,b).$$

This inequality, which ensures that $M(a,b)$ lies between a and b, is the essential feature of means. In the following discussion we assume that means satisfy these three properties.

The arithmetic, geometric and harmonic means were studied by the early Greeks largely within the broader context of the theory of ratio and proportion, which in turn was motivated in part by its application to the theory of music. Pythagoras is generally regarded as the first Greek mathematician to investigate how these three means relate to musical intervals. Pappus is credited as the first to associate these three means with Figure 3.1. By manipulating the picture, one can conclude that $H(a,b) \leq G(a,b) \leq A(a,b)$.

Since then, both for their beauty and importance, the study of means has become a cornerstone of the theory of inequalities. There are numerous new proofs, refinements and

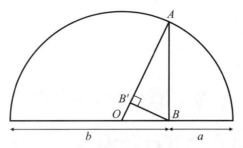

Figure 3.1. Pappus's construction: $OA = A(a,b), AB = G(a,b), BB' = H(a,b)$

variants on these means [4, pp. 21–48]. Howard Eves in [3] showed how some of these means occur in certain geometrical figures; for example, besides demonstrating three appearances of the harmonic mean, he also ranked most of the foregoing means in terms of the lengths of vertical segments of a trapezoid in Figure 3.2. Shannon Patton created an animated version of this figure using *Geometer's Sketchpad*. The detail is available at *Mathematical Magazine*'s website:

www.maa.org/pubs/mm_supplements/index.html.

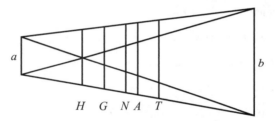

Figure 3.2. The means in a trapezoid

In this chapter, we show how to find functions that are associated to these means. To see this, we introduce [2]

$$f(t) = \frac{\int_a^b x^{t+1}\,dx}{\int_a^b x^t\,dx}. \tag{3.1}$$

This encompasses all the means at the beginning of the chapter: particular values of t in (3.1) give each of the means on our list. Indeed, it is easy to verify that

$$f(-3) = H(a,b), \quad f(-\tfrac{3}{2}) = G(a,b), \quad f(-1) = L(a,b),$$

$$f(-\tfrac{1}{2}) = N(a,b), \quad f(0) = A(a,b), \quad f(1) = T(a,b).$$

Moreover, upon showing that $f(t)$ is strictly increasing, as an immediate consequence we deduce that

$$H(a,b) \le G(a,b) \le L(a,b) \le N(a,b) \le A(a,b) \le T(a,b) \tag{3.2}$$

with equality if and only if $a = b$.

3. Means Generated by an Integral

To prove that $f(t)$ is strictly increasing for $0 < a < b$, we show that $f'(t) > 0$. By the quotient rule,

$$f'(t) = \frac{\int_a^b x^{t+1} \ln x \, dx \int_a^b x^t \, dx - \int_a^b x^{t+1} \, dx \int_a^b x^t \ln x \, dx}{(\int_a^b x^t \, dx)^2}. \tag{3.3}$$

Thus, it suffices to prove that the numerator for this quotient is positive. Since the bounds of the definite integrals are constants, the numerator for this quotient can be written as

$$\int_a^b x^{t+1} \ln x \, dx \int_a^b x^t \, dx - \int_a^b x^{t+1} \, dx \int_a^b x^t \ln x \, dx$$

$$= \int_a^b x^{t+1} \ln x \, dx \int_a^b y^t \, dy - \int_a^b y^{t+1} \, dx \int_a^b x^t \ln x \, dx$$

$$= \int_a^b \int_a^b x^t y^t (x - y) \ln x \, dx dy.$$

Substituting in a different manner, we can write the same numerator as

$$\int_a^b x^{t+1} \ln x \, dx \int_a^b x^t \, dx - \int_a^b x^{t+1} \, dx \int_a^b x^t \ln x \, dx$$

$$= \int_a^b y^{t+1} \ln y \, dy \int_a^b x^t \, dx - \int_a^b x^{t+1} \, dx \int_a^b y^t \ln y \, dy$$

$$= \int_a^b \int_a^b x^t y^t (y - x) \ln y \, dx dy.$$

Averaging the two equivalent expressions shows that the numerator equals

$$\frac{1}{2} \int_a^b \int_a^b x^t y^t (x - y)(\ln x - \ln y) \, dx dy > 0.$$

With expression (3.3), this implies that $f'(t) > 0$. Therefore, $f(t)$ is strictly increasing as desired.

Note that the above proof is based on double integrals. A related single variable proof for the positivity of the numerator starts with

$$F(u) = \int_a^u x^{t+1} \ln x \, dx \int_a^u x^t \, dx - \int_a^u x^{t+1} \, dx \int_a^u x^t \ln x \, dx.$$

Clearly $F(a) = 0$ and for $u \geq a$,

$$F'(u) = u^t \int_a^u x^t (u - x)(\ln u - \ln x) \, dx \geq 0,$$

so $F(b) \geq 0$ as long as $b > a$.

Next, notice that $f(t)$ is a continuous monotonic increasing mean. By comparing with the classic means given in (3.2), we can sharpen some inequalities. For example, since

$$f(-2) = \frac{ab(\ln b - \ln a)}{b - a} = \frac{G^2(a, b)}{L(a, b)},$$

the monotonicity of $f(t)$ implies that $f(-2)$ separates H and G which results in the following well-known interpolation inequality:

$$H(a,b) \leq \frac{G^2(a,b)}{L(a,b)} < G(a,b).$$

Furthermore, defining the power mean by

$$M_p(a,b) = \left(\frac{a^p + b^p}{2}\right)^{1/p},$$

we get the equalities

$$M_1(a,b) = A(a,b), M_0(a,b) = \lim_{p \to 0} M_p(a,b) = G(a,b), M_{-1}(a,b) = H(a,b).$$

Observing that

$$M_{1/2}(a,b) = \frac{1}{2}(G(a,b) + A(a,b)),$$

$$N(a,b) = \frac{1}{3}(G(a,b) + 2A(a,b)),$$

we challenge the reader to choose values of t to show that

$$L(a,b) < M_{1/3}(a,b) < \frac{1}{3}(2G(a,b) + A(a,b)) < M_{1/2}(a,b) < N(a,b) < M_{2/3}(a,b).$$

Now, we have seen a good deal of the power that $f(t)$ possesses. Naturally, one may rightly wonder where $f(t)$ came from and why $f(t)$ is a mean. Is there some larger principle afoot here that might reveal the reason for its effectiveness? There is maybe more than one answer to these questions, but our experience with $L(a,b)$ suggests a starting point. To see why $L(a,b)$ is a mean, applying the *mean value theorem* to $\ln x$ on $[a,b]$, we have

$$\frac{\ln b - \ln a}{b - a} = \frac{1}{c}$$

or

$$c = \frac{b-a}{\ln b - \ln a} = L(a,b),$$

where $a < c < b$. The c chosen is clearly positively homogeneous and symmetric.

Having this in mind, recall the *extended mean value theorem*, which asserts that

$$\frac{F'(c)}{G'(c)} = \frac{F(b) - F(a)}{G(b) - G(a)},$$

where $a < c < b$. Let $F(x) = x^\alpha, G(x) = x^\beta$ with $\alpha > \beta$. Then

$$c(a,b;\alpha,\beta) = \left(\frac{\beta(b^\alpha - a^\alpha)}{\alpha(b^\beta - a^\beta)}\right)^{1/(\alpha-\beta)} = \left(\frac{\int_a^b x^{\alpha-1}\,dx}{\int_a^b x^{\beta-1}\,dx}\right)^{1/(\alpha-\beta)}. \quad (3.4)$$

3. Means Generated by an Integral

Since $a < c(a, b; \alpha, \beta) < b$, and positively homogeneous and symmetry are also obvious, it follows that c constitutes a continuous mean. Thus,

$$f(t) = c(a, b; t+2, t+1),$$

is just one member of the general mean family of (3.4). In particular, setting $\beta = 1$ in (3.4), we obtain the Stolarsky mean

$$S_\alpha(a, b) = \left(\frac{b^\alpha - a^\alpha}{\alpha(b-a)} \right)^{1/(\alpha-1)}, \quad (\alpha \neq 0, 1)$$

which is monotonic increasing in α. Note that

$$S_{-1}(a, b) = G(a, b), \quad S_2(a, b) = A(a, b).$$

The limiting cases give

$$S_0(a, b) = \lim_{\alpha \to 0} S_\alpha(a, b) = L(a, b)$$

and

$$S_1(a, b) = \lim_{\alpha \to 1} S_\alpha(a, b) = e^{-1}(b^b/a^a)^{1/(b-a)} = I(a, b).$$

Here $I(a, b)$ is called the *identric* mean and also arises from the Putnam Problem 1979-B2. Like the Heronian mean, the identric mean is trapped between L and A as well. Thus

$$L(a, b) \leq I(a, b) \leq A(a, b).$$

It is also interesting to see that

$$\ln\left(\frac{I(a, b)}{G(a, b)} \right) = \frac{A(a, b) - L(a, b)}{L(a, b)}.$$

Finally, to gain a more general perspective on the topic of means, we state a set of axioms, which, if satisfied by a class of functions, entitle those functions to be called "means". The axioms will be chosen by abstracting the most important properties of $f(t)$ in (3.1).

We say function $F(a, b)$ defines a mean for $a, b > 0$ when

1. $F(a, b)$ is continuous in each variable,

2. $F(a, b)$ is strictly increasing in each variable,

3. $F(a, b) = F(b, a)$,

4. $F(ta, tb) = tF(a, b)$ for all $t > 0$,

5. $a < F(a, b) < b$ for $0 < a < b$.

We leave to the reader to show that a necessary and sufficient condition for a mean is that for $0 < a \leq b$,

$$F(a, b) = b\, f(a/b)$$

where $f(s)$ is positive, continuous and strictly increasing for $0 < s \leq 1$, and satisfies

$$s < f(s) \leq 1, \quad \text{for } 0 < s < 1.$$

In particular, if ϕ is a positive continuous function on $(0, 1]$ and we let

$$f(s) = f_\phi(s) = \frac{\int_s^1 x\phi(x)\, dx}{\int_s^1 \phi(x)\, dx},$$

then f satisfies these conditions and

$$F(a, b) = bf(a/b) = \frac{\int_a^b x\phi(x/b)\, dx}{\int_a^b \phi(x/b)\, dx}$$

defines a mean. Moreover, if ψ is positive continuous on $(0, 1]$ and ψ/ϕ is strictly increasing, then $f_\phi < f_\psi$ on $(0, 1)$.

The means generated by (3.1) craft continuous analogs to some discrete means, thereby providing a powerful method to study the means systematically. As the chapters progress, we will often translate between discrete and continuous mathematics in search of the effective approach to solve problems.

Exercises

1. Let $f(t)$ be defined by (3.1). Find another proof of its monotonicity.

2. **Putnam Problem 1957-B3.** For $f(x)$ a positive, monotonic decreasing function defined in $0 \leq x \leq 1$ prove that

$$\frac{\int_0^1 xf^2(x)\, dx}{\int_0^1 xf(x)\, dx} \leq \frac{\int_0^1 f^2(x)\, dx}{\int_0^1 f(x)\, dx}.$$

3. **Putnam Problem 1979-B2.** Let $0 < a < b$. Evaluate

$$\lim_{t \to 0} \left\{ \int_0^1 [bx + a(1-x)]^t\, dx \right\}^{1/t}.$$

4. **Slope mean:** Define $S(a, b)$ as the slope of the line which bisects the angle formed by the lines $y = ax$ and $y = bx$. Find an explicit form for $S(a, b)$ and then prove that $S(a, b)$ is a mean and satisfies the inequality

$$H(a, b) \leq S(a, b) \leq A(a, b).$$

5. We see that $M_{2/3}(a, b) \leq I(a, b)$. Is it possible to find $2/3 < t < 1$ such that $I(a, b) < M_t(a, b)$?

6. For $-1 < \alpha < 1/2$ or $2 < \alpha$, show that

$$S_\alpha(a, b) \leq M_{(\alpha+1)/3}(a, b).$$

3. Means Generated by an Integral

7. Show that
$$\frac{1}{3}(2A(a,b) + G(a,b)) \leq I(a,b) \leq \frac{2}{e} A(a,b) + \left(1 - \frac{2}{e}\right) G(a,b).$$

8. **Monthly Problem 11347** [2008, 167]. Determine all ordered 4-tuples $(\alpha, \beta, \gamma, \delta)$ of positive numbers with $\alpha > \beta$ and $\gamma > \delta$ such that for all distinct positive a and b,
$$I > \frac{\alpha A + \beta G}{\alpha + \beta} > (A^\gamma G^\delta)^{1/(\gamma+\delta)} > \sqrt{AG}.$$

9. Let agm(a,b) be the *Gauss arithmetic-geometric mean*. Prove that
$$L(a,b) \leq \text{agm}(a,b) \leq \frac{A(a,b) + G(a,b)}{2} \leq N(a,b).$$

10. Let $0 < a < b$ and $x, y \geq 1$. Prove that
$$\frac{b^{x+y} - a^{x+y}}{b^x - a^x} \geq \frac{x+y}{x} A(a,b)^y$$
where equality holds if and only if $x = y = 1$.

11. Let $0 < a < b$. Applying the midpoint and trapezoidal approximations to $\int_{\ln a}^{\ln b} e^x \, dx$ (see Figure 3.3), prove that $G(a,b) < L(a,b) < A(a,b)$. Furthermore, prove that $L(a,b) < M_{1/3}(a,b)$ via *Simpson's Three-Eighths Rule* [1, p. 195].

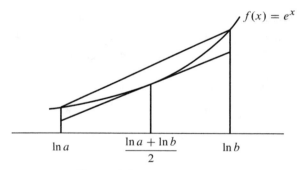

Figure 3.3. Areas in the figure

12. (**Some series involving means**).Let $x = (b-a)/(b+a)$. Prove that
    (a) $\ln(A/G) = \sum_{k=1}^{\infty} \frac{1}{2k} x^{2k}$,
    (b) $\ln(I/G) = \sum_{k=1}^{\infty} \frac{1}{2k+1} x^{2k}$,
    (c) $\ln(A/I) = \sum_{k=1}^{\infty} \frac{1}{2k(2k+1)} x^{2k}$.

13. **Open Problem**. Notice that
$$\ln I(a,b) = \int_0^1 \ln[ta + (1-t)b] \, dt.$$

Using various numerical methods to approximate the integral enables us to obtain some new inequalities. For example,

(a) For $a \neq b$, show that
$$\left(A^2(a,b)G(a,b)\right)^{1/3} < I(a,b),$$

(b) For $a \neq b$, show that
$$I^2(a,b) < \frac{2}{3} A^2(a,b) + \frac{1}{3} G^2(a,b).$$

Along these lines, one may discover more new interesting inequalities.

14. **Open Problem**. The logarithmic mean, somewhat surprisingly, has applications to the economic index analysis. Further applications, however, require a natural generalization to a logarithmic mean of n variables. Try to define a logarithmic mean L of n variables which satisfies the inequality $G \leq L \leq A$. Since the form of $L(a,b)$ does not suggest a reasonable generalization, you may start with

$$\frac{1}{L(a,b)} = \int_0^1 \frac{dx}{xb + (1-x)a}$$

as Stolarsky suggested in [5]. Similarly, try to define an identric mean I of n variables such that $G \leq L \leq I \leq A$.

References

[1] R. L. Burden and J. D. Faires, *Numerical Analysis*, 8th edition, Brooks/Cole, Belmont, 2005.

[2] H. Chen, Means generated by an integral, *Math. Magazine*, **78** (2005) 397–399.

[3] H. Eves, Means appearing in geometric figures, *Math. Magazine*, **76** (2003) 292–294.

[4] D. S. Mitrinovic, J. E. Pecaric and A. E. Fink, *Classical and New Inequalities in Analysis*, Kluwer Academic Publishers, Boston, 1993.

[5] K. Stolarsky, Generalizations of the logarithmic means, *Math. Magazine*, **48** (1975) 87-92.

4

The L'Hôpital Monotone Rule

In elementary calculus, we learn that if $f(x)$ is continuous and has a positive derivative on (a, b) then $f(x)$ is increasing on (a, b). However, if we are trying to prove the monotonicity of

$$f(x) = \frac{x^p - 1}{x^q - 1}, \quad (p > q > 0, x \geq 1) \tag{4.1}$$

in this way, we see the numerator of the derivative is

$$(p-q)x^{p+q-1} - px^{p-1} + qx^{q-1},$$

where the positivity is not immediately evident. In general, the derivative of a quotient is quite messy and the process to show monotonicity can be tedious. In this chapter, we introduce a general method for proving the monotonicity of a large class of quotients. Because of the similarity to the hypotheses to those of l'Hôpital's rule, we refer to the following theorem as the "L'Hôpital Monotone Rule."

L'Hôpital's Monotone Rule (LMR) Let $f, g : [a, b] \longrightarrow \mathbb{R}$ be continuous functions that are differentiable on (a, b) with $g' \neq 0$ on (a, b). If f'/g' is increasing (decreasing) on (a, b), then the functions

$$\frac{f(x) - f(b)}{g(x) - g(b)} \quad \text{and} \quad \frac{f(x) - f(a)}{g(x) - g(a)}$$

are likewise increasing (decreasing) on (a, b).

Proof. We may assume that $g'(x) > 0$ and $f'(x)/g'(x)$ is increasing for all $x \in (a, b)$. By the mean value theorem, for a given $x \in (a, b)$ there exists $c \in (a, x)$ such that

$$\frac{f(x) - f(a)}{g(x) - g(a)} = \frac{f'(c)}{g'(c)} \leq \frac{f'(x)}{g'(x)},$$

and so

$$f'(x)\big(g(x) - g(a)\big) - g'(x)\big(f(x) - f(a)\big) \geq 0.$$

Therefore,
$$\left(\frac{f(x)-f(a)}{g(x)-g(a)}\right)' = \frac{f'(x)(g(x)-g(a)) - g'(x)(f(x)-f(a))}{(g(x)-g(a))^2} \geq 0.$$

This shows that $(f(x)-f(a))/(g(x)-g(a))$ is increasing on (a,b) as desired. Along the same lines, we can show that $(f(x)-f(b))/(g(x)-g(b))$ is increasing.

Here we assume that a and b are finite. Along the same path, the rule can be extended easily to the case where a or b is infinity. LMR first appeared in Gromov's work [1] for volume estimation in differential geometry. Since then, LMR and its variants have been used in approximation theory, quasi-conformal theory and probability (see [3] and the references cited therein). But, for most analysis students, LMR is not as well known as it should be. To popularize and promote this monotonicity rule, we present four examples within the realm of elementary analysis. The reader can see the wide applicability of LMR and find more applications in the exercises.

Our first example is to confirm the monotonicity of (4.1) by using LMR. To see this, let $f(x) = x^p - 1$ and $g(x) = x^q - 1$. Consider

$$G(x) = \begin{cases} f(x)/g(x), & \text{if } x \neq 1, \\ p/q, & \text{if } x = 1. \end{cases}$$

We have $f'(x)/g'(x) = (p/q)x^{p-q}$, which is increasing on $(1, \infty)$ as long as $p > q > 0$. Hence, by LMR, $G(x)$ is increasing on $(1, \infty)$. In particular, $G(x) > G(1) = p/q$. This yields the following bonus inequality:

$$\frac{x^p - 1}{p} > \frac{x^q - 1}{q}, \quad \text{for } x > 1, p > q > 0.$$

Our second example comes from Wilker's inequality, which appears in [5]. Wilker asked for a proof that

$$\left(\frac{\sin x}{x}\right)^2 + \frac{\tan x}{x} > 2, \quad (0 < x < \frac{\pi}{2}). \tag{4.2}$$

Sumner et. al. in [4] provided an elementary but long solution. Based on LMR, we give a short alternative solution for (4.2). Our solution is almost algorithmic in nature. Set

$$F(x) = \begin{cases} \left(\frac{\sin x}{x}\right)^2 + \frac{\tan x}{x}, & \text{if } x \neq 0, \\ 2, & \text{if } x = 0. \end{cases}$$

Let $f(x) = \sin^2 x + x \tan x$, $g(x) = x^2$. Now

$$f'(x) = 2 \sin x \cos x + x \sec^2 x + \tan x$$

and

$$\frac{f''(x)}{g''(x)} = \cos^2 x - \sin^2 x + \sec^2 x + x \tan x \sec^2 x.$$

4. The L'Hôpital Monotone Rule

Since

$$\left(\frac{f''(x)}{g''(x)}\right)' = 3\tan x \sec^2 x (1 - \cos^4 x) + \sin x \left(\frac{x}{\sin x}\frac{1}{\cos^4 x} - \cos x\right) + 2x \tan^2 x \sec^2 x$$
$$> 0,$$

$f''(x)/g''(x)$ is increasing on $(0, \pi/2)$. Thus, using LMR twice and noticing that $f(0) = f'(0) = g(0) = g'(0) = 0$, we deduce that $f'(x)/g'(x)$ and so $F(x) = f(x)/g(x)$ is increasing on $(0, \pi/2)$.

Our third example is related to the well-known Jordan's inequality:

$$\frac{2}{\pi}x \leq \sin x \leq x, \quad (0 \leq x \leq \frac{\pi}{2}).$$

Now, by using LMR, we improve Jordan's inequality as follows.

If $x \in [0, \pi/2]$, then

$$\frac{2}{\pi}x + \frac{\pi-2}{\pi^2}x(\pi - 2x) \leq \sin x \leq \frac{2}{\pi}x + \frac{2}{\pi^2}x(\pi - 2x),$$

$$\frac{2}{\pi}x + \frac{1}{\pi^3}x(\pi^2 - 4x^2) \leq \sin x \leq \frac{2}{\pi}x + \frac{\pi-2}{\pi^2}x(\pi^2 - 4x^2),$$

$$\frac{2}{\pi}x + \frac{1}{2\pi^5}x(\pi^4 - 16x^4) \leq \sin x \leq \frac{2}{\pi}x + \frac{\pi-2}{\pi^5}x(\pi^4 - 16x^4),$$

where the coefficients are all the best possible.

We show the third inequality in the list. The other two can be proven similarly. It is easy to see that the third inequality is equivalent to

$$\frac{1}{2\pi^5} \leq \frac{\frac{\sin x}{x} - \frac{2}{\pi}}{\pi^4 - 16x^4} \leq \frac{\pi - 2}{\pi^5}. \tag{4.3}$$

Let $f_1(x) = \sin x/x$, $g_1(x) = -16x^4$, $f_2(x) = \sin x - x\cos x$, $g_2(x) = x^5$. Then

$$\frac{f_1'(x)}{g_1'(x)} = \frac{1}{64}\frac{\sin x - x\cos x}{x^5} = \frac{1}{64}\frac{f_2(x)}{g_2(x)},$$

$$\frac{f_2'(x)}{g_2'(x)} = \frac{1}{5}\frac{\sin x}{x^3}.$$

Since $\sin x/x^3$ is decreasing on $(0, \pi/2)$, we have $f_2'(x)/g_2'(x)$ is decreasing on $(0, \pi/2)$. By LMR, $f_2(x)/g_2(x)$ is decreasing on $(0, \pi/2)$ and so is $f_1'(x)/g_1'(x)$. Applying LMR once more, we have

$$\frac{f_1(x) - f_1(\pi/2)}{g_1(x) - g_1(\pi/2)} = \frac{\frac{\sin x}{x} - \frac{\pi}{2}}{\pi^4 - 16x^4}$$

decreasing on $(0, \pi/2)$. Thus, the desired inequality (4.3) follows from

$$\lim_{x \to 0} \frac{\frac{\sin x}{x} - \frac{\pi}{2}}{\pi^4 - 16x^4} = \frac{\pi - 2}{\pi^5}, \quad \lim_{x \to \pi/2} \frac{\frac{\sin x}{x} - \frac{\pi}{2}}{\pi^4 - 16x^4} = \frac{1}{2\pi^5}.$$

Our final example is Monthly Problem 11369 [2], which was recently posed by Donald Knuth: Prove that for all real t, and all $\alpha \geq 2$,

$$e^{\alpha t} + e^{-\alpha t} - 2 \leq (e^t + e^{-t})^\alpha - 2^\alpha. \tag{4.4}$$

First, observe that the inequality is invariant when t is replaced by $-t$. Hence, it is sufficient to show the inequality holds for $t \geq 0$.

Next, set $x = e^t$. Hence (4.4) is equivalent to
$$\frac{(x^2+1)^\alpha - x^{2\alpha} - 1}{x^\alpha} \geq 2^\alpha - 2. \tag{4.5}$$

To apply LMR, let $f(x) = (x^2+1)^\alpha - x^{2\alpha} - 1$ and $g(x) = x^\alpha$. Then
$$\frac{f'(x)}{g'(x)} = \frac{2x(x^2+1)^{\alpha-1} - 2x^{2\alpha-1}}{x^{\alpha-1}}.$$

Define
$$F(x) = \frac{f'(x)}{g'(x)} = 2x\left(x + \frac{1}{x}\right)^{\alpha-1} - 2x^\alpha.$$

To show that $f'(x)/g'(x)$ is increasing for $x > 1$, it suffices to prove that $F'(x) \geq 0$ for $x > 1$. Indeed,
$$F'(x) = 2\left(x + \frac{1}{x}\right)^{\alpha-1}\left[\alpha - \frac{2(\alpha-1)}{x^2+1}\right] - 2\alpha x^{\alpha-1}.$$

Using Bernoulli's inequality
$$(1+t)^p \geq 1 + pt \qquad \text{for } t \geq 0, \, p \geq 1,$$
which can be proved via LMR by considering $(1+t)^p/(1+pt)$ as well, we have
$$\left(x + \frac{1}{x}\right)^{\alpha-1} = x^{\alpha-1}\left(1 + \frac{1}{x^2}\right)^{\alpha-1} \geq x^{\alpha-1} + (\alpha-1)x^{\alpha-3}.$$

Therefore, for $x > 1$ and $\alpha \geq 2$,
$$F'(x) \geq \frac{2(\alpha-1)(\alpha-2)x^{\alpha-3}(x^2-1)}{x^2+1} \geq 0.$$

Now, LMR deduces that
$$\frac{f(x) - f(1)}{g(x) - g(1)} = \frac{(x^2+1)^\alpha - x^{2\alpha} - 2^\alpha + 1}{x^\alpha - 1}$$
is increasing for $x > 1$. In particular, applying l'Hôpital's rule yields
$$\lim_{x \to 1} \frac{f(x) - f(1)}{g(x) - g(1)} = \lim_{x \to 1} \frac{(x^2+1)^\alpha - x^{2\alpha} - 2^\alpha + 1}{x^\alpha - 1} = 2^\alpha - 2.$$

Thus,
$$\frac{f(x) - f(1)}{g(x) - g(1)} = \frac{(x^2+1)^\alpha - x^{2\alpha} - 2^\alpha + 1}{x^\alpha - 1} \geq 2^\alpha - 2.$$

This implies the desired (4.5) and therefore proves (4.4).

To end this chapter, we give an example to show that $f'(x)/g'(x)$ is not monotonic but $f(x)/g(x)$ is still monotonic. Consider
$$f(x) = x \int_0^x (1 + \cos(1/t))\,dt, \quad g(x) = x.$$

Then $f(x)/g(x)$ is increasing on $(0, \infty)$, whereas $f'(x)/g'(x)$ is not monotonic on $(0, \infty)$. This also shows that the converse of LMR is false.

4. The L'Hôpital Monotone Rule

Exercises

1. For $x > -1$, prove that
$$F(x) = \frac{x \ln(1+x)}{x - \ln(1+x)}$$
is increasing.

2. Prove that $f(x) = (1+x)^{1/x}$ is decreasing but $g(x) = (1+x)^{1+1/x}$ is increasing on $(0, \infty)$.

3. For $0 < x < \pi/2$, prove that
   (a) $\left(\frac{\sin x}{x}\right)^3 > \cos x$.
   (b) $\frac{2x}{\sin x} + \frac{x}{\tan x} > 3$.

   Remark. Using $t^2 \geq 2t - 1$ with $t = x/\sin x$ in the second inequality yields
   $$\left(\frac{x}{\sin x}\right)^2 + \frac{x}{\tan x} > 2.$$

4. **Monthly Problem 11009** [2003, 341; 2005, 92–93]. Let $a > b \geq c > d > 0$. Prove that
$$f(x) = \frac{a^x - b^x}{c^x - d^x}$$
is increasing and convex for all x.

   Remark. One may show that $f(x)$ is increasing based on LMR. Without loss of generality, assume that $d = 1$, otherwise divide the ratio by d^x. Now, let $y = c^x, \alpha = \ln b / \ln c, \beta = \ln a / \ln c$. Then $\beta > \alpha \geq 1$. One can rewrite $f(x)$ as
$$F(y) := \frac{y^\beta - y^\alpha}{y - 1} \qquad y \in (0, 1) \cup (1, \infty).$$

5. In Wilker's inequality, prove that
$$\frac{\left(\frac{\sin x}{x}\right)^2 + \frac{\tan x}{x} - 2}{x^3 \tan x} \geq \frac{16}{\pi^4}, \quad (0 < x < \pi/2).$$

6. Kober's inequality claims that $\cos x \geq 1 - 2x/\pi$ for $0 < x < \pi/2$. Prove the following refinement:
$$1 - \frac{2}{\pi}x + \frac{\pi - 2}{\pi^2}x(\pi - 2x) \leq \cos x \leq 1 - \frac{2}{\pi}x + \frac{2}{\pi^2}x(\pi - 2x).$$

7. For $0 < x < 1$, prove that
$$\frac{4}{\pi}\frac{x}{1-x^2} < \tan\left(\frac{\pi x}{2}\right) < \frac{\pi}{2}\frac{x}{1-x^2}.$$

8. **Monthly Problem 11308** [2007, 640; 2009, 183–184]. Let n be a positive integer. For $1 \leq i \leq n$, let x_i be a real number in $(0, \pi/2)$ and a_i be a real number in $[1, \infty)$. Prove that

$$\prod_{i=1}^{n} \left(\frac{x_i}{\sin x_i}\right)^{2a_i} + \prod_{i=1}^{n} \left(\frac{x_i}{\tan x_i}\right)^{a_i} > 2.$$

9. For $\pi/n \leq x \leq \pi - \pi/n$, prove that

$$\frac{\sin nx}{n \sin x} \geq -\frac{1}{3}.$$

10. Prove that
    (a) $\sum_{k=1}^{n} \frac{\sin kx}{k} > 0$, for all $0 < x < \pi, n = 1, 2, \ldots$.
    (b) $\sum_{k=1}^{n} \frac{\sin kx}{k} \geq \frac{1}{6} \sin x$, for all $\pi/n < x < \pi - \pi/n, n = 1, 2, \ldots$.

References

[1] M. Gromov and M. Taylor, Finite propagation speed, kernel estimates for functions of the Laplace operator, and the geometry of complete Riemannian manifolds, *J. Diff. Geom.*, **17** (1982) 15–53.

[2] D. Knuth, Problem 11369, *Amer. Math. Monthly*, **115** (2008) 567.

[3] I. Pinelis, On L'Hôpital-type rules for monotonicity, *J. Inequal. Pure Appl. Math.* **7** (2006), Article 40. Available at jipam.vu.edu.au/article.php?sid=657

[4] J. S. Sumner, A. A. Jagers and J. Anglesio, Inequalities involving trigonometric functions, *Amer. Math. Monthly*, **98** (1991) 264–267.

[5] J. B. Wilker, Problem E3306, *Amer. Math. Monthly*, **96** (1989) 55.

5

Trigonometric Identities via Complex Numbers

> *The shortest path between two truths in the real domain passes through the complex domain.*
> — J. Hadamard

Trigonometric identities often arise in a variety of branches of mathematics, including geometry, analysis, number theory and applied mathematics. The classical standard tables [4] and [5] have collected many of these identities, such as

$$\prod_{k=1}^{n-1} \sin\left(\frac{k\pi}{n}\right) = \frac{n}{2^{n-1}}. \tag{5.1}$$

However, the proofs for these identities are widely scattered throughout books and journals. In this chapter, we present a comprehensive study on proving trigonometric identities via complex numbers. To keep the chapter to a reasonable page limit, we only examine several families of identities in [4, 5]. But, we emphasize that the methods used here can be applied to prove many identities. As applications of these identities, we explicitly evaluate the *Poisson integral* as well as establish some more sophisticated finite trigonometric identities of the sort

$$\sum_{k=1}^{n-1} \csc^2\left(\frac{k\pi}{n}\right) = \frac{(n-1)(n+1)}{3}. \tag{5.2}$$

5.1 A Primer of complex numbers

Historically, complex numbers were first introduced by Cardano in the course of solving quadratic and cubic equations. He noticed that if the "new numbers" $-1 \pm \sqrt{-1}$ were treated as ordinary numbers with the added rule that $\sqrt{-1} \cdot \sqrt{-1} = -1$, they did solve the equation $x^2 + 2x + 2 = 0$. However, for a long period of time, it was felt that no meaning could actually be assigned to $i = \sqrt{-1}$ and it was therefore termed "imaginary". In 1748, Euler discovered the amazing identity

$$e^{i\theta} = \cos\theta + i\sin\theta. \tag{5.3}$$

Such a close connection between i and trigonometric functions was quite startling and perhaps indicated that some meaning should be attached to these "imaginary" numbers.

Finally, the work of Gauss and others clarified the meaning of complex numbers. In the complex domain, Gauss eventually established the *fundamental theorem of algebra*, which claims that any polynomial equation has at least one solution. In the middle of the nineteenth century, starting to extend the methods of the differential and integral calculus by including complex variables, Cauchy and others founded the branch of mathematics called the *Theory of Complex Analysis*. It has been acclaimed as one of the most harmonious theories in the abstract sciences.

In the following, we review some selected basic facts on complex numbers.

5.1.1 Multiplication and De Moivre's Formula

Recall the geometric representation of a complex number. Let θ denote the angle between the vector $z = x + i y$ and the positive real axis, where $0 \leq \theta < 2\pi$. The angle θ is called the *argument* and is denoted $\theta = \arg z$. Thus, we have z in a *polar coordinate representation*:

$$z = r(\cos\theta + i\,\sin\theta), \tag{5.4}$$

where

$$r = |z| = \sqrt{x^2 + y^2}, \quad \tan\theta = y/x.$$

(5.4) gives us a geometric method for performing complex multiplication: for any complex numbers z_1 and z_2,

$$|z_1 z_2| = |z_1||z_2|$$

and

$$\arg(z_1 z_2) = \arg z_1 + \arg z_2 \quad (\text{mod } 2\pi). \tag{5.5}$$

In general, this leads to the well-known *De Moivre's formula*: for a positive integer n,

$$(\cos\theta + i\,\sin\theta)^n = \cos n\theta + i\,\sin n\theta. \tag{5.6}$$

Now, appealing to Euler's formula (5.3), we get the following two popular identities:

$$1 + e^{ix} = 1 + (e^{ix/2})^2 = 2\cos^2(x/2) + 2i\,\sin(x/2)\cos(x/2) = 2\cos(x/2)e^{ix/2} \tag{5.7}$$

and

$$1 - e^{ix} = 1 - (e^{ix/2})^2 = 2\sin^2(x/2) - 2i\,\sin(x/2)\cos(x/2) = -2i\,\sin(x/2)e^{ix/2}. \tag{5.8}$$

5.1.2 Roots of Unity

The solutions of the equation $z^n - 1 = 0$ are called the *n*th *roots of unity*. De Moivre's formula can be reversed to find all roots of unity, namely

$$\{1, \omega, \omega^2, \ldots, \omega^{n-1}\},$$

where $\omega = e^{i(2\pi/n)}$ is called the *n*th *primitive root of unity*. Geometrically, these numbers are the vertices of a regular n-gon inscribed in the unit circle $|z| = 1$, and have many nice properties that interconnect algebra, geometry, combinatorics and number theory. Since

$$z^n - 1 = (z - 1)(z^{n-1} + z^{n-2} + \cdots + 1), \tag{5.9}$$

5. Trigonometric Identities via Complex Numbers

clearly, the factor $z - 1$ accounts for the root 1, and so the other roots satisfy

$$z^{n-1} + z^{n-2} + \cdots + 1 = 0, \tag{5.10}$$

which yields

$$z^{n-1} + z^{n-2} + \cdots + 1 = (z - \omega) \cdots (z - \omega^{n-1}). \tag{5.11}$$

Now with the stage appropriately set, we show how several families of trigonometric identities that appeared in [4, 5] fall into place.

5.2 Finite Product Identities

We begin with identity (5.11). Setting $z = 1$ and then taking modulus gives

$$n = \prod_{k=1}^{n-1} |1 - \omega^k|. \tag{5.12}$$

In the complex plane, this shows that the product of the distance from one vertex on the unit circle to other vertices is the constant n.

In view of (5.8),

$$1 - \omega^k = 1 - e^{i(2k\pi/n)} = -2i \sin\left(\frac{k\pi}{n}\right) e^{ik\pi/n},$$

we see that (5.1) follows from (5.12) immediately. Next, for $z = -1$, (5.11) gives

$$\frac{1}{2}[1 - (-1)^n] = (-1)^{n-1} \prod_{k=1}^{n-1} (1 + \omega^k).$$

By (5.7), we have

$$1 + \omega^k = 1 + e^{i(2k\pi/n)} = 2\cos\left(\frac{k\pi}{n}\right) e^{ik\pi/n},$$

and so

$$\prod_{k=1}^{n-1} \left|\cos\left(\frac{k\pi}{n}\right)\right| = \frac{1 - (-1)^n}{2^n}. \tag{5.13}$$

Since

$$\sin\left(\frac{k\pi}{n}\right) = \sin\left(\frac{(n-k)\pi}{n}\right), \quad \cos\left(\frac{k\pi}{n}\right) = -\cos\left(\frac{(n-k)\pi}{n}\right),$$

appealing to (5.12) and (5.13), we find that

$$\prod_{k=1}^{n-1} \sin\left(\frac{k\pi}{2n}\right) = \frac{\sqrt{n}}{2^{n-1}},$$

$$\prod_{k=1}^{n} \sin\left(\frac{k\pi}{2n+1}\right) = \frac{\sqrt{2n+1}}{2^n},$$

$$\prod_{k=1}^{n} \cos\left(\frac{k\pi}{2n+1}\right) = \frac{1}{2^n}.$$

Next, notice that

$$\prod_{k=0}^{n-1} \sin\left(x + \frac{k\pi}{n}\right) = \prod_{k=0}^{n-1} \frac{1}{2i}\left(e^{i(x+k\pi/n)} - e^{-i(x+k\pi/n)}\right)$$

$$= \frac{1}{(2i)^n} \prod_{k=0}^{n-1} e^{ik\pi/n - ix}\left(e^{2xi} - e^{i(-2k\pi/n)}\right)$$

$$= \frac{1}{(2i)^n} e^{i(n-1)\pi/2 - inx} \prod_{k=0}^{n-1}\left(e^{2xi} - e^{i(-2k\pi/n)}\right).$$

Since $\omega^{-k} = \omega^{n-k}$, setting $z = e^{2xi}$ in (5.9) we have

$$\prod_{k=0}^{n-1}\left(e^{2xi} - e^{i(-2k\pi/n)}\right) = e^{2nxi} - 1,$$

so that

$$\prod_{k=0}^{n-1} \sin\left(x + \frac{k\pi}{n}\right) = \frac{1}{(2i)^n}(e^{\pi i/2})^{n-1}(e^{nxi} - e^{-nxi}) = \frac{1}{2^{n-1}} \sin nx$$

or equivalently

$$\sin nx = 2^{n-1} \prod_{k=0}^{n-1} \sin\left(x + \frac{k\pi}{n}\right). \tag{5.14}$$

Replacing x by $x + \pi/2$ in (5.14) yields

$$\sin nx = (-1)^{n/2} 2^{n-1} \prod_{k=0}^{n-1} \cos\left(x + \frac{k\pi}{n}\right), \quad \text{when } n \text{ is even,}$$

$$\cos nx = (-1)^{(n-1)/2} 2^{n-1} \prod_{k=0}^{n-1} \cos\left(x + \frac{k\pi}{n}\right), \quad \text{when } n \text{ is odd.}$$

Finally, notice that the equation $z^{2n} - 1 = 0$ has $2n$ roots evenly spaced in angle around the unit circle, with a separation of π/n radians. Setting $\xi = \exp(i\pi/n)$, the factorization of $z^{2n} - 1$ yields

$$z^{2n} - 1 = \prod_{k=-n}^{n-1}(z - \xi^k).$$

Combining the conjugate factors gives

$$z^{2n} - 1 = (z^2 - 1)\prod_{k=1}^{n-1}\left(z^2 - 2z\cos\left(\frac{k\pi}{n}\right) + 1\right).$$

Dividing by z^n on both sides yields

$$z^n - z^{-n} = (z - z^{-1})\prod_{k=1}^{n-1}\left(z - 2\cos\left(\frac{k\pi}{n}\right) + z^{-1}\right).$$

5. Trigonometric Identities via Complex Numbers

Setting $z = e^{ix}$ and appealing to De Moivre's formula, we obtain

$$\frac{\sin nx}{\sin x} = 2^{n-1} \prod_{k=1}^{n-1} \left(\cos x - \cos\left(\frac{k\pi}{n}\right) \right). \tag{5.15}$$

Similarly, using the factorizations of $z^{2n+1} - 1$ and $z^{2n} + 1$ respectively, we can establish other product identities:

$$(z-1) \prod_{k=1}^{n} \left(z^2 - 2z \cos\left(\frac{2k\pi}{2n+1}\right) + 1 \right) = z^{2n+1} - 1,$$

$$(z+1) \prod_{k=1}^{n} \left(z^2 + 2z \cos\left(\frac{2k\pi}{2n+1}\right) + 1 \right) = z^{2n+1} - 1,$$

$$\prod_{k=0}^{n-1} \left(z^2 - 2z \cos\left(\frac{(2k+1)\pi}{2n}\right) + 1 \right) = z^{2n} + 1.$$

5.3 Finite Summation Identities

We have illustrated how to obtain the finite product identities from the factorizations. In the following, we show how to establish the finite summation identities based on partial fractions. Since $\omega^{-k} = \omega^{n-k}$, we see that

$$\{1, \omega^{-1}, \omega^{-2}, \ldots, \omega^{-(n-1)}\}$$

are also roots of $z^n - 1$. Thus,

$$1 - z^n = \prod_{k=0}^{n-1} (1 - z\omega^k), \tag{5.16}$$

from which the partial fraction decomposition of $1 - z^n$ has the form of

$$\frac{1}{1 - z^n} = \sum_{k=0}^{n-1} \frac{A_k}{1 - z\omega^k},$$

where A_k are constants to be determined. Multiplying by $1 - z^n$ on both sides yields

$$\sum_{k=0}^{n-1} A_k \frac{1 - z^n}{1 - z\omega^k} = 1.$$

Since

$$\lim_{z \to \omega^{-k}} \frac{1 - z^n}{1 - z\omega^i} = \begin{cases} 0, & \text{if } i \neq k, \\ n, & \text{if } i = k, \end{cases}$$

we obtain $A_k = 1/n$ and so

$$\frac{1}{1 - z^n} = \frac{1}{n} \sum_{k=0}^{n-1} \frac{1}{1 - z\omega^k}.$$

or equivalently
$$\frac{n}{1-z^n} = \sum_{k=0}^{n-1} \frac{1}{1-z\omega^k}. \tag{5.17}$$

In view of the identity
$$\frac{1}{a+ib} = \frac{a-ib}{(a+ib)(a-ib)} = \frac{a-ib}{a^2+b^2},$$

setting $z = xe^{i\alpha}$ in (5.17) and then extracting the real part gives
$$\frac{n(1-x^n \cos n\alpha)}{1-2x^n \cos n\alpha + x^{2n}} = \sum_{k=0}^{n-1} \frac{1-x\cos(\alpha+2k\pi/n)}{1-2x\cos(\alpha+2k\pi/n)+x^2}. \tag{5.18}$$

Replacing x by $1/x$ in (5.18) yields
$$\frac{nx^n(x^n - \cos n\alpha)}{1-2x^n \cos n\alpha + x^{2n}} = \sum_{k=0}^{n-1} \frac{x^2 - x\cos(\alpha+2k\pi/n)}{1-2x\cos(\alpha+2k\pi/n)+x^2}. \tag{5.19}$$

Now, subtracting (5.19) from (5.18) we get
$$\frac{n(1-x^{2n})}{1-2x^n \cos n\alpha + x^{2n}} = \sum_{k=0}^{n-1} \frac{1-x^2}{1-2x\cos(\alpha+2k\pi/n)+x^2}. \tag{5.20}$$

Alternately, adding (5.19) and (5.20) we have
$$\frac{n(1-x^n \cos n\alpha)}{1-2x^n \cos n\alpha + x^{2n}} = \sum_{k=0}^{n-1} \frac{1-x\cos(\alpha+2k\pi/n)}{1-2x\cos(\alpha+2k\pi/n)+x^2}. \tag{5.21}$$

For $\alpha = 0$, we may specify (5.20) and (5.21), respectively, as
$$\sum_{k=0}^{n-1} \frac{1}{1-2x\cos(2k\pi/n)+x^2} = \frac{n(1+x^n)}{(1-x^n)(1-x^2)}, \tag{5.22}$$

and
$$\sum_{k=0}^{n-1} \frac{1-x\cos(2k\pi/n)}{1-2x\cos(2k\pi/n)+x^2} = \frac{n}{1-x^n}. \tag{5.23}$$

Another way of deriving a finite summation identity is to manipulate an existing finite product identity. For example, if we take the logarithm of both sides of (5.14), we obtain
$$\ln \sin nx = (n-1)\ln 2 + \sum_{k=0}^{n-1} \ln \sin\left(x + \frac{k\pi}{n}\right),$$

and then differentiate, the result is seen to be
$$\sum_{k=0}^{n-1} \cot\left(x + \frac{k\pi}{n}\right) = n \cot nx. \tag{5.24}$$

5. Trigonometric Identities via Complex Numbers

Replacing x by $x + \pi/2$ in (5.24) gives

$$\sum_{k=0}^{n-1} \tan\left(x + \frac{k\pi}{n}\right) = n \tan\left(nx + \frac{n-1}{2}\pi\right). \tag{5.25}$$

Differentiating (5.24) yields

$$\sum_{k=0}^{n-1} \csc^2\left(x + \frac{k\pi}{n}\right) = n^2 \csc^2 nx.$$

This establishes (5.2) as follows:

$$\sum_{k=1}^{n-1} \csc^2\left(\frac{k\pi}{n}\right) = \lim_{x \to 0}\left(n^2 \csc^2 nx - \csc^2 x\right) = \frac{(n-1)(n+1)}{3}.$$

Similarly, differentiating (5.25) gives

$$\sum_{k=0}^{n-1} \sec^2\left(x + \frac{k\pi}{n}\right) = n^2 \sec^2\left(nx + \frac{n-1}{2}\pi\right).$$

This identity will allow us to derive some more sophisticated trigonometric identities (see Exercise 14).

5.4 Euler's Infinite Product

One of the highlights in Euler's 1748 work, *Introduction to Analysis of the Infinite*, is the representation of the sine as an infinite product

$$\sin x = x \prod_{k=1}^{\infty}\left(1 - \frac{x^2}{k^2 \pi^2}\right). \tag{5.26}$$

This result is the key to many other expansions and identities. It opens up the theory of infinite products and partial fraction decompositions of transcendental functions. We will see some of its remarkable corollaries in the next chapter, which includes the partial fractions of trigonometric functions and the summation of the Riemann zeta function for all even positive integers. We present here an elementary derivation of (5.26).

The first step is to establish the following fact: for an odd natural number n,

$$\sin nx = P_n(\sin x),$$

where $P_n(t)$ is an nth degree polynomial.

For $n = 3$, we see the required polynomial is just $P_3(t) = 3t - 4t^3$ since $\sin 3x = 3 \sin x - 4 \sin^3 x$. In general, for any positive integer m, we start with De Moivre's formula (5.6)

$$\cos mx + i \sin mx = (\cos x + i \sin x)^m$$

and take its imaginary part, which is

$$\sin mx = \sin x \left(\binom{m}{1} \cos^{m-1} x - \binom{m}{3} \sin^2 x \cos^{m-3} x + \binom{m}{5} \sin^4 x \cos^{m-5} x + \cdots\right).$$

In particular, if $m = 2n + 1$, appealing to $\cos^2 x = 1 - \sin^2 x$, we arrive at

$$\sin(2n+1)x = (2n+1)\sin x \left(1 - \frac{(2n+1)^2 - 1^2}{3!}\sin^2 x \right.$$
$$\left. + \frac{[(2n+1)^2 - 1^2][(2n+1)^2 - 3^2]}{5!}\sin^4 x + \cdots \right).$$

That is,
$$\sin(2n+1)x = (2n+1)\sin x \cdot P(\sin^2 x), \tag{5.27}$$

where $P(t)$ is an nth degree polynomial with $P(0) = 1$. Note that if $r_1, r_2, \ldots, r_n$ are roots of $P(t)$, then in factored form

$$P(t) = \prod_{k=1}^{n} \left(1 - \frac{t}{r_k}\right).$$

This is self evident, since substituting $t = 0$ gives $P(0) = 1$, just as substituting $t = r_k$ yields $P(r_k) = 0$ for $k = 1, 2, \ldots, n$.

Next, due to (5.27), if x is a solution of $\sin(2n+1)x = 0$, but $\sin x \neq 0$, then $\sin^2 x$ is a solution of $P(t) = 0$. Consider

$$x = \frac{k\pi}{2n+1}, \quad \text{for } k = 1, 2, \ldots, n.$$

For each of these values we have $(2n+1)x = k\pi$, and thus $\sin(2n+1)x = 0$, while $0 < x < \pi/2$ implies that we get n distinct positive roots for $P(t)$, namely

$$r_1 = \sin^2(\pi/(2n+1)), r_2 = \sin^2(2\pi/(2n+1)), \ldots, r_n = \sin^2(n\pi/(2n+1)).$$

Thus, in factored form,

$$\sin(2n+1)x = (2n+1)\sin x \prod_{k=1}^{n}\left(1 - \frac{\sin^2 x}{\sin^2(k\pi/(2n+1))}\right).$$

Replacing x by $x/(2n+1)$ yields

$$\sin x = (2n+1)\sin\left(\frac{x}{2n+1}\right) \prod_{k=1}^{n}\left(1 - \frac{\sin^2(x/(2n+1))}{\sin^2(k\pi/(2n+1))}\right), \tag{5.28}$$

which is true for all natural numbers n.

For a reader who is familiar with the uniformly convergence theorem for infinite products, which justifies that interchanging the product and the limit is valid, it is easy to see that (5.26) follows from (5.28) by letting $n \to \infty$. To avoid using such an advanced result, we rewrite $\sin x$ as $A_K^n B_K^n$, where $(K+1)\pi > |x|, n > K$ and

$$A_K^n = (2n+1)\sin\left(\frac{x}{2n+1}\right) \prod_{k=1}^{K}\left(1 - \frac{\sin^2(x/(2n+1))}{\sin^2(k\pi/(2n+1))}\right),$$

$$B_K^n = \prod_{k=K+1}^{n}\left(1 - \frac{\sin^2(x/(2n+1))}{\sin^2(k\pi/(2n+1))}\right).$$

5. Trigonometric Identities via Complex Numbers

For fixed K, $1 \leq k \leq K$, we have
$$\lim_{n\to\infty} (2n+1) \sin\left(\frac{x}{2n+1}\right) = x$$
and
$$\lim_{n\to\infty} \frac{\sin^2(x/(2n+1))}{\sin^2(k\pi/(2n+1))} = \frac{x^2}{k^2\pi^2}.$$

Thus,
$$\lim_{n\to\infty} A_K^n = x \prod_{k=1}^{K}\left(1 - \frac{x^2}{k^2\pi^2}\right) \stackrel{\text{def}}{=} A_K.$$

Since $B_K^n = \sin x / A_K^n$, we see that $\lim_{n\to\infty} B_K^n = B_K$ exists. To estimate B_K, using the well-known inequality
$$\frac{2}{\pi}\theta < \sin\theta < \theta, \quad \text{for } 0 < \theta < \pi/2,$$
we have
$$\sin^2 \frac{x}{2n+1} < \frac{x^2}{(2n+1)^2}$$
and
$$\sin^2 \frac{k\pi}{2n+1} > \frac{4}{\pi^2} \cdot \frac{k^2\pi^2}{(2n+1)^2} \quad (k = K+1, \ldots, n).$$

Thus,
$$1 - \frac{\sin^2(x/(2n+1))}{\sin^2(k\pi/(2n+1))} > 1 - \frac{x^2}{4k^2}$$
and
$$1 > B_K^n > \prod_{k=K+1}^{n}\left(1 - \frac{x^2}{4k^2}\right).$$

Since $\sum x^2/(4k^2)$ converges, we see that the infinite product
$$\prod_{k=K+1}^{\infty}\left(1 - \frac{x^2}{4k^2}\right)$$
converges to 1 as K goes to infinity and so
$$\lim_{K\to\infty} B_K = 1.$$

Thus, we finally establish Euler's infinite product (5.26) by
$$\sin x = \lim_{K\to\infty} A_K B_K = x \prod_{k=1}^{\infty}\left(1 - \frac{x^2}{k^2\pi^2}\right).$$

As a corollary, from $\cos x = \sin 2x/(2\sin x)$, we find that
$$\cos x = \prod_{k=1}^{\infty}\left(1 - \frac{4x^2}{(2k-1)^2\pi^2}\right). \tag{5.29}$$

Moreover, for $x = \pi/2$, (5.26) leads to the famous Wallis's formula, namely
$$\frac{2}{\pi} = \prod_{k=1}^{\infty}\left(1 - \frac{1}{4k^2}\right) = \frac{1\cdot 3\cdot 3\cdot 5\cdot 5\cdot 7\cdot 7\cdots}{2\cdot 2\cdot 4\cdot 4\cdot 6\cdot 6\cdot 8\cdots}. \tag{5.30}$$

5.5 Sums of inverse tangents

The evaluation of inverse tangent sums of the form

$$\sum_{k=1}^{\infty} \arctan f(k)$$

for a real function f has appeared in the Mathematical Association of America journal problem sections from time to time. For example, in a brief search in the Monthly one will find Raisbeck's E930 back in 1950 and Borwein and Baily's 11438 in 2009. There are very good chances that such problems can be solved by using the addition and difference formulas for inverse tangents. But the *principle of the argument* for complex multiplication provides an equally simple method.

Let $z_k = a_k + i\, b_k$ $(1 \le k \le n)$ with $a_k, b_k \in \mathbb{R}$. Consider $P_n = \prod_{k=1}^n z_k$. Appealing to (5.5), we have

$$\arg(P_n) = \sum_{k=1}^n \arg(z_k) = \sum_{k=1}^n \arctan\left(\frac{b_k}{a_k}\right) \quad (\text{mod } 2\pi). \tag{5.31}$$

Similarly, for a convergent infinite product $P = \prod_{k=1}^{\infty} z_k$,

$$\arg(P) = \sum_{k=1}^{\infty} \arctan\left(\frac{b_k}{a_k}\right) \quad (\text{mod } 2\pi). \tag{5.32}$$

Of course, throughout $\arctan x$ will always denote the *principal value*.

It seems that (5.31) and (5.32) are not as widely known as they deserve to be. To help spread the word, we demonstrate two examples.

Example 1 (Monthly Problem E930 [1950,483; 1951, 262–263]). If

$$\prod_{k=1}^n (x + r_k) = \sum_{j=0}^n a_j x^{n-j}, \tag{5.33}$$

show that

$$\sum_{k=1}^n \arctan r_k = \arctan \frac{a_1 - a_3 + a_5 - \cdots}{a_0 - a_2 + a_4 - \cdots}.$$

The published solution by Apostol [8] involves the addition formula for $\arctan x$, the elementary symmetric function of the n variables and mathematical induction. Here we present a short proof based on (5.31). For $x = 1/i$, (5.33) becomes

$$\prod_{k=1}^n (1 + i\, r_k)/i = \sum_{j=0}^n a_j i^{j-n},$$

which is equivalent to

$$\prod_{k=1}^n (1 + i\, r_k) = \sum_{j=0}^n a_j i^j = (a_0 - a_2 + a_4 - \cdots) + i\,(a_1 - a_3 + a_5 - \cdots).$$

Now, the required identity follows from applying (5.31) to this equation.

5. Trigonometric Identities via Complex Numbers

Example 2. Replacing x by πz in (5.26) yields

$$\sin \pi z = \pi z \prod_{k=1}^{\infty} \left(1 - \frac{z^2}{k^2}\right). \tag{5.34}$$

For $z = x + i y$,

$$\sin \pi z = \sin \pi x \cosh \pi y + i \cos \pi x \sinh \pi y.$$

Appealing to (5.32), we recover an elegant formula in [3] by Glasser and Klamkin, namely

$$\sum_{k=1}^{\infty} \arctan\left(\frac{2xy}{k^2 - x^2 + y^2}\right) = \arctan\left(\frac{y}{x}\right) - \tan^{-1}\left(\frac{\tanh \pi y}{\tan \pi x}\right) \pmod{2\pi},$$

which, for $x = y = 1/\sqrt{2}$, becomes

$$\sum_{k=1}^{\infty} \arctan\left(\frac{1}{k^2}\right) = \frac{\pi}{4} - \arctan\left(\frac{\tanh \pi/\sqrt{2}}{\tan \pi/\sqrt{2}}\right) \pmod{2\pi}. \tag{5.35}$$

This solves the Monthly Problem 3379, 1990 posed by Chapman.

5.6 Two Applications

In contrast to Fourier series, identities such as (5.20) and (5.22) are finite sums over the angles equally dividing the plane. They often crop up in discrete Fourier series and number theory. We conclude this chapter with two applications of (5.20) and (5.22). Our first application shows how to use (5.22) to evaluate the following Poisson integral

$$I = \int_0^{2\pi} \frac{d\theta}{1 - 2x \cos \theta + x^2}, \quad (|x| < 1)$$

via the limit of a Riemann sum. This integral is usually evaluated by using either the residue theorem or Fourier series.

Since

$$1 - 2x \cos \theta + x^2 \geq (1 - |x|)^2 > 0 \quad \text{for } |x| < 1,$$

the integrand is continuous and integrable. Partition the interval $[0, 2\pi]$ into n equal subintervals by the partition points

$$\left\{\theta_k = \frac{2k\pi}{n} : \quad 1 \leq k \leq n\right\}.$$

By (5.22), we get

$$I = \lim_{n \to \infty} \frac{2\pi}{n} \left(\sum_{k=0}^{n-1} \frac{1}{1 - 2x \cos\left(\frac{2k\pi}{n}\right) + x^2}\right)$$

$$= \lim_{n \to \infty} \frac{2\pi}{n} \frac{n(1 + x^n)}{(1 - x^n)(1 - x^2)}$$

$$= \frac{2\pi}{1 - x^2}.$$

The second application uses (5.20) and (5.22) to deduce some more sophisticated finite trigonometric identities like (5.2). Regrouping (5.22) as

$$\sum_{k=1}^{n-1} \frac{1}{1 - 2x \cos\left(\frac{2k\pi}{n}\right) + x^2} = \frac{n(1 + x^n)}{(1 - x^n)(1 - x^2)} - \frac{1}{(1 - x)^2},$$

and letting $x \to 1$, we re-prove (5.2) as

$$\sum_{k=1}^{n-1} \csc^2\left(\frac{k\pi}{n}\right) = 4 \lim_{x \to 1} \left(\frac{n(1 + x^n)}{(1 - x^n)(1 - x^2)} - \frac{1}{(1 - x)^2}\right) = \frac{(n - 1)(n + 1)}{3},$$

where l'Hôpital's rule has been used to calculate the limit.

Similarly, we assume that n is odd first. Setting $\alpha = \pi$ in (5.20), we have

$$\sum_{k=0}^{n-1} \frac{1}{1 + 2x \cos\left(\frac{2k\pi}{n}\right) + x^2} = \frac{n(1 - x^{2n})}{(1 + x^n)^2 (1 - x^2)}.$$

Letting $x \to 1$, we obtain

$$\sum_{k=0}^{n-1} \sec^2\left(\frac{k\pi}{n}\right) = 4 \lim_{x \to 1} \frac{n(1 - x^{2n})}{(1 + x^n)^2 (1 - x^2)} = n^2, \qquad (5.36)$$

where l'Hôpital's rule has been used again.

If n is even, setting $\alpha = \pi$ in (5.20) and then regrouping yields

$$\sum_{k=0, k \neq n/2}^{n-1} \frac{1}{1 + 2x \cos\left(\frac{2k\pi}{n}\right) + x^2} = \frac{n(1 - x^{2n})}{(1 + x^n)^2 (1 - x^2)} - \frac{1}{(1 - x)^2}.$$

Letting $x \to 1$, we have

$$\sum_{k=0, k \neq n/2}^{n-1} \sec^2\left(\frac{k\pi}{n}\right) = \frac{(n - 1)(n + 1)}{3}. \qquad (5.37)$$

Appealing to the trigonometric identities $1 + \tan^2 x = \sec^2 x$ and $1 + \cot^2 x = \csc^2 x$, by (5.35)–(5.37), we find that

$$\sum_{k=1}^{n-1} \cot^2\left(\frac{k\pi}{n}\right) = \frac{(n - 1)(n - 2)}{3}$$

and

$$\sum_{k=0}^{n-1} \tan^2\left(\frac{k\pi}{n}\right) = \begin{cases} n(n - 1), & \text{if } n \text{ is odd}; \\ \frac{(n-1)(n-2)}{3} & \text{if } n \text{ is even, } k \neq n/2. \end{cases} \qquad (5.38)$$

If we replace the power 2 in these identities by an arbitrary positive even power, finding an explicit closed form becomes more difficult. We will investigate these problems systematically by using the generating functions in Chapter 14. Remarkably, the identity (5.22) also provides us an experimental method to search for more trigonometric identities (see Exercises 15 and 16).

5. Trigonometric Identities via Complex Numbers

Exercises

1. Let $\omega = e^{i(2\pi/n)}$ and n be a positive integer. Prove that

   (a) $\prod_{k=1}^{n-1} |1 - \omega^k|^{2k} = n^n$.

   (b) $\prod_{k=1}^{n-1} \left(\sin\left(\frac{k\pi}{n}\right)\right)^k = \left(\frac{1}{2}\right)^{n(n+1)/2} n^{n/2}$.

   (c) $\prod_{k=1}^{n-1} \left|\cos\left(\frac{k\pi}{n}\right)\right|^k = \left(\frac{1}{2}\right)^{n(n+1)/2} [1 - (-1)^n]^{n/2}$.

2. Prove that the sum of the squared distance from any point P on the unit circumcircle of a regular n-gon to its vertices is $2n$.

3. Determine the closed forms of

   $$\sum_{k=1}^{n} r^k \sin(kx) \quad \text{and} \quad \sum_{k=0}^{n} r^k \cos(kx).$$

   Find their limits as $n \to \infty$ if $|r| < 1$.

4. Let $n \geq 3$. Find

   $$\prod_{k=1}^{n-1} (1 + 2\cos(k\pi/n)).$$

5. Define $P_n(x)$ by

   $$P_n(x) = \sum_{k=0}^{n} \binom{n}{k}^2 x^{2k}(1-x)^{2(n-k)}.$$

   Show that for $0 \leq x \leq 1$,

   $$P_n(x) \geq \frac{1}{2^{2n}} \binom{2n}{n}.$$

6. Prove that

   (a) $\prod_{k=1}^{n-1} \sin\left(\frac{(2k+1)\pi}{2n}\right) = \left(\frac{1}{2}\right)^{n-1}$.

   (b) $\prod_{k=1}^{n-1} \sin\left(\frac{(3k+1)\pi}{3n}\right) = \frac{\sqrt{3}}{2^n}$.

7. Prove that

   (a) $\frac{\sin x}{x} = \prod_{k=1}^{\infty} \cos\left(\frac{x}{2^k}\right)$.

   (b) $\frac{\sin x}{x} = \cos^2(x/2) + \sum_{k=1}^{\infty} \sin^2(x/2^{k+1}) \prod_{m=1}^{k} \cos(x/2^m)$.

8. **Putnam Problem 1949-A6.** For every x, prove that

   $$\frac{\sin x}{x} = \prod_{n=1}^{\infty} \frac{1 + 2\cos(2x/3^n)}{3}.$$

Comment. A new approach to this problem is to apply the Fourier transform to the product expansion Exercise 7(a), then use the convolution and delta distributions. Based on this method, one can generate a family of similar identities. For example,

$$\frac{\sin x}{x} = \prod_{k=1}^{\infty} \frac{1}{6}\left(2\cos\frac{x}{6^k} + 2\cos\frac{3x}{6^k} + 2\cos\frac{5x}{6^k}\right).$$

The details and further examples can be found in [7].

9. **Monthly Problem 11383** [2008, 757]. Show that

$$\sum_{n=1}^{\infty} \arccos\left(\frac{1 + \sqrt{n^2 + 2n}\sqrt{n^2 + 4n + 3}}{(n+1)(n+2)}\right) = \frac{\pi}{3}.$$

Remark. The answer seems to be $\pi/6$.

10. **Monthly Problem 5486** [1967, 447; 1968, 421–422]. Prove that

$$\sum_{k=0}^{n} \csc^2\left(\frac{(2k+1)\pi}{2n}\right) = n^2 \quad \text{and} \quad \sum_{k=1}^{2n} \csc^2\left(\frac{k\pi}{2n+1}\right) = \frac{4}{3}n(n+1).$$

11. Prove that

$$\sinh x = x \prod_{n=1}^{\infty}\left(1 + \frac{x^2}{n^2\pi^2}\right), \quad \cosh x = \prod_{n=1}^{\infty}\left(1 + \frac{4x^2}{(2n-1)^2\pi^2}\right).$$

Use the latter result to prove that

$$\lim_{n\to\infty} n \prod_{k=1}^{n}\left(1 - \frac{1}{k} + \frac{5}{4k^2}\right) = \frac{\cosh(\pi)}{\pi}.$$

12. Prove that

$$\sum_{k=1}^{\infty} \arctan\left(\frac{2xy}{(2k-1)^2 - x^2 + y^2}\right) = \arctan\left(\tan\frac{\pi x}{2} \tanh\frac{\pi y}{2}\right) \pmod{2\pi}.$$

13. Prove that

$$\sum_{k=1}^{\infty}(-1)^k \arctan\left(\frac{x}{2n+1}\right) = \arctan\left(\tanh\frac{\pi x}{4}\right),$$

where it is assumed that $-\pi/2 \le \arctan t \le \pi/2$.

14. **Putnam Problem 1986-A3.** Evaluate

$$\sum_{k=1}^{\infty} \text{arccot}(k^2 + k + 1),$$

where arccot x for $x \ge 0$ denotes the number θ in the interval $0 < \theta \le \pi/2$ with $\cot\theta = x$.

5. Trigonometric Identities via Complex Numbers

15. Prove that
$$\sum_{k=0}^{n-1} \tan^2\left(x + \frac{k\pi}{n}\right) = n^2 \sec^2\left(nx + \frac{n-1}{2}\pi\right) - n.$$

In general, let
$$S_m(x) = \sum_{k=0}^{n-1} \tan^m\left(x + \frac{k\pi}{n}\right).$$

Prove that
$$S_{m+1}(x) = \frac{1}{m} S'_m(x) - S_{m-1}, \quad \text{for } m = 2, 3, \ldots.$$

16. In view of (5.22), define
$$F_p(x) = \sum_{k=0}^{n-1} \frac{1 - x^2}{(1 - 2x\cos(2k\pi/n) + x^2)^p}.$$

If $n \not\equiv 0 \pmod{4}$, show that
$$\sum_{k=0}^{n-1} \sec^p(2k\pi/n) = 2^{p-1}(-i)^p(F_p(i)). \tag{5.39}$$

Moreover, prove the following recursive relation
$$F_{1+p}(x) = \frac{F_p(x)}{1 - x^2} + \frac{x}{p}\frac{d}{dx}\left(\frac{F_p(x)}{1 - x^2}\right).$$

This enables us to compute $F_p(x)$ in succession.

Comment. Using the sampling theorems with second-order discrete eigenvalue problems, Hassan in [6] derives some trigonometric identities such as (5.39). With the help of Mathematica or Maple, Exercises 15 and 16 provide us an elementary experimental method to search for this kind of trigonometric identities. In the course of doing this, one should not miss

$$\sum_{k=0}^{n-1} \sec^2\left(\frac{(2k+1)\pi}{4n}\right) = 2n^2, \quad \sum_{k=0}^{n-1} \sec^2\left(\frac{(2k+1)\pi}{4n+2}\right) = \frac{2n(n+1)}{3}.$$

References

[1] H. Chen, On a new trigonometric identity, *Internat. J. Math. Ed. Sci. Tech.*, **37** (2006) 215–223.

[2] W. Dunham, *Euler: The Master of Us All*, The Mathematical Association of America, 1990.

[3] M. Glasser and M. Klamkin, On some inverse tangent summations, *Fibonacci Quarterly*, **14** (1976) 385–388.

[4] I. Gradshteyn and I. Ryzhik, *Table of Integrals, Series, and Product*, 6th edition, edited by A. Jeffrey and D. Zwillinger, Academic Press, New York, 2000.

[5] E. Hansen, *A Table of Series and Products*, Prentice Hall, New Jersey, 1975.

[6] H. Hassan, New trigonometric sums by sampling theorem, *J. Math. Anal. Appl.*, **339** (2008) 811–827.

[7] K. Morrison, Cosine products, Fourier transforms, and random sums, *Amer. Math. Monthly*, **102** (1995) 716–724.

[8] G. Raisbeck, Problem E930, *Amer. Math. Monthly*, **58** (1951) 262–263.

6
Special Numbers

Numbers constitute the only universal language.   — N. West

All numbers are not created equal. Some sequences of numbers in analysis seem quite complicated and mysterious, yet they have played an important role and have made many unexpected appearances in analysis and number theory. A large body of literature exists on these special numbers but much of it in widely scattered books and journals. In this chapter we investigate three of these special sets of numbers, the Fibonacci, harmonic and Bernoulli numbers. We derive some basic results that are readily accessible to those with a knowledge of elementary analysis. These materials will serve as a brief primer on these special numbers and lay the foundation for many of the latter chapters. We start with the Fibonacci numbers, which are defined by a recursive relation. Then we examine the harmonic numbers and some of their important features. Finally, we investigate the Bernoulli numbers, which are attributed to Jacob Bernoulli in the discussion of closed forms for the sums of kth powers of consecutive integers.

6.1 Generating Functions

Special numbers are often defined by recurrences. But applications may well require explicit closed forms of the general terms. Generating functions provide one of the best routes to finding these formulas. In the preface of his fascinating book [10], H. Wilf writes "The full beauty of the subject of generating functions emerges from tuning in on both channels: the discrete and the continuous." He further observes that "A generating function is a clothesline on which we hang up a sequence of numbers for display." Thus, roughly speaking, generating functions will transform problems about *sequences* into problems about *functions*. This is useful because we have piles of mathematical machinery for manipulating functions.

Definition 1 *The ordinary generating function for the infinite sequence* $\{a_0, a_1, a_2, \ldots\}$ *is the power series*
$$G(x) = a_0 + a_1 x + a_2 x^2 + \cdots.$$

Now it is clear that the unknown numbers a_n are arranged neatly on the clothesline: indexing from 0, a_n is the coefficient of x^n in the generating function. Therefore, once we identify the function $G(x)$, we will be able to find the explicit formula for the a_n's by expanding $G(x)$ as a power series, and vice versa.

Recall that the sum of an infinite geometric series, for $|x| < 1$,

$$1 + x + x^2 + x^3 + \cdots = \frac{1}{1-x}.$$

This indicates that the generating function of the sequence $\{1, 1, 1, \ldots\}$ is $1/(1-x)$. On the other hand, if $G(x) = -\ln(1-x)$, we have

$$\frac{G^{(n)}(0)}{n!} = \frac{1}{n}.$$

Therefore $-\ln(1-x)$ generates the sequence $\{1, 1/2, 1/3, \ldots\}$.

Sometimes a sequence $\{a_n\}$ has a generating function whose properties are quite complicated, while the related sequence $\{a_n/n!\}$ has a generating function that is quite simple. This leads to beneath definition:

Definition 2 *The exponential generating function for the infinite sequence $\{a_0, a_1, a_2, \ldots\}$ is the power series*

$$G(x) = a_0 + \frac{a_1}{1!}x + \frac{a_2}{2!}x^2 + \frac{a_3}{3!}x^3 + \cdots.$$

Thus the series for e^x is the exponential generating function for $\{1, 1, 1, \ldots\}$, and the series for $\cos x$ and $\sin x$, for the periodic sequences $\{1, 0, -1, 0 \ldots\}$ and $\{0, 1, 0, -1, \ldots\}$, respectively.

The magic of generating functions is that we can carry out all sorts of manipulations on sequences by performing mathematical operations on their associated generating functions. For more detailed theories and applications of generating functions, please refer to [10]. In the following, using Fibonacci, harmonic and Bernoulli numbers as case studies, we explore how generating functions lead to simple, direct proofs of basic properties of these sets of numbers.

6.2 Fibonacci Numbers

Fibonacci numbers have a long and glorious mathematical history. They are named after the Italian mathematician Leonardo of Pisa, who was known as Fibonacci (a contraction of *filius Bonaccio*, "son of Bonaccio."). In 1202, Fibonacci published *Liber Abbaci*, or "Book of Calculation," a highly influential book of practical mathematics. In this book, he introduced and promoted the mighty Hindu-Arabic numerals to Europeans, who were still doing calculations with Roman numbers. His book also contained the following curious problem: "How many pairs of rabbits can be produced from a single pair in a year if every month each begets a new pair which from the second month on becomes productive?" In the course of answering this rabbit problem, the *Fibonacci numbers*

$$1, 1, 2, 3, 5, 8, 13, 21, \ldots$$

and the formula

$$F_{n+2} = F_n + F_{n+1} \quad \text{for } n \geq 1, \tag{6.1}$$

with $F_1 = F_2 = 1$ are revealed, where F_n denotes the nth Fibonacci number.

6. Special Numbers

The formula (6.1), which is often called a *recursive formula*, does not directly give the value of F_n. Instead it gives a rule telling us how to compute F_n from the previous numbers. To find the exact value of F_n, we now turn to the method of generating functions. Let the generating function be

$$G(x) = \sum_{n=1}^{\infty} F_n x^n.$$

In view of (5.1), then

$$\begin{aligned} G(x) &= x + x^2 + F_3 x^3 + F_4 x^4 + F_5 x^5 + \cdots \\ &= x + x^2 + (F_1 + F_2)x^3 + (F_2 + F_3)x^4 + (F_3 + F_4)x^5 + \cdots \\ &= x + x(F_1 x + F_2 x^2 + F_3 x^3 + \cdots) + x^2(F_1 x + F_2 x^2 + F_3 x^3 + \cdots) \\ &= x + xG(x) + x^2 G(x) \end{aligned}$$

and so

$$G(x) = \frac{x}{1 - x - x^2}.$$

To determine the coefficient of x^n in $G(x)$, we factor

$$1 - x - x^2 = (1 - \alpha x)(1 - \beta x),$$

where $\alpha = (1 + \sqrt{5})/2$, $\beta = (1 - \sqrt{5})/2$. Using the method of partial fractions and the geometric series expansion yields

$$\begin{aligned} G(x) &= \frac{x}{(1 - \alpha x)(1 - \beta x)} \\ &= \frac{1}{\sqrt{5}} \left(\frac{1}{1 - \alpha x} - \frac{1}{1 - \beta x} \right) \\ &= \frac{1}{\sqrt{5}} \sum_{k=1}^{\infty} (\alpha^n - \beta^n) x^n. \end{aligned}$$

Thus, we obtain *Binet's formula*

$$F_n = \frac{1}{\sqrt{5}} (\alpha^n - \beta^n) \tag{6.2}$$

as an explicit expression for the Fibonacci number F_n.

The formula (6.2) also can be obtained from the exponential generating function. To see this, let

$$G_e(x) = \sum_{k=1}^{\infty} \frac{F_n}{n!} x^n.$$

Differentiating $G_e(x)$ successively gives

$$G_e'(x) = F_1 + F_2 x + \frac{F_3}{2!} x^2 + \frac{F_4}{3!} x^3 + \frac{F_5}{4!} x^4 + \frac{F_6}{5!} x^5 + \cdots,$$

$$G_e''(x) = F_2 + F_3 x + \frac{F_4}{2!} x^2 + \frac{F_5}{3!} x^3 + \frac{F_6}{4!} x^4 + \frac{F_7}{7!} x^5 + \cdots.$$

It follows that, from (6.1) and $F_1 = F_2 = 1$, $G_e(x)$ satisfies the differential equation

$$G_e''(x) - G_e'(x) - G_e(x) = 0.$$

This, together with $G_e(0) = 0, G_e'(0) = F_1 = 1$, yields

$$G_e(x) = \frac{1}{\sqrt{5}} (e^{\alpha x} - e^{\beta x}).$$

Finally, the formula (6.2) is an immediate consequence of the power series expansion of e^{ax}.

In light of (6.2), Fibonacci numbers grow very fast. In fact, $F_{45} = 1{,}134{,}903{,}170$. That is, in less than four years, one will have more than one billion pairs of rabbits! Moreover, one can determine exactly how fast Fibonacci numbers grow. In particular, considering the ratios of successive Fibonacci numbers yields a particularly striking result:

$$\lim_{n\to\infty} \frac{F_{n+1}}{F_n} = \lim_{n\to\infty} (F_n)^{1/n} = \phi, \tag{6.3}$$

where $\phi = (1 + \sqrt{5})/2 = 1.618034\ldots$ was called the golden ratio by the ancient Greeks. Recently, Viswanath in [8] took a fresh look at Fibonacci numbers by studying the Fibonacci-like recurrence:

$$\mathcal{F}_{n+2} = \pm \mathcal{F}_{n+1} \pm \mathcal{F}_n, \quad \mathcal{F}_0 = 0, \mathcal{F}_1 = 1,$$

where each $\pm$ sign is chosen randomly with probability $1/2$. He unexpectedly found that

$$\lim_{n\to\infty} (|\mathcal{F}_n|)^{1/n} = 1.13198824\ldots.$$

This constant may explain the case of rabbits randomly allowed to prey on each other, and enables us closer to simulate the real world.

Since the thirteenth century, Fibonacci numbers have intrigued us to the present day. There is a specialized journal entitled *Fibonacci Quarterly* that was founded in 1962 and is devoted to Fibonacci numbers and their generalizations. Recently, Fibonacci numbers even appeared in a popular novel, Dan Brown's bestseller *The Da Vinci Code*. Although Fibonacci numbers were originally represented as an unrealistic model of a population of rabbits, scientists did find them turning up in many models of the natural world. For example, the flowers of surprisingly many plants have a Fibonacci number of petals. Lilies have 3 petals, buttercups have 5, sunflowers often have 55, 89 or 144. Similar numerical patterns involving successive Fibonacci numbers also occur in pine cones and pineapples. However, the precise reasons for Fibonacci numerology in plant life are still unknown. There's still plenty of room for mathematical exploration and experimentation in a problem that began centuries ago.

6.3 Harmonic numbers

The *harmonic numbers* $\{H_n\}$ are defined by

$$H_n = 1 + \frac{1}{2} + \frac{1}{3} + \cdots + \frac{1}{n}. \tag{6.4}$$

6. Special Numbers

These numbers satisfy the recurrence relation

$$H_n = H_{n-1} + \frac{1}{n}, \qquad (\text{for } n > 1)$$

and $H_1 = 1$. Hence

$$\sum_{n=1}^{\infty} H_n x^n = x + \sum_{n=2}^{\infty} H_n x^n$$

$$= x + \sum_{n=2}^{\infty} \left(H_{n-1} + \frac{1}{n} \right) x^n$$

$$= x \left(\sum_{n=1}^{\infty} H_n x^n \right) + \sum_{n=1}^{\infty} \frac{x^n}{n}.$$

Since it is valid to interchange the order of summation and integration for a convergent power series, for $|x| < 1$,

$$\sum_{n=1}^{\infty} \frac{1}{n} x^n = \sum_{n=1}^{\infty} \int_0^x t^{n-1} dt = \int_0^x \left(\sum_{n=1}^{\infty} t^{n-1} \right) dt = \int_0^x \frac{dt}{1-t} = -\ln(1-x).$$

Thus, we find that the generating function of $\{H_n\}$ is

$$\sum_{n=1}^{\infty} H_n x^n = -\frac{\ln(1-x)}{1-x}. \tag{6.5}$$

Unlike the Fibonacci numbers, a closed form of H_n for general n does not exist. Moreover, even as $\lim_{n \to \infty} 1/n = 0$, H_n itself is divergent! Modern calculus textbooks prove this fact either by establishing the estimates $H_{2^n} \geq 1 + \frac{1}{2}n$ or by comparing H_n with $\int_1^\infty dx/x$. But the following argument is equally simple and insightful. Suppose that $\lim_{n \to \infty} H_n = S$. Then

$$\frac{1}{2} + \frac{1}{4} + \cdots + \frac{1}{2n} + \cdots = \frac{1}{2} \left(1 + \frac{1}{2} + \cdots + \frac{1}{n} + \cdots \right) = \frac{1}{2} S$$

and so the sum of the odd numbered terms,

$$1 + \frac{1}{3} + \cdots + \frac{1}{2n+1} + \cdots$$

must be the other half of S. But this is impossible since

$$\frac{1}{2n-1} > \frac{1}{2n}$$

for every positive integer n.

On the other hand, define the alternate signed harmonic numbers by

$$H'_n = 1 - \frac{1}{2} + \frac{1}{3} + \cdots + \frac{(-1)^{n+1}}{n}.$$

H'_{2n} exhibits the following particular beautiful form — *Catalan's identity*:

$$H'_{2n} = 1 - \frac{1}{2} + \frac{1}{3} - \cdots + \frac{1}{2n-1} - \frac{1}{2n}$$
$$= H_{2n} - 2\left(\frac{1}{2} + \frac{1}{4} + \cdots + \frac{1}{2n}\right)$$
$$= H_{2n} - H_n = \frac{1}{n+1} + \frac{1}{n+2} + \cdots + \frac{1}{2n},$$

and

$$\lim_{n \to \infty} H'_{2n} = \lim_{n \to \infty} \frac{1}{n} \sum_{k=1}^{n} \frac{1}{1 + k/n} = \int_0^1 \frac{dx}{1+x} = \ln 2.$$

Now we study the asymptotic behavior of H_n as $n \to \infty$. Define

$$a_n = \frac{1}{n} - \ln\left(\frac{n+1}{n}\right).$$

Since

$$a_n = \int_0^1 \left(\frac{1}{n} - \frac{1}{n+x}\right) dx = \int_0^1 \frac{x\,dx}{n(n+x)},$$

a simple estimation on the integral yields

$$0 < a_n < \int_0^1 \frac{dx}{n^2} = \frac{1}{n^2}$$

and so $\sum_{n=1}^{\infty} a_n$ converges by the series comparison test. Therefore,

$$\lim_{n \to \infty} (H_n - \ln n) = \lim_{n \to \infty} \left(\sum_{k=1}^{n} a_k + \ln\left(\frac{n+1}{n}\right)\right) = \sum_{n=1}^{\infty} a_n \qquad (6.6)$$

exists. Euler denoted this limit as γ, namely

$$\gamma = \lim_{n \to \infty} (H_n - \ln n) = 0.5772156649\ldots, \qquad (6.7)$$

where γ is called *Euler's constant*. A careful analysis of the series remainder in (6.6) will yield nice estimates on γ. Indeed, a closed form of the remainder can be established and various asymptotic estimates for the remainder, which imply the well-known Young and De Temple inequalities, are then derived. See Exercise 9. Although there is compelling evidence that γ is not rational, a proof has not yet been found.

Next, due to the symmetry, we have

$$\sum_{k=1}^{n} \frac{H_k}{k} = \sum_{k=1}^{n} \sum_{i=1}^{k} \frac{1}{i \cdot k}$$
$$= \frac{1}{2}\left(\sum_{1 \leq i \leq k \leq n} \frac{1}{i \cdot k} + \sum_{i=k} \frac{1}{i \cdot k}\right)$$
$$= \frac{1}{2}\left(H_n^2 + \sum_{k=1}^{n} \frac{1}{k^2}\right).$$

6. Special Numbers

This elegant identity is a special case of what today is known as an "Euler sum", which we will explore more rigorously in Chapter 17.

Finally, we turn to the link between the *Riemann hypothesis* and the harmonic numbers. It is well known that the distribution of prime numbers among all positive integers does not display any regular pattern. However, in 1859, Riemann observed that the frequency of prime numbers is very closely related to the behavior of an elaborate function

$$\zeta(z) = 1 + \frac{1}{2^z} + \frac{1}{3^z} + \cdots$$

called the *Riemann zeta function*. In an eight page paper Riemann made his bold conjecture about the zeros of his Zeta function: except for negative even integers, all other zeros of $\zeta(z)$ lie on the vertical line $\text{Re}(z) = 1/2$. Using computers, mathematicians have managed to show that Riemann's hypothesis is true for the first 1.5 billion zeros of the zeta function. A proof of the Riemann hypothesis will shed light on many of the mysteries surrounding the distribution of prime numbers, and satisfy the curiosity of many generations of mathematicians. It will also have major consequence for what has come to be a crucial component of modern life — internet security. In 2000, the Clay Mathematics Institute (claymath.org) offered a $1 million prize for proof of the Riemann hypothesis. Recently, Jeffrey Lagarias in [3] made a surprising observation: the Riemann hypothesis is equivalent to the following elementary statement: for all $n \geq 1$,

$$\sigma(n) \leq H_n + \ln(H_n)e^{H_n},$$

with equality only for $n = 1$, where $\sigma(n) = \sum_{d|n} d$ denotes the sum of the divisors of n; for example, $\sigma(6) = 12$. A weaker version of this inequality appears in Exercise 10.

6.4 Bernoulli Numbers

There are many methods for introducing the Bernoulli numbers. The most versatile way was conceived by Euler. Instead of using recurrences, as we did in the previous sections, he observed that the Bernoulli numbers, B_n, possess an exponential generating function

$$\frac{x}{e^x - 1} = \sum_{n=0}^{\infty} \frac{B_n}{n!} x^n. \tag{6.8}$$

Rewrite the left-hand side in the alternative form

$$\frac{x}{e^x - 1} = \frac{xe^{-x/2}}{e^{x/2} - e^{-x/2}} = \frac{x}{2} \frac{\cosh \frac{x}{2} - \sinh \frac{x}{2}}{\sinh \frac{x}{2}} = \frac{x}{2} \coth \frac{x}{2} - \frac{x}{2}. \tag{6.9}$$

Notice that $(x/2) \coth(x/2)$ is an even function. Therefore we find that all but one of the Bernoulli numbers of odd terms are zero:

$$B_1 = -\frac{1}{2}, \quad B_{2k+1} = 0, \quad k = 1, 2, 3, \ldots. \tag{6.10}$$

In general, to obtain the B_n's, we invoke the Cauchy product

$$\left(B_0 + B_1 x + \frac{B_2}{2!} x^2 + \frac{B_3}{3!} x^3 + \cdots \right) \left(x + \frac{1}{2!} x^2 + \frac{1}{3!} x^3 + \cdots \right) = x.$$

Recalling that the *Cauchy product* of two convergent power series is given by

$$\left(\sum_{i=0}^{\infty} a_i x^i\right) \cdot \left(\sum_{j=0}^{\infty} b_j x^j\right) = \sum_{k=0}^{\infty} \left(\sum_{i+j=k} a_i b_j\right) x^k, \qquad (6.11)$$

we obtain $B_0 = 1$ and the recursive formula

$$\frac{B_0}{1!} \cdot \frac{1}{n!} + \frac{B_1}{1!} \cdot \frac{1}{(n-1)!} + \cdots + \frac{B_{n-1}}{(n-1)!} \cdot \frac{1}{1!} = 0, \quad n \geq 2.$$

Multiplying through by $n!$ and then appealing to the binomial coefficients yields a briefer form,

$$\sum_{k=0}^{n-1} \binom{n}{k} B_k = 0, \quad (\text{for } n \geq 2). \qquad (6.12)$$

If B^k is interpreted as the Bernoulli number B_k, in view of the binomial theorem, (6.12) is abbreviated to

$$(1 + B)^n = B^n.$$

By taking $n = 2, 4, 6, 8$ in (6.12), we find that

$$B_2 = \frac{1}{6}, \ B_4 = -\frac{1}{30}, \ B_6 = \frac{1}{42}, \ B_8 = -\frac{1}{30}.$$

Historically, Bernoulli numbers and polynomials were introduced in Jacob Bernoulli's book *Ars Conjectandi* (The Art of Conjecturing). To find the area under the curve $y = x^k$ from 0 to 1, it comes down to evaluating

$$S_k(n) = 1^k + 2^k + \cdots + (n-1)^k$$

as Archimedes did in the case of $k = 2$. Jacob Bernoulli gave an elegant description of $S_k(n)$. In the course of doing this, he discovered these special numbers and polynomials. Let the generating function of $\{S_k(n)\}$ be

$$G(n, x) = \sum_{k=0}^{\infty} \frac{S_k(n)}{k!} x^k.$$

Interchanging the summation order gives

$$G(n, x) = \sum_{i=1}^{n-1} \sum_{k=0}^{\infty} \frac{i^k}{k!} x^k$$

$$= \sum_{i=1}^{n-1} e^{ix} = \frac{e^{nx} - 1}{e^x - 1} = \frac{e^{nx} - 1}{x} \cdot \frac{x}{e^x - 1},$$

6. Special Numbers

where the second term is the expected Bernoulli numbers which were defined in (6.8). On the other hand, using the power series expansion of e^{nx} and (6.11) yields

$$G(n,x) = \left(\sum_{i=0}^{\infty} \frac{n^{i+1} x^i}{(i+1)!}\right) \left(\sum_{j=0}^{\infty} \frac{B_j}{j!} x^j\right)$$

$$= \sum_{k=0}^{\infty} \left(\sum_{i+j=k} \frac{n^{i+1} B_j}{(i+1)! j!}\right) x^k$$

$$= \sum_{k=0}^{\infty} \left(\frac{1}{k+1} \sum_{j=0}^{k} \binom{k+1}{j} B_j n^{k+1-j}\right) \frac{x^k}{k!}.$$

Define the kth degree *Bernoulli polynomial* by

$$B_k(x) = \sum_{j=0}^{k} \binom{k}{j} B_j x^{k-j}. \tag{6.13}$$

Equating the coefficients in those two expressions of $G(n,x)$ gives

$$S_k(n) = \frac{1}{k+1}(B_{k+1}(n) - B_{k+1}).$$

In the following, we summarize some basic properties and formulas regarding $B_k(x)$ and B_k.

6.4.1 The generating function of $B_k(x)$

From (6.8) and (6.10), we have the generating function of $B_k(x)$,

$$\sum_{k=0}^{\infty} B_k(x) \frac{t^k}{k!} = \sum_{k=0}^{\infty} \left(\sum_{j=0}^{k} \binom{k}{j} B_j x^{k-j}\right) \frac{t^k}{k!}$$

$$= \sum_{k=0}^{\infty} \left(\sum_{j=0}^{k} \frac{t^j B_j (xt)^{k-j}}{j!(k-j)!}\right)$$

$$= \left(\sum_{k=0}^{\infty} \frac{B_k}{k!} t^k\right) \left(\sum_{i=0}^{\infty} \frac{(xt)^k}{k!}\right)$$

$$= \frac{t e^{xt}}{e^t - 1}.$$

In particular, in terms of the given Bernoulli numbers, the first six Bernoulli polynomials are

$$B_0(x) = 1,$$
$$B_1(x) = x - \frac{1}{2},$$
$$B_2(x) = x^2 - x + \frac{1}{6},$$
$$B_3(x) = x^3 - \frac{3}{2}x^2 + \frac{1}{2}x,$$
$$B_4(x) = x^4 - 2x^3 + x^2 - \frac{1}{30},$$
$$B_5(x) = x^5 - \frac{5}{2}x^4 + \frac{5}{3}x^3 - \frac{1}{6}x.$$

6.4.2 Difference, Differentiation and Integration

Let $G_B(x,t)$ denote the generating function of $B_k(x)$:

$$G_B(x,t) = \frac{te^{xt}}{e^t - 1} = \sum_{k=0}^{\infty} B_k(x) \frac{t^k}{k!},$$

from which the following three properties of $B_k(x)$ are derived.

1. Note that

$$G_B(x+1,t) = \frac{te^{(x+1)t}}{e^t - 1} = te^{xt} + \frac{te^{xt}}{e^t - 1} = te^{xt} + G_B(x,t).$$

Equating coefficients of t^k on both sides gives

$$B_0(x+1) = B_0(0), \quad B_k(x+1) = B_k(x) + kx^{k-1} \quad (k \geq 1). \tag{6.14}$$

2. Next, differentiating $G_B(x,t)$ with respect to x yields

$$\frac{\partial}{\partial x} G_B(x,t) = \frac{t^2 e^{xt}}{e^t - 1} = \sum_{k=0}^{\infty} B_k'(x) \frac{t^k}{k!}.$$

But clearly

$$\frac{t^2 e^{xt}}{e^t - 1} = \sum_{k=0}^{\infty} B_k(x) \frac{t^{k+1}}{k!},$$

therefore equating coefficients of $t^k/k!$ gives

$$B_k'(x) = k\, B_{k-1}(x). \tag{6.15}$$

Repeating this process leads to

$$B_k^{(n)}(x) = \frac{k!}{(k-n)!} B_{k-n}(x).$$

6. Special Numbers

3. Replacing k by $k+1$ in (6.15) and then integrating from a to x gives

$$\int_a^x B_k(x)\,dx = \frac{1}{k+1}(B_{k+1}(x) - B_{k+1}(a)). \tag{6.16}$$

This, appealing to (6.14), implies

$$\int_t^{t+1} B_k(x)\,dx = t^k,$$

which, when $t = 0$, gives

$$\int_0^1 B_k(x)\,dx = 0 \quad (k \geq 1). \tag{6.17}$$

Moreover, (6.16) leads to a recursive relation

$$B_k(x) = k \int_0^x B_{k-1}(t)\,dt + B_k, \quad k \geq 1. \tag{6.18}$$

This equation and (6.17), together with $B_0 = 1$, $B_0(x) = 1$, provide another method for defining the Bernoulli polynomials and Bernoulli numbers recursively. See Exercise 12. Indeed, Lehmer observed that, beginning with Bernoulli's discovery, there have been six approaches for defining the Bernoulli polynomials. See [4] for details.

6.4.3 Euler-Maclaurin summation formula

As an application of (6.15), we derive a remarkable formula known as the *Euler-Maclaurin summation formula*, which precisely relates a discrete sum with a continuous integral. Along the way we also show how Bernoulli polynomials and Bernoulli numbers arise naturally.

Start with

$$\int_0^1 f(x)\,dx = \int_0^1 f(x) B_0(x)\,dx.$$

Appealing to $B_1'(x) = B_0(x)$, integration by parts gives

$$\int_0^1 f(x)\,dx = \int_0^1 B_1'(x) f(x)\,dx$$

$$= B_1(x) f(x)\Big|_0^1 - \int_0^1 B_1(x) f'(x)\,dx$$

$$= \frac{f(0) + f(1)}{2} - \int_0^1 B_1(x) f'(x)\,dx.$$

In view of $B_1(x) = (1/2) B_2'(x)$, integration by parts again gives

$$\int_0^1 f(x)\,dx = \frac{1}{2}(f(0) + f(1)) - \frac{1}{2!}(B_2(1) f'(1) - B_2(0) f'(0)) + \frac{1}{2}\int_0^1 B_2(x) f''(x)\,dx$$

$$= \frac{1}{2}(f(0) + f(1)) - \frac{1}{2!} B_2(f'(1) - f'(0)) + \frac{1}{2}\int_0^1 B_2(x) f''(x)\,dx.$$

Thus, by (6.10) and (6.12), repeating integration by parts yields

$$\int_0^1 f(x)\,dx = \frac{1}{2}(f(0) + f(1)) - \sum_{i=1}^{k} \frac{B_{2i}}{(2i)!}(f^{(2i-1)}(1)$$

$$- f^{(2i-1)}(0)) + \frac{1}{(2k)!} \int_0^1 B_{2k}(x) f^{(2k)}(x)\,dx.$$

Replacing $f(x)$ by $f(x + j)$ for any positive integer j gives

$$\int_0^1 f(x + j)\,dx = \frac{1}{2}(f(j) + f(j + 1)) - \sum_{i=1}^{k} \frac{B_{2i}}{(2i)!}(f^{(2i-1)}(j + 1)$$

$$- f^{(2i-1)}(j)) + \frac{1}{(2k)!} \int_0^1 B_{2k}(x) f^{(2k)}(x + j)\,dx.$$

Next, for any positive integer $N > 1$,

$$\int_0^N f(x)\,dx = \sum_{j=0}^{N-1} \int_j^{j+1} f(x)\,dx = \sum_{j=0}^{N-1} \int_0^1 f(x + j)\,dx,$$

$$\sum_{j=0}^{N-1} \frac{1}{2}(f(j) + f(j + 1)) = \sum_{j=0}^{N} f(j) - \frac{1}{2}(f(0) + f(N)),$$

$$\sum_{j=0}^{N-1} (f^{(2i-1)}(j + 1) - f^{(2i-1)}(j)) = f^{(2i-1)}(N) - f^{(2i-1)}(0).$$

Thus we establish the *Euler-Maclaurin summation formula*:

$$\int_0^N f(x)\,dx = \sum_{j=0}^{N} f(j) - \frac{1}{2}(f(0) + f(N)) - \sum_{i=1}^{k} \frac{B_{2i}}{(2i)!}(f^{(2i-1)}(N)$$

$$- f^{(2i-1)}(0)) + \frac{1}{(2k)!} \int_0^1 \sum_{j=0}^{N-1} B_{2k}(x) f^{(2k)}(x + j)\,dx. \quad (6.19)$$

Equivalently, we can write this as a relation between a finite sum and an integral

$$\sum_{j=0}^{N} f(j) = \int_0^N f(x)\,dx + \frac{1}{2}(f(0) + f(N))$$

$$+ \sum_{i=1}^{k} \frac{B_{2i}}{(2i)!}(f^{(2i-1)}(N) - f^{(2i-1)}(0)) + R_N, \quad (6.20)$$

where

$$R_N = -\frac{1}{(2k)!} \int_0^1 \sum_{j=0}^{N-1} B_{2k}(x) f^{(2k)}(x + j)\,dx$$

$$= -\frac{1}{(2k)!} \int_0^N B_{2k}(x - \lfloor x \rfloor) f^{(2k)}(x)\,dx.$$

6. Special Numbers

Here $\lfloor x \rfloor$ denotes the largest integer $\leq x$. The formulas (6.19) and (6.20) enable us to approximate integrals by sums or, conversely, to estimate the values of certain sums by means of integrals. For example, let $f(x) = \ln x$. Since $f^{(n)}(x) = (-1)^{n-1}(n-1)!x^{-n}$, by (6.20), we get

$$\sum_{i=1}^{n} \ln k = \int_0^n \ln x \, dx + \frac{1}{2}(\ln 1 + \ln n)$$
$$+ \frac{1}{2!} B_2 \left(\frac{1}{n} - \frac{1}{1}\right) + \frac{1}{4!} B_4 \left(\frac{2}{n^3} - \frac{2}{1^3}\right) + \frac{1}{6!} B_6 \left(\frac{24}{n^5} - \frac{24}{1^5}\right) + \cdots.$$

By using the known Bernoulli numbers, we have

$$\ln n! = n \ln n - n + \frac{1}{2} \ln n + \frac{1}{12n} - \frac{1}{360n^3} + \frac{1}{1250n^5} + R_n.$$

Now exponentiating both sides yields

$$n! = n^n e^{-n} \sqrt{n} e^{R_n} \exp\left(\frac{1}{12n} - \frac{1}{360n^3} + \frac{1}{1250n^5} + \cdots\right).$$

Appealing to the power series of e^x, we obtain

$$n! = n^n e^{-n} \sqrt{n} e^{R_n} \left(1 + \frac{1}{12n} + \frac{1}{288n^2} + \cdots\right). \tag{6.21}$$

A careful analysis shows that (see Exercise 13)

$$e^{R_n} \to \sqrt{2\pi} \quad \text{as } n \to \infty.$$

This, together with (6.21), yields the famous *Stirling's formula*

$$n! = \sqrt{2\pi n} \, n^n e^{-n} \left(1 + \frac{1}{12n} + \frac{1}{288n^2} + \cdots\right). \tag{6.22}$$

6.4.4 Power series in terms of B_k

Appealing to (6.10), we rewrite (6.9) as

$$\frac{x}{2} \coth \frac{x}{2} = \sum_{n=0}^{\infty} \frac{B_{2n}}{(2n)!} x^{2n}.$$

Substituting x by $2xi$ and noticing that $\coth(ix) = -i \cot x$, we find that

$$x \cot x = \sum_{n=0}^{\infty} (-1)^n \frac{2^{2n} B_{2n}}{(2n)!} x^{2n} = 1 - \frac{1}{3}x^2 - \frac{1}{45}x^4 - \frac{2}{945}x^6 - \cdots. \tag{6.23}$$

Furthermore, by the trigonometric identities

$$\tan x = \cot x - 2 \cot(2x), \quad \csc x = \cot x + \tan(x/2),$$

we obtain

$$\tan x = \sum_{n=1}^{\infty} (-1)^{n+1} \frac{2^{2n}(2^{2n}-1) B_{2n}}{(2n)!} x^{2n-1}, \qquad (6.24)$$

$$x \csc x = 1 + \sum_{n=1}^{\infty} (-1)^{n+1} \frac{2(2^{2n-1}-1) B_{2n}}{(2n)!} x^{2n}. \qquad (6.25)$$

There are many applications for these power series expansions, the most impressive of which links to the values of the Riemann zeta function at even positive integers. It is well known that $\zeta(2) = \pi^2/6$. Now, we conclude this chapter by deriving Euler's amazing formula: for any positive integer k,

$$\zeta(2k) = \sum_{n=1}^{\infty} \frac{1}{n^{2k}} = \frac{(-1)^{k-1} 2^{2k-1} B_{2k}}{(2k)!} \pi^{2k}. \qquad (6.26)$$

Recall Euler's infinite product (5.26), namely

$$\frac{\sin x}{x} = \prod_{n=1}^{\infty} \left(1 - \frac{x^2}{n^2 \pi^2}\right).$$

Taking the logarithm of both sides and then differentiating yields

$$\cot x - \frac{1}{x} = \sum_{n=1}^{\infty} \frac{-2x}{\pi^2 n^2 - x^2},$$

which results in Euler's elegant partial fraction expansion:

$$\cot x = \frac{1}{x} - \sum_{n=1}^{\infty} \frac{2x}{\pi^2 n^2 - x^2} = \frac{1}{x} - \sum_{n=1}^{\infty} \left(\frac{1}{n\pi - x} - \frac{1}{n\pi + x}\right). \qquad (6.27)$$

With the previous established power series expansion (6.23), we now have proven that

$$1 - 2 \sum_{n=1}^{\infty} \frac{x^2}{\pi^2 n^2 - x^2} = 1 + \sum_{k=1}^{\infty} \frac{(-1)^k 2^{2k} B_{2k}}{(2k)!} x^{2k}. \qquad (6.28)$$

In terms of a geometric series expansion,

$$\frac{x^2}{\pi^2 n^2 - x^2} = \frac{x^2/(\pi^2 n^2)}{1 - x^2/(\pi^2 n^2)} = \sum_{k=1}^{\infty} \frac{x^{2k}}{\pi^{2k} n^{2k}}.$$

Thus (6.28) becomes

$$1 - 2 \sum_{n=1}^{\infty} \left(\sum_{k=1}^{\infty} \frac{x^{2k}}{\pi^{2k} n^{2k}}\right) = 1 + \sum_{k=1}^{\infty} \frac{(-1)^k 2^{2k} B_{2k}}{(2k)!} x^{2k}.$$

Interchanging the order of summation on the left yields

$$1 + \sum_{k=1}^{\infty} \left(-2 \sum_{n=1}^{\infty} \frac{1}{\pi^{2k} n^{2k}}\right) x^{2k} = 1 + \sum_{k=1}^{\infty} \frac{(-1)^k 2^{2k} B_{2k}}{(2k)!} x^{2k}.$$

6. Special Numbers

Now (6.26) follows from equating the coefficients of x^{2k} on both sides. In particular, for $k = 2, 3, 4$, we get

$$\sum_{n=1}^{\infty} \frac{1}{n^4} = \frac{\pi^4}{90}, \quad \sum_{n=1}^{\infty} \frac{1}{n^6} = \frac{\pi^6}{945}, \quad \sum_{n=1}^{\infty} \frac{1}{n^8} = \frac{\pi^8}{9450}.$$

In light of (6.26), we see that $\zeta(2k)$ is a rational multiple of π^{2k}, and hence irrational. In spite of such success, Euler's original proof, as it emerged from (6.28), was geared toward even powers of x only, and thus unable to attain a formula for $\zeta(2k+1)$ analogous to (6.26). The question whether $\zeta(2k+1)$ is irrational has been asked since the time of Euler. For almost 200 years there had been no progress until 1978 when R. Apery proved the irrationality of $\zeta(3)$ by using only results known at the time of Euler. Despite considerable effort we still know very little about the irrationality of $\zeta(2k+1)$ for $k \geq 2$.

Exercises

1. Let F_n be the Fibonacci numbers and

$$a_n = F_0 + F_1 + F_2 + \cdots + F_n.$$

   Find a generating function for a_n and then prove that

$$a_n = F_{n+2} - 1.$$

2. Let $a_0 = 1$. For $n \geq 1$, define a_n by

$$a_n = \sum_{i=1}^{n-1} (n-i)a_i + 1.$$

   Show that $a_n = F_{2n-1}$.

3. **Monthly Problem 11258** [2006, 939; 2008, 949]. Let F_n denote the nth Fibonacci number, and let i denote $\sqrt{-1}$. Prove that

$$\sum_{k=0}^{\infty} \frac{F_{3^k} - 2F_{1+3^k}}{F_{3^k} + i\, F_{2 \cdot 3^k}} = i + \frac{1}{2}(1 - \sqrt{5}).$$

4. Let H_n be the nth harmonic number. For all $n \geq 2$, prove that

$$\sum_{k=1}^{n-1} H_k = nH_n - n$$

   and

$$H_n^2 \geq 2\left(\frac{H_2}{2} + \frac{H_3}{3} + \cdots + \frac{H_n}{n}\right).$$

5. Prove that

$$\ln^2(1-x) = 2 \sum_{n=1}^{\infty} \frac{H_n}{n+1} x^n$$

and
$$\ln^3(1-x) = -6\sum_{n=1}^{\infty}\left(\frac{1}{n+2}\sum_{k=1}^{n}\frac{H_k}{k+1}\right)x^{n+2}.$$

6. Let
$$a_n = \sum_{k=1}^{n}\left(1 - \frac{1}{2} + \frac{1}{3} - \cdots + \frac{(-1)^{k-1}}{k} - \ln 2\right).$$
Does the sequence $\{a_n\}$ converge? If so, to what value?

7. **Monthly Problem 11382** [2008, 665]. Show that if p is prime and $p > 5$, then
$$\sum_{k=1}^{p-1}\frac{H_k^2}{k} \equiv \sum_{k=1}^{p-1}\frac{H_k}{k^2} \pmod{p^2}.$$

Remark. Two rational numbers are congruent modulo d if their difference can be expressed as a reduced fraction of the form da/b with b relatively prime to a and d. A good entrance to this problem is Terence Tao's "Solving Mathematical Problems," Section 2.3.

8. Prove that if $f(x) = \sum_{n=1}^{\infty} a_n x^n$ converges, then
$$\sum_{n=1}^{\infty} a_n H_n x^n = \int_0^1 \frac{f(x) - f(tx)}{1-t}\,dt.$$
In particular, show that
$$\sum_{n=1}^{\infty}\frac{1}{n2^{n-1}}H_n = \zeta(2), \qquad \sum_{n=1}^{\infty}\frac{1}{n^2}H_{n-1} = \zeta(3).$$

9. Let $R_n = H_n - \ln(n+1)$ and γ be Euler's constant. Prove that
$$\gamma - R_n = \sum_{n=1}^{\infty}\frac{a_k}{(n+1)(n+2)\cdots(n+k)},$$
where
$$a_1 = \frac{1}{2},\ a_k = \frac{1}{k}\int_0^1 x(1-x)\cdots(k-1-x)dx \quad (k > 1).$$
For the reader familiar with *Stirling numbers of the first kind* $s(k, i)$, explicitly,
$$a_k = \frac{(-1)^{k+1}}{k}\sum_{i=1}^{k}\frac{s(k,i)}{i+1}.$$

10. **Monthly Problem 10949** [2002, 569; 2004, 264–265]. Show that, for each positive integer n,
$$\sigma(n) \leq H_n + 2\ln(H_n)e^{H_n},$$
with equality only for $n = 1$, where $\sigma(n) = \sum_{d|n} d$ denotes the sum of the divisors of n.

6. Special Numbers

11. Prove that

    (a) $\pi \tan \pi x = \sum_{n=-\infty}^{\infty} \frac{1}{n+1/2-x}$.
    (b) $\pi \sec \pi x = \sum_{n=-\infty}^{\infty} \frac{(-1)^n}{n+1/2+x}$.
    (c) $\frac{\pi^2}{\sin^2 \pi x} = \sum_{n=-\infty}^{\infty} \frac{1}{(x+n)^2}$.

12. Define $b_0 = 1, b_0(x) = 1$ and

    $$b_n(x) = n \int_0^x b_{n-1}(t)\, dt + b_n,$$

    where the constant b_n is chosen so that $\int_0^1 b_n(t)\, dt = 0$. Show that $b_n(x) = B_n(x)$ and $b_n = B_n$.

13. Show that

    $$\lim_{n \to \infty} \frac{n!}{n^n \sqrt{n} e^{-n}} = \sqrt{2\pi}.$$

14. Show that

    $$\gamma = \sum_{k=1}^{n} \frac{1}{k} - \ln n - \frac{1}{2n} + \sum_{k=1}^{m} \frac{B_{2k}}{2k} \frac{1}{n^{2k}} + \cdots.$$

15. Show that, for $n \geq 1$,

    $$\sum_{k=0}^{n-1} \binom{2n}{2k} \frac{B_{2k+2}}{(2k+2)} = \frac{1}{(2n+1)(2n+2)}.$$

16. **Monthly Problem E 3237** [1987, 995; 1989, 364–365]. Prove that

    $$B_m = \frac{1}{n(1-n^m)} \sum_{k=0}^{m-1} n^k \binom{m}{k} B_k \sum_{j=1}^{n-1} j^{m-j},$$

    for any positive integer $m, n > 1$.

17. Prove that

    $$\sum_{n=0}^{\infty} \frac{(-1)^n}{(2n+1)^{2k+1}} = (-1)^{k+1} \frac{(2\pi)^{2k+1}}{2(2k+1)!} B_{2k+1}(1/4).$$

18. Prove *Raabe's multiplication identity*

    $$\frac{1}{m} \sum_{k=0}^{m-1} B_n\left(x + \frac{k}{m}\right) = m^{-n} B_n(mx).$$

    Comment. Lehmer in [4] proves that the nth Bernoulli polynomial $B_n(x)$ is the unique monic polynomial of degree n which satisfies this identity. Therefore this identity provides another definition of the Bernoulli polynomials.

References

[1] P. Chandra and E. Weisstein, "Fibonacci Number." From MathWorld—A Wolfram Web Resource. mathworld.wolfram.com/FibonacciNumber.html

[2] T. Koshy, *Fibonacci and Lucas Numbers with Applications*, New York, John Wiley, 2001

[3] J. Lagarias, An elementary problem equivalent to the Riemann Hypothesis, *Amer. Math. Monthly*, **109** (2002) 534–543.

[4] D. H. Lehmer, A new approach to Bernoulli polynomials, *Amer. Math. Monthly*, **95** (1988) 905–911.

[5] N. J. A. Sloane, The On-Line Encyclopedia of Integer Sequences, research.att.com/~njas/sequences/

[6] Sondow, Jonathan and Weisstein, Eric W. "Harmonic Number." From MathWorld—A Wolfram Web Resource. mathworld.wolfram.com/HarmonicNumber.html

[7] K. R. Stromberg, *Introduction to classical real analysis,* Wadsworth, Belmont, California, 1981.

[8] D. Viswanath, Random Fibonacci sequences and the number 1.13198824..., *Math. Comp.*, **69** (2000) 1131–1155.

[9] Wikipedia, "Bernoulli Numbers," available at en.wikipedia.org/wiki/Bernoulli numbers

[10] H. Wilf, *Generatingfunctionology*, A. K. Peters, Ltd., 1994. It can be downloaded from math.upenn.edu/~wilf/DownldGF.html

7
On a Sum of Cosecants

Mathematics is the study of analogies between analogies. —G. C. Rota

Undergraduate research has been a longstanding practice in the experimental sciences. Only recently, however, have significant numbers of undergraduates begun to participate in mathematical research. While many math students are interested in research, they often don't know what problems are open or how to get started. Finding appropriate problems becomes of fundamental importance. Joseph Gallian has run an excellent undergraduate mathematics research program [4] at the University of Minnesota, Duluth since 1977. He offers the following criteria for successful undergraduate research problems:

- not much background reading is required;
- partial results are probable;
- the problem has been posed recently;
- new results will likely be publishable.

How does one actually find an appropriate problem? One popular way is to generalize or specialize an existing result and see what develops. As Zeilberger in [6] suggested: at first, leave the exact form of the generalization or specialization blank, and as you move forward, see what kind of generalization or specialization would be required to make the proof work. Keep 'guessing and erasing' until you get it done, just like doing a crossword puzzle.

This chapter is intended to illustrate this strategy by studying a trigonometric sum in the form of $\sum_{k=1}^{n-1} \csc\left(\frac{k\pi}{n}\right)$.

7.1 A well-known sum and its generalization

The well-known sum we have in mind is

$$\sum_{k=1}^{n-1} \sin\left(\frac{k\pi}{n}\right) = \cot\left(\frac{\pi}{2n}\right). \tag{7.1}$$

There are many ways to derive (7.1). Here we present a simple proof using complex numbers. Indeed, we derive a more general formula:

$$\sum_{k=1}^{n-1} \sin(kx) = \frac{\sin[(n-1)x/2]\sin(nx/2)}{\sin(x/2)}. \tag{7.2}$$

To see this, appealing to

$$1 + r + r^2 + \cdots + r^{n-1} = \frac{r^n - 1}{r - 1} \tag{7.3}$$

and De Moivre's formula, we get

$$\sum_{k=1}^{n-1} \sin(kx) = \text{Im}\left(\sum_{k=1}^{n-1} e^{ikx}\right) = \text{Im}\left(e^{ix}\frac{e^{i(n-1)x} - 1}{e^{ix} - 1}\right)$$

$$= \text{Im}\left(e^{inx/2}\frac{e^{i(n-1)x/2} - e^{-i(n-1)x/2}}{e^{ix/2} - e^{-ix/2}}\right)$$

$$= \frac{\sin[(n-1)x/2]}{\sin(x/2)}\text{Im}(e^{inx/2})$$

$$= \frac{\sin[(n-1)x/2]\sin(nx/2)}{\sin(x/2)},$$

as proposed. Now (7.1) follows from (7.2) by setting $x = \pi/n$.

In analogy to (7.1) we seek to compute in closed form

$$I_n = \sum_{k=1}^{n-1} \frac{1}{\sin(k\pi/n)} = \sum_{k=1}^{n-1} \csc\left(\frac{k\pi}{n}\right). \tag{7.4}$$

To begin with, based on the identity (5.2) and other favorable results [2] on the power sums $\sum_{k=1}^{n-1} \csc^{2p}(k\pi/n)$, it seemed to me that there ought to be a closed form for I_n. I spent many hours reading related journal papers, using Mathematica to explore and test potential results, and even consulting experts in the field via email. As a matter of fact, no closed form of I_n exists [1]. In lieu of this, could we have a 'nice' estimate for I_n and eventually establish an asymptotic formula for (7.4)? Well, searching for these answers offers a 'nice' problem. As we will see, this problem is hard enough that the results will be publishable yet accessible enough to be understood by analysis students. The investigation also indicates that a certain generalization of an existing result can produce quite a different problem, one of interest on its own. To show how to proceed in such an investigation, we take the reader through some rough estimates before deriving the best possible one. This demonstrates how special cases, that often are easier to solve, lead to insights and improvements of the existing results.

7.2 Rough estimates

Some rough estimates come from playing with the expression of (7.4) directly. We start with collecting some appropriate tools. The selection of the following results is pretty straightforward.

7. On a Sum of Cosecants

Jordan's inequality: For $0 < x \leq \pi/2$,

$$\frac{2x}{\pi} \leq \sin x < x.$$

Clearly, this is equivalent to

$$\frac{1}{x} < \csc x \leq \frac{\pi}{2x}. \tag{7.5}$$

Young's inequality: Let H_n be the nth harmonic number defined by (6.4) and let γ be Euler's constant defined by (6.7). Then

$$\frac{1}{2(n+1)} < H_n - \ln n - \gamma < \frac{1}{2n}. \tag{7.6}$$

In addition, for convenience, we introduce the *big O notation*: given two sequences $\{a_n\}$ and $\{b_n\}$ such that $b_n \geq 0$ for all n. We write

$$a_n = O(b_n)$$

if there exists a constant $M > 0$ such that $|a_n| \leq M b_n$ for all n.

Now we proceed to the estimates for I_n. Since

$$\csc\left(\frac{k\pi}{n}\right) = \csc\left(\frac{(n-k)\pi}{n}\right),$$

the second inequality of (7.5) implies

$$I_n \leq 2 \sum_{k=1}^{\lfloor n/2 \rfloor} \csc\left(\frac{k\pi}{n}\right) \leq n \sum_{k=1}^{\lfloor n/2 \rfloor} \frac{1}{k}.$$

Notice that

$$n \frac{1}{2\lfloor n/2 \rfloor} = \begin{cases} 1, & \text{if } n = 2m, \\ 1 + \frac{1}{2m}, & \text{if } n = 2m+1. \end{cases}$$

This, together with the second inequality of (7.6), yields that, for $n \geq 3$,

$$I_n \leq n(\ln n + \gamma - \ln 2) + 1 + O\left(\frac{1}{n}\right).$$

Next, to estimate I_n from below, for $n \geq 3$, we rewrite

$$I_n = \begin{cases} 2 \sum_{k=1}^{\lfloor (n-1)/2 \rfloor} \csc\left(\frac{k\pi}{n}\right), & \text{if } n \text{ is odd}, \\ 1 + 2 \sum_{k=1}^{\lfloor (n-1)/2 \rfloor} \csc\left(\frac{k\pi}{n}\right), & \text{if } n \text{ is even}. \end{cases}$$

By the first inequality of (7.5) we have

$$I_n \geq 2 \sum_{k=1}^{\lfloor (n-1)/2 \rfloor} \csc\left(\frac{k\pi}{n}\right) \geq \frac{2n}{\pi} \sum_{k=1}^{\lfloor (n-1)/2 \rfloor} \frac{1}{k}. \tag{7.7}$$

Clearly, if $n = 2m$,
$$\ln\left\lfloor\frac{n-1}{2}\right\rfloor = \ln(n-2) - \ln 2; \quad \frac{n}{2(\lfloor(n-1)/2\rfloor + 1)} = 1,$$
while if $n = 2m + 1$,
$$\ln\left\lfloor\frac{n-2}{2}\right\rfloor = \ln(n-1) - \ln 2; \quad \frac{n}{2(\lfloor(n-1)/2\rfloor + 1)} = 1 - \frac{1}{n+1}.$$

Recall that for $0 < x < 1$,
$$\ln(1-x) = -\sum_{k=1}^{\infty} \frac{x^k}{k}.$$

This yields
$$\ln(n-2) = \ln n + \ln\left(1 - \frac{2}{n}\right) = \ln n - \frac{2}{n} + O\left(\frac{1}{n^2}\right).$$

Finally, using the first inequality of (7.5) in (7.7) we have
$$I_n \geq \frac{2n}{\pi}(\ln n + \gamma - \ln 2) - \frac{2}{\pi} + O\left(\frac{1}{n}\right).$$

In summary, for $n \geq 3$, we have established the following estimate:
$$\frac{2n}{\pi}(\ln n + \gamma - \ln 2) - \frac{2}{\pi} + O\left(\frac{1}{n}\right) \leq I_n \leq n(\ln n + \gamma - \ln 2) + 1 + O\left(\frac{1}{n}\right). \quad (7.8)$$

7.3 Tying up the loose bounds

A careful look back of the derivation of (7.8), we see that these vague inequalities (7.5)–(7.6) make for a big gap between the upper and lower bounds. In order to obtain better estimates, it requires us to make the preceding proofs more precise. This is not an isolated case. In general, whenever we apply some underlying inequality to a new problem, the success or failure of the application will depend on our ability to reframe the problem so that the applied inequality is the best possible. To fix the bounds in (7.8) more firmly, we now turn to the precise analysis of the equality in (7.4) and take advantage of two expansions established in Chapter 6.

Rewriting the partial fraction expansion (6.27) as
$$\cot \pi x = \frac{1}{\pi x} + \frac{1}{\pi}\sum_{i=1}^{\infty}\left(\frac{1}{x-i} + \frac{1}{x+i}\right),$$
from $\csc \alpha = \cot(\alpha/2) - \cot \alpha$, we find
$$\csc \pi x = \frac{1}{\pi x} + \frac{1}{\pi}\sum_{i=1}^{\infty}(-1)^i\left(\frac{1}{x-i} + \frac{1}{x+i}\right),$$
thereby implying that
$$\csc\left(\frac{k\pi}{n}\right) = \frac{n}{\pi}\left(\frac{1}{k} + \sum_{i=1}^{\infty}(-1)^i\left(\frac{1}{k+ni} + \frac{1}{k-ni}\right)\right).$$

7. On a Sum of Cosecants

Since
$$\sum_{i=1}^{\infty} \frac{(-1)^i}{k-ni} = \frac{1}{n-k} + \sum_{i=2}^{\infty} \frac{(-1)^i}{k-ni},$$
shifting the index i to $i+1$ on the right-hand side gives
$$\sum_{i=1}^{\infty} \frac{(-1)^i}{k-ni} = \frac{1}{n-k} + \sum_{i=1}^{\infty} \frac{(-1)^i}{(n-k)+ni}.$$

Noticing
$$\sum_{k=1}^{n-1} \frac{1}{n-k} = \sum_{k=1}^{n-1} \frac{1}{k}, \quad \sum_{k=1}^{n-1} \frac{1}{(n-k)+ni} = \sum_{k=1}^{n-1} \frac{1}{k+ni},$$
and interchanging the summations yields
$$I_n = \frac{n}{\pi} \sum_{k=1}^{n-1} \left(\frac{1}{k} + \sum_{i=1}^{\infty} (-1)^i \left(\frac{1}{k+ni} + \frac{1}{k-ni} \right) \right)$$
$$= \frac{2n}{\pi} \sum_{k=1}^{n-1} \frac{1}{k} + \frac{2n}{\pi} \sum_{i=1}^{\infty} (-1)^i \sum_{k=1}^{n-1} \frac{1}{k+ni}. \tag{7.9}$$

Next, recall
$$H_n = \ln n + \gamma + \frac{1}{2n} + O\left(\frac{1}{n^2}\right).$$

This leads to
$$\sum_{k=1}^{n-1} \frac{1}{k+ni} = H_{n(i+1)} - H_{ni} - \frac{1}{n(i+1)}$$
$$= \ln\left(1 + \frac{1}{i}\right) - \frac{1}{n(i+1)} + \frac{1}{n} O(i^{-2}),$$

so that
$$\sum_{i=1}^{\infty} (-1)^i \sum_{k=1}^{n-1} \frac{1}{k+ni} = \sum_{i=1}^{\infty} (-1)^i \ln(1+1/i) - \frac{1}{n} \sum_{i=1}^{\infty} \frac{(-1)^i}{i+1} + \frac{O(\sum_{i=1}^{\infty} i^{-2})}{n}.$$

Appealing to Wallis' formula (5.30), which asserts that
$$\prod_{n=1}^{\infty} \frac{2n}{2n-1} \cdot \frac{2n}{2n+1} = \frac{\pi}{2},$$
we obtain
$$\sum_{i=1}^{\infty} (-1)^i \ln(1+1/i) = -\ln\left(\prod_{n=1}^{\infty} \frac{2n}{2n-1} \cdot \frac{2n}{2n+1}\right) = -\ln(\pi/2).$$

Furthermore, in view of the well-known facts that
$$\sum_{i=1}^{\infty} \frac{(-1)^i}{i+1} = \ln 2 - 1 \quad \text{and} \quad \sum_{i=1}^{\infty} \frac{1}{i^2} = \frac{\pi^2}{6},$$

we have
$$\sum_{i=1}^{\infty}(-1)^i \sum_{k=1}^{n-1}\frac{1}{k+ni} = -\ln(\pi/2) + O\left(\frac{1}{n}\right).$$
This, in conjunction with (7.9), yields
$$I_n = \frac{2n}{\pi}(\ln n + \gamma - \ln(\pi/2)) + O(1). \tag{7.10}$$
To sharpen the preceding term $O(1)$, we use (6.25) in the form of
$$\csc x = \sum_{i=0}^{\infty} a_i x^{2i-1} \tag{7.11}$$
with $a_0 = 1$ and $a_i > 0$. Since
$$(\ln(\tan(x/2)))' = \csc x, \tag{7.12}$$
from (7.11), integrating (7.12) provides
$$\ln(\tan(x/2)) - \ln(x/2) = \sum_{i=1}^{\infty} \frac{a_i}{2i} x^{2i}. \tag{7.13}$$
On the other hand, by (7.11) we have
$$I_n \le \frac{2n}{\pi}\sum_{k=1}^{[n/2]}\frac{1}{k} + 2\sum_{i=1}^{\infty} a_i (\pi/n)^{2i-1} \sum_{k=1}^{[n/2]} k^{2i-1}. \tag{7.14}$$
Since $n/[n/2] \le 3$, along the same lines we used in the rough estimate, we get
$$\frac{2n}{\pi}\sum_{k=1}^{[n/2]}\frac{1}{k} < \frac{2n}{\pi}(\ln n + \gamma + \ln 2) + \frac{3}{\pi}.$$
To estimate the double sum in (7.14), noticing
$$\sum_{k=1}^{[n/2]} k^{2i-1} < \int_0^{n/2} t^{2i-1}\, dt + (n/2)^{2i-1},$$
and using the equations (7.11) and (7.13), we have
$$2\sum_{i=1}^{\infty} a_i(\pi/n)^{2i-1}\sum_{k=1}^{[n/2]} k^{2i-1} < \frac{2n}{\pi}\sum_{i=1}^{\infty}\frac{a_i}{2i}(\pi/2)^{2i} + 2\sum_{i=1}^{\infty} a_i(\pi/2)^{2i-1}$$
$$= \frac{2n}{\pi}\left(\ln(\tan(\pi/4)) - \ln(\pi/4)\right) + 2\left(\csc(\pi/2) - \frac{2}{\pi}\right)$$
$$= -\frac{2n}{\pi}\ln(\pi/4) + 2 - \frac{4}{\pi}.$$
Collecting all this together, finally, we have established
$$I_n < \frac{2n}{\pi}(\ln n + \gamma - \ln(\pi/2)) + 2 - \frac{1}{\pi}. \tag{7.15}$$

7. On a Sum of Cosecants

7.4 Final Remarks

In the course of searching for the optimal bounds for I_n, we see the process evolve in the following basic pattern: we start with a few familiar tools, which obtain relatively simple results; we wish to improve these results and find that the familiar tools no longer serve for this purpose but reveal the direction of the improvement; then we appeal to either deeper mathematical tools or new ideas to reach a better solution of the problem at hand. As students begin to understand and work with this kind of methodology, they may develop their own internal monitors and become their own questioners. In this regard, I_n again provides a good example.

Let
$$J_n = \frac{2n}{\pi}(\ln n + \gamma - \ln(\pi/2)).$$

Mathematica 6.0 tabulates some selected differences of $J_n - I_n$:

n	$J_n - I_n$
3	0.0287351170881755128
4	0.0216641822966619335
5	0.0173744361336004932
6	0.0144985127666732041
7	0.0124376321763192829
8	0.0108888316916454100
9	0.0096825731521195440
10	0.0087166472057682210
100	0.0008726545781646500
1000	0.0000872664525524908
10000	0.0000087266462638730
100000	0.0000008726648900160

The numerical evidence now suggests that I_n is bounded by J_n along. We now have another new problem worthy of further investigation:

Open Problem: For $n \geq 3$, prove that
$$I_n \leq \frac{2n}{\pi}(\ln n + \gamma - \ln(\pi/2)).$$

Exercises

1. Show that
$$\sum_{k=0}^{n} \csc\left(\frac{x}{2^k}\right) = \cot\left(\frac{x}{2^{n+1}}\right) - \cot x.$$

2. Evaluate the limit
$$L = \lim_{n\to\infty} \frac{n}{2^n} \sum_{k=1}^{n} \frac{2^k}{k}.$$

3. Let $n \geq 2$. Prove that
$$\sum_{k=1}^{n-1} \csc\left(\frac{k\pi}{n}\right) = -\frac{1}{n} \sum_{k=0}^{n-1} (2k+1) \cot\left(\frac{(2k+1)\pi}{2n}\right).$$

4. Let $A_n = \sum_{k=n+1}^{\infty} (-1)^{k-1}/k$. Consider the inequalities
$$\frac{1}{2n+a} < |A_n| < \frac{1}{2n+b}.$$

   Kazarinoff (*Amer. Math. Monthly*, **62** (1955) 726–727) proved this estimate with $a = 2$ and $b = -2$. Find the smallest a and the largest b such that this estimate holds for every $n \geq 1$ or for every $n \geq n_0$, some positive integer. If replacing A_n by $\mathcal{A}_n = \sum_{k=n+1}^{\infty} (-1)^{k-1}/(2k-1)$, establish the corresponding estimates.

5. Determine the best possible constants α and β such that inequalities
$$\frac{e}{2n+\alpha} \leq e - \left(1 + \frac{1}{n}\right)^n \leq \frac{e}{2n+\beta}$$

   hold for every $n \geq 1$.

6. **Monthly Problem E 3432** [1991, 264; 1992, 684–685]. Prove that for every positive integer n we have
$$\frac{1}{2n+2/5} < H_n - \ln n - \gamma < \frac{1}{2n+1/3}.$$

   Show that $2/5$ can be replaced by a slightly smaller number, but $1/3$ cannot be replaced by a slightly large number.

   Remark. The best possible constant to replace $2/5$ is $1/(1-\gamma) - 2$.

7. Prove (7.10) by applying the trapezoidal rule to the integral
$$\int_{1/2n}^{1-1/2n} \left(\frac{1}{\sin(\pi x)} - \frac{1}{\pi x(1-x)}\right) dx,$$

   This approach is due to Michael Renardy. See
   www.siam.org/journals/problems/downloadfiles/05-001s.pdf

8. Prove the complete asymptotic expansion
$$I_n \sim \frac{2n}{\pi}(\ln n + \gamma - \ln(\pi/2)) + \frac{2n}{\pi} \sum_{k=1}^{\infty} \frac{2^{2k-1}-1}{k(2k)!} B_{2k}^2 \left(-\frac{\pi^2}{n^2}\right)^k,$$

   where B_{2k} are Bernoulli numbers.

9. **Open Problem.** Can one establish (7.10) via the Euler-Maclaurin summation formula?

10. **Open Problem.** In 1890 Mathieu defined $S(x)$, now called *Mathieu's series*, as

$$S(x) := \sum_{k=1}^{\infty} \frac{2k}{(k^2 + x^2)^2}.$$

Various inequalities have been established on $S(x)$. For example, Alzer et al. proved that

$$\frac{1}{x^2 + 1/(2\zeta(3))} < S(x) < \frac{1}{x^2 + 1/6}, \quad \text{for all } x \in \mathbb{R}.$$

Although this estimate is asymptotically sharp, it leaves a big gap for $x \approx 0$. Can one find a best possible estimate of $S(x)$ applicable on the entire $\mathbb{R}$? The following interesting observation,

$$S(x) = \sum_{k=1}^{\infty} \frac{1}{x} \int_0^{\infty} e^{-kt} t \sin(xt)\, dt = \frac{1}{x} \int_0^{\infty} \frac{t}{e^t - 1} \sin(xt)\, dt,$$

may be useful.

References

[1] J. Borwein, personal communication.

[2] H. Chen, On some trigonometric power sums, *Internat. J. Math. Math. Sci.*, **30** (2002) 185–191.

[3] ——, Bounds for a trigonometric sum — a case study of undergraduate math research, *Internat. J. Math. Ed. Sci. Tech.*, **37** (2006) 215–223.

[4] J. Gallian, The Duluth Undergraduate Research Program, available at www.d.umn.edu/~jgallian/progdesc.html

[5] Wolfram, "The best-known properties and formulas for the cosecant function," available at functions.wolfram.com/ElementaryFunctions/Csc/introductions/Csc/05/

[6] D. Zeilberger, The method of undetermined generalization and specialization, *Amer. Math. Monthly*, **103** (1996) 233–239.

8

The Gamma Products in Simple Closed Forms

> *The art of doing mathematics consists in finding that special case which contains all the germs of generality.* — D. Hilbert

To be a good problem solver, it is not just what you know, but how you use what you know. As we have seen in previous chapters, many formulas and theorems emerge from the analysis of specific calculations and special cases. Halmos once said "the source of all great mathematics is the special case, the concrete example. It is frequent in mathematics that every instance of a concept of seemingly great generality is in essence the same as a small and concrete special case." Thus, to solve a difficult problem, one should initially make the problem concrete and physical whenever possible. Often, building a base case will give us understanding and insight. After recognizing the similarities between the base case and the original problem, one may eventually extract the idea or technique that obtains the more general result. This principle of "understanding special cases deeply" often gives one the feeling 'oh, I see now' and often offers a route to solving a problem. In this chapter, we demonstrate how to use this principle by studying products involving the gamma function.

Of the so-called "special functions", the gamma function is one of the most fundamental (see Chapter 1 of [1]). This function is closely related to factorials and crops up in many unexpected places in analysis and statistics, especially in systematically building integration formulas, summing infinite series, and defining probability distributions. In the wake of these processes, many remarkable identities appeared. These identities are simple enough to introduce to students in elementary analysis but deep enough to have called forth contributions from the finest mathematicians. As a first example we choose a particularly striking identity — *Gauss's multiplication formula*. It states that for any real number $x > 0$ and any positive integer n,

$$\prod_{k=1}^{n-1} \Gamma\left(x + \frac{k}{n}\right) = (2\pi)^{(n-1)/2} n^{1/2-nx} \Gamma(nx), \qquad (8.1)$$

where $\Gamma(x)$ denotes the usual gamma function, defined by

$$\Gamma(x) = \int_0^\infty t^{x-1} e^{-t}\, dt.$$

Integration by parts establishes the recursion

$$\Gamma(x+1) = x\Gamma(x). \tag{8.2}$$

Letting $x = 1$ in (8.1), using (8.2), and simplifying, we get

$$\prod_{k=1}^{n-1} \Gamma\left(\frac{k}{n}\right) = (2\pi)^{(n-1)/2} n^{-1/2}. \tag{8.3}$$

The gamma products we have in mind are variations of (8.3). We are interested in finding simple values for products of fewer factors. Let A denote a proper subset of $\{1, 2, \ldots, n-1\}$ and (k, n) be the greatest common divisor of k and n. Consider the gamma product

$$\prod_{k \in A, (k,n)=1} \Gamma\left(\frac{k}{n}\right). \tag{8.4}$$

Emboldened by Gauss's success, we may ask two things:

I. Does (8.4) still retain a closed form?

II. Is it possible to identify those sets A such that (8.4) possesses a value analogous to (8.3)?

There are countless interesting answers for Problem I from the well-known *reflection formula*

$$\Gamma(x)\Gamma(1-x) = \frac{\pi}{\sin \pi x} \tag{8.5}$$

to the more involved Chowla-Selberg formulas in [4] such as

$$\Gamma^2(1/4) = 4\sqrt{\pi} K(k_1), \quad \Gamma(1/8)\Gamma(3/8) = (\sqrt{2}-1)^{1/2} 2^{13/4} \pi^{1/2} K(k_2),$$

where $K(k)$ is a *complete elliptic integral of the first kind* and k_i is the ith *elliptic integral singular values* (see Chapter 1 of [2] and [3, 6, 7] for details). However, Problem II apparently has not received much attention. Most existing results are mere consequences of (8.3) such as

$$\Gamma(1/4)\Gamma(3/4) = \sqrt{2}\pi, \quad \Gamma(1/8)\Gamma(3/8)\Gamma(5/8)\Gamma(7/8) = 2\sqrt{2}\pi^2.$$

In the following, motivated by a recently posed Monthly problem, we first investigate a base case and solve it by two different methods. The kernels of sophisticated ideas in both methods enable us to identity a large class of nontrivial gamma products that solve Problem II. Our study here, solely using a special case of (8.1), (8.3) and (8.5), is particularly accessible to undergraduate analysis students. Surprisingly, in contrast to the usual derivations in [3, 6, 7], elliptical integrals, elliptic integral singular values and zeta functions do not manifest at all.

Recently, Monthly Problem 11426 in [5], proposed by Glasser, asks the reader to find

$$\frac{\Gamma(1/14)\,\Gamma(9/14)\,\Gamma(11/14)}{\Gamma(3/14)\,\Gamma(5/14)\,\Gamma(13/14)}.$$

8. The Gamma Products in Simple Closed Forms

In the course of solving this problem, we actually find

$$\Gamma(1/14)\,\Gamma(9/14)\,\Gamma(11/14) = 4\pi^{3/2}, \tag{8.6}$$

$$\Gamma(3/14)\,\Gamma(5/14)\,\Gamma(13/14) = 2\pi^{3/2}. \tag{8.7}$$

These will serve as our base case for investigating Problem II. We will give two proofs. Although one proof is enough to establish the truth of these identities, we will see the ideas used in each proof are insightful for determining the general sets of A that solve Problem II.

First Proof. Start with *Legendre's duplication formula*, which is the special case of (8.1) with $n = 2$, namely

$$\Gamma(2x) = \pi^{-1/2} 2^{2x-1}\,\Gamma(x)\,\Gamma(x + 1/2). \tag{8.8}$$

Setting $x = 1/14, 9/14$ and $11/14$ in (8.8) successively, we have

$$\Gamma(1/7) = \pi^{-1/2} 2^{-6/7}\,\Gamma(1/14)\,\Gamma(4/7), \tag{8.9}$$

$$\Gamma(9/7) = \pi^{-1/2} 2^{2/7}\,\Gamma(9/14)\,\Gamma(8/7), \tag{8.10}$$

$$\Gamma(11/7) = \pi^{-1/2} 2^{4/7}\,\Gamma(11/14)\,\Gamma(9/7). \tag{8.11}$$

Applying (8.2) on both sides of (8.10) and (8.11) respectively yields

$$\Gamma(2/7) = \pi^{-1/2} 2^{-5/7}\,\Gamma(9/14)\,\Gamma(1/7), \tag{8.12}$$

$$\Gamma(4/7) = \pi^{-1/2} 2^{-3/7}\,\Gamma(11/14)\,\Gamma(2/7). \tag{8.13}$$

The product of the identities (8.9), (8.12) and (8.13) gives

$$\Gamma(1/7)\Gamma(2/7)\Gamma(4/7) = \pi^{-3/2} 2^{-2}\,\Gamma(1/14)\Gamma(4/7)\Gamma(9/14)\Gamma(1/7)\Gamma(11/14)\Gamma(2/7).$$

This proves (8.6) as desired after removal of the common factors. The identity (8.7) follows in a similar fashion.

The above proof relies on the cancellation, but there is no apparent general principle to show us how to group the gamma products to make the cancellation work. In the following, we present another proof by iterating (8.8) with different selected values. As a result, the gamma product is nearly always ready for cancellation.

Second Proof. Setting $x = 1/14, 2/14$ and $4/14$ in (8.8) recursively yields

$$\Gamma(2/14) = \pi^{-1/2} 2^{-6/7}\,\Gamma(1/14)\,\Gamma(1/14 + 1/2), \tag{8.14}$$

$$\Gamma(4/14) = \pi^{-1/2} 2^{-5/7}\,\Gamma(2/14)\,\Gamma(2/14 + 1/2), \tag{8.15}$$

$$\Gamma(8/14) = \pi^{-1/2} 2^{-3/7}\,\Gamma(4/14)\,\Gamma(4/14 + 1/2). \tag{8.16}$$

Since

$$\Gamma(8/14) = \Gamma(1/14 + 1/2),$$

multiplying the identities (8.14)–(8.16) and then eliminating the common factors yields

$$\Gamma(1/14)\Gamma(2/14 + 1/2)\Gamma(4/14 + 1/2) = 4\pi^{3/2}, \tag{8.17}$$

which is equivalent to (8.6).

Next, we prove (8.7) using (8.6). Multiplying both sides of (8.6) by the corresponding reflection part of each argument yields

$$\Gamma(1/14)\Gamma(13/14)\Gamma(9/14)\Gamma(5/14)\Gamma(11/14)\Gamma(3/14)$$
$$= 4(\pi)^{3/2}\Gamma(3/14)\Gamma(5/14)\Gamma(13/14). \tag{8.18}$$

By (8.5), the left-hand side of (8.18) is reduced to

$$\frac{\pi^3}{\sin(\pi/14)\sin(3\pi/14)\sin(5\pi/14)}.$$

So to show that (8.7) is true, it is enough to show

$$8\sin(\pi/14)\sin(3\pi/14)\sin(5\pi/14) = 1.$$

Since $\sin\alpha = \cos(\pi/2 - \alpha)$, appealing to the double angle formula for sine, we get

$$8\sin(\pi/14)\sin(3\pi/14)\sin(5\pi/14) = 8\sin(\pi/14)\cos(2\pi/14)\cos(4\pi/14)$$
$$= 4\sin(2\pi/14)\cos(2\pi/14)\cos(4\pi/14)/\cos(\pi/14)$$
$$= \sin(8\pi/14)/\cos(\pi/14)$$
$$= \sin(\pi/2 + \pi/14)/\cos(\pi/14)$$
$$= 1,$$

as required.

It is interesting to notice that (8.7) can be rewritten as

$$\Gamma(1/2 - 4/14)\Gamma(1/2 - 2/14)\Gamma(13/14) = 2\pi^{3/2}. \tag{8.19}$$

When we look back on this special case, the successive selection of the values $2^k/14$ offers us a convenient way to cancel common factors. Indeed, the final product on the left-hand side will leave only one gamma value. Thus, once this value on the left-hand side is removed, we obtain a gamma product which solves Problem II as expected. These are not isolated problems. It is remarkable that the ideas used in the special case are strong enough to extend to several entire families of gamma products. We first extend the special case to all even positive integers. Observe that, in the special case, the product of any two terms of (8.9)–(8.11) is not sufficient to cancel unnecessary terms thereby leading to a simple closed form; meanwhile, except the triples of (8.6) and (8.7), other combinations of three products do not have simple closed form either. In general, this forces us to take the gamma product over all odd terms and results in

Theorem 8.1. *For any positive integer k,*

$$\prod_{i=1}^{2k} \Gamma\left(\frac{2i-1}{2(2k)}\right) = 2^{(2k-1)/2}\pi^k, \tag{8.20}$$

$$\prod_{i=1}^{2k+1} \Gamma\left(\frac{2i-1}{2(2k+1)}\right) = 2^{(2k+1)/2}\pi^k. \tag{8.21}$$

8. The Gamma Products in Simple Closed Forms

Proof. We prove (8.20) only. (8.21) can be treated similarly. By (8.3) we have

$$\prod_{j=1}^{4k-1} \Gamma\left(\frac{j}{2(2k)}\right) = (2\pi)^{(4k-1)/2}(4k)^{-1/2}.$$

Splitting the even and odd indices j yields

$$\prod_{j=1}^{2k} \Gamma\left(\frac{2j-1}{2(2k)}\right) \cdot \prod_{j=1}^{2k-1} \Gamma\left(\frac{2j}{2(2k)}\right) = (2\pi)^{(4k-1)/2}(4k)^{-1/2}. \qquad (8.22)$$

Using (8.3) once more gives

$$\prod_{j=1}^{2k-1} \Gamma\left(\frac{2j}{2(2k)}\right) = \prod_{j=1}^{2k-1} \Gamma\left(\frac{j}{2k}\right) = (2\pi)^{(2k-1)/2}(2k)^{-1/2}. \qquad (8.23)$$

(8.20) now follows from (8.22) and (8.23).

The use of the recursive values of $2^i/14$ in (8.8) enables us to remove inductively the common factors. Thus, selection of the recursive values and cancelation help us reframe our problem so that an underlying product, whether it solves Problem II or not, could be found more efficiently. Indeed, setting $x = 1/n, 2/n, \ldots, 2^{m-1}/n$ in (8.8) successively, we find

$$\Gamma(2/n) = \pi^{-1/2} 2^{2/n-1} \Gamma(1/n) \Gamma(1/n + 1/2),$$
$$\Gamma(2^2/n) = \pi^{-1/2} 2^{2^2/n-1} \Gamma(2/n) \Gamma(2/n + 1/2),$$
$$\ldots\ldots\ldots\ldots\ldots$$
$$\Gamma(2^{m-1}/n) = \pi^{-1/2} 2^{2^{m-1}/n-1} \Gamma(2^{m-2}/n) \Gamma(2^{mk-2}/n + 1/2),$$
$$\Gamma(2^m/n) = \pi^{-1/2} 2^{2^m/n-1} \Gamma(2^{m-1}/n) \Gamma(2^{m-1}/n + 1/2).$$

Multiplying these identities and then canceling the common factors we have

$$\Gamma\left(\frac{2^m}{n}\right) = \pi^{-m/2} 2^{2(2^m-1)/n - m} \Gamma\left(\frac{1}{n}\right) \prod_{k=0}^{m-1} \Gamma\left(\frac{2^k}{n} + \frac{1}{2}\right). \qquad (8.24)$$

Thus, (8.24) will admit a simple closed form as long as one can eliminate $\Gamma(2^m/n)$. To this end, set $2^m/n = 1/n + 1/2$, which is equivalent to $n = 2(2^m - 1)$. In this case, appealing to (8.24) we actually have proved

Theorem 8.2. *Let m be a positive integer and $n = 2(2^m - 1)$. Then*

$$\Gamma\left(\frac{1}{n}\right) \prod_{k=1}^{m-1} \Gamma\left(\frac{2^k}{n} + \frac{1}{2}\right) = 2^{m-1} \pi^{m/2}. \qquad (8.25)$$

Next, we see (8.19) is the base case for the following generalization.

Theorem 8.3. *Let m be a positive integer and $n = 2(2^m - 1)$. Then*

$$\Gamma\left(\frac{n-1}{n}\right) \prod_{k=1}^{m-1} \Gamma\left(\frac{1}{2} - \frac{2^k}{n}\right) = 2\pi^{m/2}. \qquad (8.26)$$

Proof. Let S denote the left-hand side of (8.26). Multiplying both sides of (8.25) by S yields

$$\Gamma\left(\frac{1}{n}\right)\Gamma\left(\frac{n-1}{n}\right)\prod_{k=1}^{m-1}\Gamma\left(\frac{2^k}{n}+\frac{1}{2}\right)\Gamma\left(\frac{1}{2}-\frac{2^k}{n}\right) = 2^{m-1}\pi^{m/2}S.$$

By (8.5) and its equivalent

$$\Gamma\left(\frac{1}{2}-x\right)\Gamma\left(\frac{1}{2}+x\right) = \frac{\pi}{\cos\pi x},$$

we have

$$S = \frac{2\pi^{m/2}}{2^m \sin(\pi/n) \prod_{k=1}^{m-1}\cos(2^k\pi/n)}.$$

Now, to establish (8.25), it remains to show that for all $m \geq 2$,

$$2^m \sin(\pi/n) \prod_{k=1}^{m-1} \cos(2^k\pi/n) = 1.$$

Repeatedly using the double angle formula and in view of $2^m/n = 1/n + 1/2$, we have

$$2^m \sin\left(\frac{\pi}{n}\right) \prod_{k=1}^{m-1}\cos\left(\frac{2^k\pi}{n}\right) = \frac{2^{m-1}\sin(2\pi/n)}{\cos(\pi/n)} \prod_{k=1}^{m-1}\cos\left(\frac{2^k\pi}{n}\right)$$

$$= \frac{\sin(2^m\pi/n)}{\cos(\pi/n)} = 1,$$

as desired.

Setting $m = 3$, we recover (8.6)–(8.7) from (8.25)–(8.26), respectively. Moreover, we have

$$\Gamma(1/30)\,\Gamma(17/30)\,\Gamma(19/30)\Gamma(23/30) = 8\pi^2,$$
$$\Gamma(7/30)\,\Gamma(11/30)\,\Gamma(13/30)\Gamma(29/30) = 2\pi^2,$$
$$\Gamma(1/62)\,\Gamma(33/62)\,\Gamma(35/62)\Gamma(39/62)\Gamma(47/62) = 16\pi^{5/2}, \quad (8.27)$$
$$\Gamma(15/62)\,\Gamma(23/62)\,\Gamma(27/62)\Gamma(29/62)\Gamma(61/62) = 2\pi^{5/2}. \quad (8.28)$$

It appears that $\Gamma(1/n)$ has no simple closed form for $n > 3$. So the three theorems above characterize some families of gamma products with simple closed forms. We do not yet know whether these are the only families possessing such simple closed forms.

If we dispense with the restriction $(k,n) = 1$ in Problem II, (8.24) may yield more simple closed forms. For example, let $2^m/n = 2/n + 1/2$, i.e., $n = 2(2^m - 2)$. (8.24) becomes

$$\Gamma\left(\frac{1}{n}\right)\prod_{k=0,k\neq 1}^{m-1}\Gamma\left(\frac{2^k}{n}+\frac{1}{2}\right) = \pi^{m/2}2^{m-1-2/n}.$$

8. The Gamma Products in Simple Closed Forms

In particular,

$$\Gamma(1/12)\Gamma(7/12)\Gamma(5/6) = 2^{11/6}\pi^{3/2},$$
$$\Gamma(1/28)\Gamma(15/28)\Gamma(9/14)\Gamma(11/14) = 2^{41/14}\pi^{3/2},$$
$$\Gamma(1/60)\Gamma(31/60)\Gamma(17/30)\Gamma(19/30)\Gamma(23/30) = 2^{119/30}\pi^{5/2}.$$

Along these lines, (8.24) will generate many more gamma products in simple closed forms. The details are left to the reader.

To conclude this chapter, it is worth noticing that for a given problem there might be different ways to form a base case. For each base case, there may be more than one way to the solution. Some base cases might be more helpful than others in solving the problem at hand. Formulating good base questions may be just as important as posing general questions. In my personal experience, most of the Monthly problems could be good base problems due to the quality of the problems and richness in mathematical ideas. Indeed, many posed problems emerge from special cases in the proposer's research. Trying to generalize a Monthly problem often leads to a good undergraduate research problem.

Exercises

1. Prove that
$$\int_0^\infty \ln \Gamma(x)\, dx = \frac{1}{2} \ln(2\pi).$$

2. Prove that
$$\int_x^{x+1} \ln \Gamma(t)\, dt = x \ln x - x + C$$
and determine the value of C.

3. Recall that
$$\mathrm{agm}(a,b) = \left(\frac{2}{\pi} \int_0^{\pi/2} \frac{dx}{\sqrt{a^2 \sin^2 x + b^2 \cos^2 x}} \right)^{-1}.$$

   Prove that
$$\mathrm{agm}(1, 1/\sqrt{2}) = \frac{2\pi^{3/2}}{\Gamma^2(1/4)}.$$

4. For all $0 < a_i, b_i < 1$ with $\sum_{i=1}^k (a_i - b_i) = 0$, prove that
$$\prod_{n=1}^\infty \frac{(n-a_1)(n-a_2)\cdots(n-a_k)}{(n-b_1)(n-b_2)\cdots(n-b_k)} = \prod_{i=1}^k \frac{\Gamma(1-b_i)}{\Gamma(1-a_i)}.$$

5. **Weierstrass' product formula** states that
$$\frac{1}{\Gamma(x)} = xe^{\gamma x} \prod_{n=1}^\infty \left(1 + \frac{x}{n}\right) e^{-x/n},$$

where γ is Euler's constant defined by (5.8). Using this formula, prove
$$\sin \pi x = \pi x \prod_{n=1}^{\infty}\left(1 - \frac{x^2}{n^2}\right).$$

6. For $0 < x < 1$, show that
$$\frac{\pi}{\sin \pi x} = \frac{1}{x} + 2x \sum_{n=1}^{\infty} \frac{(-1)^n}{x^2 - n^2} = \lim_{n \to \infty} \sum_{k=-n}^{n} \frac{1}{x-k}$$
directly from the reflection formula without using Euler's sine product formula.

7. **Putnam Problem 1984-A5.** Let R be the region consisting of all triples (x, y, z) of nonnegative real numbers satisfying $x + y + z \leq 1$. Let $w = 1 - x - y - z$. Express the value of the triple integral
$$\iiint_R x^1 y^9 z^8 w^4 \, dx \, dy \, dz$$
in the form $a!b!c!d!/n!$, where a, b, c, d and n are positive integers.

Note. This is the special case of Liouville's integral: for all $p_i > 0$, $(1 \leq i \leq n)$,
$$L = \int \cdots \int_{\substack{x_1,\ldots,x_n \geq 0; \\ x_1+\cdots+x_n \leq 1}} f(x_1 + x_2 + \cdots + x_n) x_1^{p_1-1} x_2^{p_2-1} \cdots x_n^{p_n-1} dx_1 dx_2 \cdots dx_n.$$

8. Prove that
$$L = \frac{\Gamma(p_1)\Gamma(p_2)\cdots\Gamma(p_n)}{\Gamma(p_1 + p_2 + \cdots + p_n)} \int_0^1 f(t) t^{p_1+p_2+\cdots+p_n-1} \, dt.$$

Use this result to derive the volume and surface area of the n-dimensional unit ball, where **Putnam Problem 1951-B7** is a special case.

9. Using the above result, prove Gauss' multiplication formula.

10. For each positive integer n, show that
$$\sum_{k=0}^{n-1}\left[\Gamma\left(\frac{2k+1}{2n}\right)\Gamma\left(\frac{2n-2k-1}{2n}\right)\right]^2 = n^2 \pi^2.$$

11. Prove that
$$-\frac{\Gamma'(x)}{\Gamma(x)} = \gamma + \sum_{n=1}^{\infty}\left(\frac{1}{x+n-1} - \frac{1}{n}\right).$$

12. **Open Problem.** By (8.1) with $n = 3$ we have
$$\Gamma(3x) = (2\pi)^{-1} 3^{3x-1/2} \Gamma(x)\Gamma(x+1/3)\Gamma(x+2/3).$$
Find some products, as general as possible, in the simple closed forms.

8. The Gamma Products in Simple Closed Forms

13. **Open Problem.** Similar to the first proof, we have

$$\Gamma(3/62)\,\Gamma(17/62)\,\Gamma(37/62)\Gamma(43/62)\Gamma(55/62) = 8\pi^{5/2},$$
$$\Gamma(5/62)\,\Gamma(9/62)\,\Gamma(41/62)\Gamma(49/62)\Gamma(51/62) = 8\pi^{5/2},$$
$$\Gamma(7/62)\,\Gamma(19/62)\,\Gamma(25/62)\Gamma(45/62)\Gamma(59/62) = 4\pi^{5/2},$$
$$\Gamma(11/62)\,\Gamma(13/62)\,\Gamma(21/62)\Gamma(53/62)\Gamma(57/62) = 4\pi^{5/2}.$$

(a) Using these results as the base cases, can you find some general formulas similar to (8.25) and (8.26)?

(b) Adding the identities (8.27) and (8.28) that derived in the text, we see the numerator set

$$\{(1, 33, 35, 39, 47), (15, 23, 27, 29, 61), (3, 17, 37, 43, 55)$$
$$(5, 9, 41, 49, 51), (7, 19, 25, 45, 59), (11, 13, 21, 53, 57)\}$$

is an exactly disjoint partition of all odd numbers less than 62, exclusive of 31, where $\Gamma(1/2) = \sqrt{\pi}$ is well known. They are the only number sets that admit the simple closed forms. Can you generalize this observation?

Remark. Recently, Nijenhuis characterized the case $n \equiv 2 \pmod 4$ via group theory approach. See arXiv:0907.1689v1.

14. Let p be prime and $n = p^m$. Show that

$$\prod_{(k,n)=1} \Gamma\left(\frac{k}{n}\right) = \frac{(2\pi)^{\phi(n)/2}}{\sqrt{p}},$$

where $\phi(n)$ is the Euler-phi function.

References

[1] G. Andrews, R. Askey & R. Roy, *Special Functions*, Encyclopedia of Mathematics and Its Applications 71, Cambridge University Press, 1999.

[2] J. Borwein and P. Borwein, *Pi and the AGM*, John Wiley, New York, 1987.

[3] J. Borwein and J. Zucker, Fast evaluation of the Gamma function for small rational fractions using complete elliptic integrals of the first kind, *IMA J. Numerical Analysis*, **12**(1992) 519–526.

[4] S. Chowla and A. Selberg, On Epstein's zeta function, *J. Reine Agnew. Math.*, **227**(1967) 86–110.

[5] M. L. Glasser, Problem 11426, *Amer. Math. Monthly*, **116**(2009) 365.

[6] Eric W. Weisstein, "Elliptic Integral Singular Value." From MathWorld—A Wolfram Web Resource. mathworld.wolfram.com/EllipticIntegralSingularValue.html

[7] ——, "Gamma Function." From MathWorld—A Wolfram Web Resource. mathworld.wolfram.com/GammaFunction.html

9

On the Telescoping Sums

In analysis, there are only a few series with partial sums that we can calculate in closed form and almost all of these come from telescoping sums. In this chapter, we apply this technique to a variety of problems, beginning with some that are arithmetic and trigonometric, moving to others that involve Apéry-like formulas and famous families of numbers, and finally ending with a class of problems that can be solved algorithmically.

We begin with Gauss's legendary method for summing arithmetic series. Consider an arithmetic series of n terms with first term a, last term b and common difference d. Write the sum twice in the form of

$$S = a + (a+d) + \cdots + (b-d) + b,$$
$$S = b + (b-d) + \cdots + (a+d) + a.$$

Summing the terms vertically in pairs yields $2S = (a+b)n$, which has a solution

$$S = \frac{(a+b)n}{2}.$$

When we turn to a geometric series $\sum_{k=1}^{\infty} ar^k$, since the general terms are not additively symmetric, Gauss's pairing trick can not be applied anymore. However, rather than writing the partial sum S twice, we look at S and rS:

$$S = a + ar + ar^2 + \cdots + ar^{n-1},$$
$$rS = ar + ar^2 + ar^3 + \cdots + ar^n.$$

Instead of adding, we subtract the two equalities yielding

$$S - rS = a - ar + ar - ar^2 + \cdots + ar^{n-1} - ar^n. \tag{9.1}$$

Now, all terms on the right cancel out except for first and last; solving for S yields

$$S = \frac{a(1-r^n)}{1-r}.$$

Since the expression (9.1) on the right contracts the way a telescope does, it is called a *telescoping* sum.

But this is not the end of the story. Following the idea above, in general, we call a sequence $\{a_k\}$ *telescoping* if there exists a sequence $\{b_k\}$ with the property that

$$a_k = b_{k+1} - b_k, \quad \text{for all } k \geq 1.$$

In this case, we have

$$\sum_{k=1}^{n} a_k = \sum_{k=1}^{n} (b_{k+1} - b_k)$$
$$= (b_2 - b_1) + (b_3 - b_2) + \cdots + (b_{n+1} - b_n)$$
$$= b_{n+1} - b_1.$$

In particular, if $\lim_{k \to \infty} b_k$ exists, we find the infinite series

$$\sum_{k=1}^{\infty} a_k = \lim_{k \to \infty} b_k - b_1.$$

This technique is useful in two ways. First, given a sequence $\{b_k\}$ it is easy to generate a sequence $\{a_k\}$ such that $\sum_{k=1}^{n} a_k = b_{n+1} - b_1$. This enables us to construct many convergent and divergent infinite series. Second, it determines the sums of the convergent series. A natural question arises, namely, how can we telescope a series? That is, how can we find such a sequence $\{b_k\}$ from the given one? As usual there are no general methods to aid the search for the sequence. In elementary calculus, this technique is usually introduced in a few examples, and students regard it as a clever but unexpected trick. Now, we are ready to explore various approaches for constructing telescoping sums, and to show how they can be systematically used to evaluate various series. More importantly, we will see that this method is not as exotic as one might think.

9.1 The sum of products of arithmetic sequences

Let $\{a_k\}$ be an arithmetic sequence with a nonzero common difference d. Thus, $a_n = a_0 + nd$. For any nonnegative integer m, we have

$$a_k a_{k+1} \cdots a_{k+m+1} - a_{k-1} a_k \cdots a_{k+m} = a_k a_{k+1} \cdots a_{k+m} (a_{k+m+1} - a_{k-1})$$
$$= (m+2)d \, a_k a_{k+1} \cdots a_{k+m},$$

and so

$$a_k a_{k+1} \cdots a_{k+m} = \frac{1}{(m+2)d} \left(a_k a_{k+1} \cdots a_{k+m+1} - a_{k-1} a_k \cdots a_{k+m} \right).$$

Therefore, the sequence $\{a_k a_{k+1} \cdots a_{k+m}\}$ is telescoping, and

$$\sum_{k=1}^{n} a_k a_{k+1} \cdots a_{k+m} = \frac{1}{(m+2)d} \left(a_n a_{n+1} \cdots a_{n+m+1} - a_0 a_1 \cdots a_{1+m} \right). \quad (9.2)$$

9. On the Telescoping Sums

In particular, setting $a_k = k$ in (9.2) yields

$$\sum_{k=1}^{n} k(k+1)\cdots(k+m) = \frac{n(n+1)\cdots(n+m+1)}{m+2}, \tag{9.3}$$

from which the closed formulas for the sums of powers of positive integers follow easily. Indeed, taking $m = 0$ and $m = 1$ in (9.3), respectively, yields

$$\sum_{k=1}^{n} k = \frac{n(n+1)}{2} \tag{9.4}$$

and

$$\sum_{k=1}^{n} k(k+1) = \frac{n(n+1)(n+2)}{3}. \tag{9.5}$$

Therefore,

$$\sum_{k=1}^{n} k^2 = \sum_{k=1}^{n} [k(k+1) - k] = \frac{n(n+1)(2n+1)}{6}. \tag{9.6}$$

Similarly, substituting $m = 2$ in (9.3) gives

$$\sum_{k=1}^{n} k(k+1)(k+2) = \frac{n(n+1)(n+2)(n+3)}{4}.$$

Combining this formula with (9.4) and (9.5) results in

$$\sum_{k=1}^{n} k^3 = \sum_{k=1}^{n} [k(k+1)(k+2) - 3k(k+1) + k] = \frac{n^2(n+1)^2}{4}. \tag{9.7}$$

In most beginning calculus textbooks, formulas (9.4), (9.6) and (9.7), which often are proved by mathematical induction, are useful in the evaluation of such integrals as $\int_0^1 x^k \, dx$ by appealing to the definition of a Riemann integral. Students often ask how to find these formulas in the first place. Using (9.3), we now have a method for their discovery.

9.2 The sum of products of reciprocals of arithmetic sequences

Under the same assumptions in the preceding on $\{a_k\}$, for any positive integer m, since

$$\frac{1}{a_{k+1}a_{k+2}\cdots a_{k+m}} - \frac{1}{a_k a_{k+1}\cdots a_{k+m-1}} = \frac{a_k - a_{k+m}}{a_k a_{k+1}\cdots a_{k+m}}$$

$$= -\frac{md}{a_k a_{k+1}\cdots a_{k+m}},$$

it follows that

$$\frac{1}{a_k a_{k+1}\cdots a_{k+m}} = \frac{1}{md}\left(\frac{1}{a_k a_{k+1}\cdots a_{k+m-1}} - \frac{1}{a_{k+1}a_{k+2}\cdots a_{k+m}}\right).$$

Therefore, the sequence $\left\{\dfrac{1}{a_k a_{k+1} \cdots a_{k+m}}\right\}$ is telescoping, and

$$\sum_{k=1}^{n} \frac{1}{a_k a_{k+1} \cdots a_{k+m}} = \frac{1}{md}\left(\frac{1}{a_1 a_2 \cdots a_m} - \frac{1}{a_{n+1} a_{n+2} \cdots a_{n+m}}\right). \qquad (9.8)$$

For $a_k = a + kd$ and $m = 1, 2$, we may specify (9.8) respectively, as

$$\sum_{k=1}^{n} \frac{1}{(a+kd)[a+(k+1)d]} = \frac{1}{d}\left(\frac{1}{a+d} - \frac{1}{a+(n+1)d}\right) \qquad (9.9)$$

and

$$\sum_{k=1}^{n} \frac{1}{(a+kd)[a+(k+1)d][a+(k+2)d]}$$
$$= \frac{1}{2d}\left(\frac{1}{(a+d)(a+2d)} - \frac{1}{[a+(n+1)d][a+(n+2)d]}\right). \qquad (9.10)$$

The well-known sums

$$\sum_{k=1}^{n} \frac{1}{k(k+1)} = 1 - \frac{1}{n+1} \qquad (9.11)$$

and

$$\sum_{k=1}^{n} \frac{1}{k(k+1)(k+2)} = \frac{1}{4} - \frac{1}{2(n+1)(n+2)} \qquad (9.12)$$

follow immediately from setting $a = 0$, $d = 1$ in (9.9) and (9.10). Sometimes it is worthwhile to manipulate a sequence into a slightly different form before applying (9.8). For example, since

$$\frac{1}{k(k+2)} = \frac{k+1}{k(k+1)(k+2)} = \frac{1}{(k+1)(k+2)} + \frac{1}{k(k+1)(k+2)},$$

with (9.11) and (9.12), this leads to

$$\sum_{k=1}^{n} \frac{1}{k(k+2)} = \frac{3}{4} - \frac{2n+3}{2(n+1)(n+2)}.$$

Generally, applying partial fractions, we get

$$\sum_{k=1}^{n} \frac{1}{k(k+m)} = \frac{n}{m} \sum_{k=1}^{m} \frac{1}{k(k+n)} \qquad (9.13)$$

and

$$\sum_{k=1}^{n} \frac{1}{(k+l)(k+m)} = \frac{n}{m-l} \sum_{k=1}^{m-l} \frac{1}{(k+l)(k+l+n)}. \qquad (9.14)$$

9. On the Telescoping Sums

Furthermore, appealing to formulas (9.8), (9.13) and (9.14), we can establish the following infinite sums:

$$\sum_{k=1}^{\infty} \frac{1}{k(k+m)} = \frac{1}{m}\left(1 + \frac{1}{2} + \cdots + \frac{1}{m}\right),$$

$$\sum_{k=1}^{\infty} \frac{1}{(k+l)(k+m)} = \frac{1}{m-l}\left(\frac{1}{l+1} + \cdots + \frac{1}{m}\right), \quad (m > l),$$

$$\sum_{k=1}^{\infty} \frac{1}{(a+kd)\cdots[a+(k+m)d]} = \frac{1}{md(a+d)(a+2d)\cdots(a+md)}. \tag{9.15}$$

9.3 Trigonometric sums

In this section, we show how to telescope various sums involving the trigonometric functions.

9.3.1 Angles in arithmetic sequences

First, similar to Gauss's pairing trick, we derive the formula

$$S := \sum_{k=0}^{n} \sin(x+k\theta) = \frac{\sin\left(x + \frac{n}{2}\theta\right)\sin\left(\frac{n+1}{2}\theta\right)}{\sin\left(\frac{\theta}{2}\right)}. \tag{9.16}$$

Write S twice in the form of

$$S = \sin x + \sin(x+\theta) + \cdots + \sin(x+n\theta),$$
$$S = \sin(x+n\theta) + \sin[x+(n-1)\theta] + \cdots + \sin x.$$

Summing the terms vertically in pairs and using the sum to product formula

$$\sin\alpha + \sin\beta = 2\sin\frac{\alpha+\beta}{2}\cos\frac{\alpha-\beta}{2},$$

we get

$$2S = 2\sin\left(x + \frac{n}{2}\theta\right)\sum_{k=0}^{n}\cos\left(k\theta - \frac{n}{2}\theta\right).$$

To eliminate the cosine sum, we multiply both sides by $\sin\theta/2$ and then apply the product to sum formula

$$\sin\alpha\cos\beta = \frac{1}{2}[\sin(\alpha+\beta) - \sin(\beta-\alpha)] \tag{9.17}$$

to obtain

$$2S\sin\theta/2 = \sin\left(x + \frac{n}{2}\theta\right)\sum_{k=0}^{n}\left(\sin\left((k+1)\theta - \frac{n+1}{2}\theta\right) - \sin\left(k\theta - \frac{n+1}{2}\theta\right)\right).$$

Telescoping the sum yields

$$2S\sin\theta/2 = 2\sin\left(x + \frac{n+1}{2}\theta\right)\sin\left(\frac{n+1}{2}\theta\right),$$

thereby implying (9.16) as required.

Next, differentiating (9.16) with respect to x gives an analogous formula for the summation of cosines, namely

$$\sum_{k=0}^{n} \cos(x + k\theta) = \frac{\cos\left(x + \frac{n}{2}\theta\right) \sin\left(\frac{n+1}{2}\theta\right)}{\sin\left(\frac{\theta}{2}\right)}. \tag{9.18}$$

Furthermore, combining (9.16) and (9.18), we get another interesting formula,

$$\frac{\sin x + \sin(x + \theta) + \cdots + \sin(x + n\theta)}{\cos x + \cos(x + \theta) + \cdots + \cos(x + n\theta)} = \tan\left(x + \frac{n}{2}\theta\right).$$

Along the same lines, one can establish the following alternating sums which are perhaps less well known:

$$\sum_{k=0}^{n} (-1)^k \sin(x + k\theta) = \frac{\sin\left(\frac{n+1}{2}(\theta + \pi)\right) \sin\left(x + \frac{n}{2}(\theta + \pi)\right)}{\cos\left(\frac{\theta}{2}\right)},$$

$$\sum_{k=0}^{n} (-1)^k \cos(x + k\theta) = \frac{\sin\left(\frac{n+1}{2}(\theta + \pi)\right) \cos\left(x + \frac{n}{2}(\theta + \pi)\right)}{\cos\left(\frac{\theta}{2}\right)}.$$

Finally, we derive a closed form for

$$T_s := \sum_{k=1}^{n-1} r^k \sin k\theta$$

in the same manner as we handled the sum in (9.1). Notice that

$$(1 - 2r \cos \theta + r^2) T_s = r \sin \theta + r^2 \sin 2\theta + \cdots + r^{n-2} \sin(n-2)\theta + r^{n-1} \sin(n-1)\theta$$
$$- 2r^2 \sin \theta \cos \theta - \cdots - 2r^{n-1} \sin(n-2)\theta \cos \theta$$
$$- 2r^n \sin(n-1)\theta \cos \theta r^3 \sin \theta + \cdots + r^{n-1} \sin(n-3)\theta$$
$$+ r^n \sin(n-2)\theta + r^{n+1} \sin(n-1)\theta.$$

Let $[r^k]$ denote the coefficient of r^k. Combining the like terms yields

$$[r^k] = \begin{cases} \sin 2\theta - 2 \sin \theta \cos \theta, & k = 2, \\ \sin(k\theta) - 2 \sin(k-1)\theta \cos \theta + \sin(k-2)\theta, & 2 < k \leq n-1, \\ -2 \sin(n-1)\theta \cos \theta + \sin(n-2)\theta, & k = n. \end{cases}$$

Appealing to (9.17), we see that all $[r^k]$ vanish for $2 \leq k \leq n-1$ except $[r^n]$, which is simplified to $-\sin n\theta$. Therefore

$$T_s = \frac{r \sin \theta - r^n \sin n\theta + r^{n+1} \sin(n-1)x}{1 - 2r \cos \theta + r^2}. \tag{9.19}$$

For $|r| < 1$, this implies

$$\sum_{k=1}^{\infty} r^k \sin k\theta = \frac{r \sin \theta}{1 - 2r \cos \theta + r^2}. \tag{9.20}$$

9.3.2 Angles in geometric sequences

We begin with the trigonometric identity

$$\sin^3 x = \frac{3}{4}\sin x - \frac{1}{4}\sin(3x).$$

Replacing x by $3^k x$, multiplying both sides by $1/3^k$, and then summing over k from 0 to n, gives

$$\sum_{k=0}^{n} \frac{\sin^3(3^k x)}{3^k} = \sum_{k=0}^{n}\left(\frac{1}{4 \cdot 3^{k-1}}\sin(3^k x) - \frac{1}{4 \cdot 3^k}\sin(3^{k+1} x)\right)$$

$$= \frac{3}{4}\sin x - \frac{\sin(3^{n+1} x)}{3^n \cdot 4}.$$

Similarly, we find

$$\sum_{k=0}^{n}(-1)^k \frac{\cos^3(3^k x)}{3^k} = \frac{3}{4}\cos x + (-1)^n \frac{\cos(3^{n+1} x)}{3^n \cdot 4}.$$

Next, it is straightforward to verify that

$$\frac{\sin(2x)}{1 + 2\cos(2x)} = \frac{1}{4}(\cot x - 3\cot(3x)).$$

Replacing $2x$ by $x/3^k$ and multiplying both sides by $1/3^k$ yields

$$\frac{\sin(x/3^k)}{3^k(1 + 2\cos(x/3^k))} = \frac{1}{4}\left(\frac{1}{3^k}\cot(x/2 \cdot 3^k) - \frac{1}{3^{k-1}}\cot(x/2 \cdot 3^{k-1})\right).$$

Summing over k from 0 to n, we now obtain

$$\sum_{k=0}^{n} \frac{\sin(x/3^k)}{3^k(1 + 2\cos(x/3^k))} = \frac{1}{4}\left(\frac{1}{3^n}\cot(x/2 \cdot 3^n) - \cot(x/2)\right).$$

Moreover, using the identities

$$\tan x = \cot x - 2\cot(2x),$$
$$\sec^2 x = 4\csc^2(2x) - \csc^2 x,$$

and

$$\csc x = \cot(x/2) - \cot x$$

respectively, the interested reader can correspondingly establish

$$\sum_{k=1}^{n-1} 2^k \tan(2^k x) = \cot x - 2^n \cot(2^n x),$$

$$\sum_{k=0}^{n-1} 2^{2k}\sec^2(2^k x) = 2^{2n}\csc^2(2^n x) - \csc^2 x,$$

$$\sum_{k=0}^{n} \csc(2^k x) = \cot(x/2) - \cot(2^n x).$$

9.3.3 Sums related to the inverse tangent function

Monthly Problem 10292, 1993 asks for the value of $\sum_{n=1}^{\infty} \arctan(2/n^2)$. Appealing to the identity

$$\arctan \alpha - \arctan \beta = \arctan\left(\frac{\alpha - \beta}{1 + \alpha\beta}\right), \qquad (9.21)$$

we have

$$\arctan\left(\frac{2}{n^2}\right) = \arctan(n+1) - \arctan(n-1).$$

The sum telescopes to

$$\lim_{n\to\infty} (\arctan(n+1) + \arctan(n) - \arctan 1) = 3\pi/4.$$

More often, the same idea can be extended to those situations in which the telescopic nature is hidden by a function. For $k \geq 1$ and any nonnegative function f, (9.21) gives

$$\arctan f(k) - \arctan f(k+1) = \arctan\left(\frac{f(k) - f(k+1)}{1 + f(k)f(k+1)}\right).$$

Thus,

$$\sum_{k=1}^{n} \arctan\left(\frac{f(k) - f(k+1)}{1 + f(k)f(k+1)}\right) = \arctan f(1) - \arctan f(n+1).$$

For instance, with $f(k) = 1/(2k-1)$ we get

$$\sum_{k=1}^{n} \arctan\left(\frac{1}{2k^2}\right) = \frac{\pi}{4} - \arctan\left(\frac{1}{2n+1}\right)$$

and with $f(k) = kx^k, x \geq 0$,

$$\sum_{k=1}^{n-1} \arctan\left(\frac{kx - k + x}{1 + k(k+1)x^{2k+1}} x^k\right) = \arctan(nx^n) - \arctan(x).$$

In general, let m be a positive integer and let f be monotonic increasing with $f(0) = 0$ and $\lim_{x\to\infty} f(x) = \infty$. We have

$$\sum_{k=0}^{\infty} \arctan\left(\frac{f(k+m) - f(k)}{1 + f(k)f(k+m)}\right) = \frac{m\pi}{2} - \sum_{i=1}^{m-1} \arctan\left(\frac{f(i) - f(i-1)}{1 + f(i)f(i-1)}\right). \qquad (9.22)$$

To see this, first notice that

$$\frac{m\pi}{2} = m\int_0^{\infty} \frac{dx}{1+x^2} = \sum_{k=0}^{\infty} \int_{f(k)}^{f(k+m)} \frac{dx}{1+x^2}$$

$$+ (m-1)\int_{f(0)}^{f(1)} \frac{dx}{1+x^2} + (m-2)\int_{f(1)}^{f(2)} \frac{dx}{1+x^2} + \cdots + \int_{f(m-2)}^{f(m-1)} \frac{dx}{1+x^2}.$$

9. On the Telescoping Sums

After carrying out all the integrations, we obtain

$$\sum_{n=0}^{\infty} \left(\arctan f(k+m) - \arctan f(k) \right)$$
$$= \frac{m\pi}{2} - (m-1)\left(\arctan f(1) - \arctan f(0) \right) - (m-2)\left(\arctan f(2) - \arctan f(1) \right)$$
$$- \cdots - \left(\arctan f(m-1) - \arctan f(m-2) \right)$$
$$= \frac{m\pi}{2} - \sum_{i=1}^{m-1} \left(\arctan f(i) - \arctan f(i-1) \right).$$

Now, (9.22) follows from (9.21) immediately. When $m = 2$, $f(x) = x$ and $m = 1$, $f(x) = 2x$, (9.22) yields

$$\sum_{k=0}^{\infty} \arctan\left(\frac{2}{(k+1)^2} \right) = \frac{3\pi}{4} \quad \text{and} \quad \sum_{k=0}^{\infty} \arctan\left(\frac{2}{(2k+1)^2} \right) = \frac{\pi}{2}.$$

After shifting the index in the first identity, we have solved the Monthly Problem 10292 again.

It is interesting to see the case where $m = 1$, $f(0) = 0$, $f(k) = 2k - 1 + a$ for $a > 0$. Here (9.22) gives

$$\arctan(a+1) + \sum_{k=1}^{\infty} \arctan\left(\frac{2}{(2k+a)^2} \right) = \frac{\pi}{2},$$

where the value is independent of a. Differentiating this with respect to a yields

$$\sum_{k=1}^{\infty} \frac{4(2k+a)}{4 + (2k+a)^4} = \frac{1}{a^2 + 2a + 2}.$$

In particular, for $a = 0$ and $a = 1$, unexpectedly, we get

$$\sum_{k=1}^{\infty} \frac{2k}{1 + 4k^4} = \frac{1}{2} \quad \text{and} \quad \sum_{k=1}^{\infty} \frac{4(2k+1)}{4 + (2k+1)^4} = \frac{1}{5}.$$

9.4 Some more telescoping sums

In this section, we explore some other methods to form telescoping sums.

9.4.1 Sums of some famous families of numbers

We first examine the sums involving the Fibonacci numbers in which the telescoping technique can be utilized. Since

$$F_k = F_{k+2} - F_{k+1},$$

the series $\{F_k\}$ is telescoping and

$$\sum_{k=1}^{n} F_k = F_{n+2} - 1.$$

Moreover, since
$$\frac{1}{F_{k-1}F_{k+1}} = \frac{1}{F_{k-1}F_k} - \frac{1}{F_k F_{k+1}},$$
we find
$$\sum_{k=2}^{n} \frac{1}{F_{k-1}F_{k+1}} = 1 - \frac{1}{F_n F_{n+1}},$$
$$\sum_{k=2}^{n} \frac{F_k}{F_{k-1}F_{k+1}} = 2 - \frac{1}{F_n} - \frac{1}{F_{n+1}}.$$

In general, for any sequence $\{a_k\}$ with
$$a_{k+1} = \alpha a_k + a_{k-1}, \quad (k \geq 3)$$
where α is a nonzero constant, we have
$$\sum_{k=2}^{n} a_k = \frac{1}{\alpha}(a_{n+1} + a_n - a_1 - a_2).$$

Based on this fact, a large class of series related to special numbers can be readily computed.

9.4.2 Apéry-like formulas

For $x \neq 0$ and any sequence $\{a_k\}$, let
$$b_1 = \frac{1}{x}, \quad b_k = \frac{a_1 a_2 \cdots a_{k-1}}{x(x+a_1)\cdots(x+a_{k-1})} \quad (k \geq 2).$$

Then
$$b_k - b_{k+1} = \frac{a_1 a_2 \cdots a_{k-1}}{(x+a_1)\cdots(x+a_k)}.$$

The telescoping sum results in
$$\sum_{k=1}^{n} \frac{a_1 a_2 \cdots a_{k-1}}{(x+a_1)\cdots(x+a_k)} = \frac{1}{x} - \frac{a_1 a_2 \cdots a_n}{x(x+a_1)\cdots(x+a_n)}. \tag{9.23}$$

This, along with
$$\zeta(3) = \sum_{k=1}^{\infty} \frac{1}{k^3} = \frac{5}{2} \sum_{k=1}^{\infty} \frac{(-1)^{k+1}}{k^3 \binom{2k}{k}},$$

has been made famous by Apéry's proof [1] of the irrationality of $\zeta(3)$. Appealing to the identities (9.23) and (9.8), we establish an analogous formula for $\zeta(2)$:
$$\zeta(2) = \sum_{k=1}^{\infty} \frac{1}{k^2} = 3 \sum_{k=1}^{\infty} \frac{1}{k^2 \binom{2k}{k}}. \tag{9.24}$$

9. On the Telescoping Sums

To prove (9.24), we set $x = n$ and $a_k = k$ in (9.23) thereby giving

$$\sum_{k=1}^{n} \frac{(k-1)!}{(n+1)(n+2)\cdots(n+k)} = \frac{1}{n} - \frac{n!}{n(n+1)\cdots(2n)}.$$

This leads to

$$\frac{1}{n^2} = \sum_{k=1}^{n} \frac{(k-1)!}{n(n+1)\cdots(n+k)} + \frac{(n-1)!}{n(n+1)\cdots(2n)}$$

$$= \sum_{k=1}^{n-1} \frac{(k-1)!}{n(n+1)\cdots(n+k)} + \frac{(n-1)!}{n^2(n+1)\cdots(2n-1)}.$$

Therefore,

$$\sum_{n=1}^{\infty} \frac{1}{n^2} = \sum_{n=1}^{\infty} \sum_{k=1}^{n} \frac{(k-1)!}{n(n+1)\cdots(n+k)} + \sum_{n=1}^{\infty} \frac{(n-1)!}{n^2(n+1)\cdots(2n-1)}. \quad (9.25)$$

For the double series in (9.20), we rearrange the series and sum in columns obtaining

$$\sum_{n=1}^{\infty} \sum_{k=1}^{n} \frac{(k-1)!}{n(n+1)\cdots(n+k)} = \sum_{n=1}^{\infty} (n-1)! \left[\sum_{k=n}^{\infty} \frac{1}{(k+1)(k+2)\cdots(n+k+1)} \right]$$

$$= \sum_{n=1}^{\infty} \frac{(n-1)!}{n(n+1)\cdots(2n)},$$

where we have used the following telescoping sum from the identity (9.8):

$$\sum_{k=n}^{\infty} \frac{1}{(k+1)(k+2)\cdots(n+k+1)}$$

$$= \frac{1}{n} \sum_{k=n}^{\infty} \left(\frac{1}{(k+1)(k+2)\cdots(n+k)} - \frac{1}{(k+2)(k+3)\cdots(n+k+1)} \right)$$

$$= \frac{1}{n(n+1)\cdots(2n)}.$$

Hence the right-hand side of (9.25) becomes

$$\sum_{n=1}^{\infty} (n-1)! \left[\frac{1}{n(n+1)\cdots(2n)} + \frac{1}{n^2(n+1)\cdots(2n-1)} \right] = 3 \sum_{n=1}^{\infty} \frac{(n-1)!}{n(n+1)\cdots(2n)}$$

$$= 3 \sum_{n=1}^{\infty} \frac{1}{n^2 \binom{2n}{n}}$$

and (9.24) follows as desired.

9.4.3 A variant of telescoping sums

Sometimes a series itself is not telescoping, but it can be decomposed into the sum of a telescoping series and a well-known series. Let us illustrate this approach by two examples.

Example 1. Monthly Problem E 3352 (1989). Show that

$$\sum_{k=0}^{\infty} \frac{1}{k!(k^4 + k^2 + 1)} = \frac{e}{2}. \tag{9.26}$$

Since

$$(k^2 - k + 1)(k^2 + k + 1) = k^4 + k^2 + 1,$$

partial fractions gives

$$\frac{2}{k^4 + k^2 + 1} = \frac{k+1}{k^2 + k + 1} - \frac{k-1}{k^2 - k + 1}.$$

Thus,

$$\frac{1}{k!(k^4 + k^2 + 1)} = \frac{1}{2}\left(\frac{k+1}{k!(k^2 + k + 1)} - \frac{k-1}{k!(k^2 - k + 1)}\right)$$

$$= \frac{1}{2}\left(\frac{k}{(k+1)!(k^2 + k + 1)} - \frac{k-1}{k!(k^2 - k + 1)} + \frac{1}{(k+1)!}\right).$$

Setting $b_k = \frac{k-1}{2[k!(k^2-k+1)]}$ yields

$$\frac{1}{k!(k^4 + k^2 + 1)} = b_{k+1} - b_k + \frac{1}{2}\frac{1}{(k+1)!}.$$

This transforms the left-hand side of (9.26) into

$$\frac{1}{2} + \frac{1}{2}\sum_{k=0}^{\infty} \frac{1}{(k+1)!} = \frac{1}{2}\sum_{k=0}^{\infty} \frac{1}{k!} = \frac{e}{2}.$$

Example 2. Transform $\zeta(3)$ into a faster converging series.

We start with

$$S = \sum_{k=2}^{\infty} \frac{1}{(k-1)k(k+1)},$$

which telescopes and, by (9.12) sums to $1/4$ after shifting the index of summation. Therefore,

$$\zeta(3) = \frac{5}{4} - \sum_{k=2}^{\infty}\left(\frac{1}{k^3 - k} - \frac{1}{k^3}\right) = \frac{5}{4} - \sum_{k=2}^{\infty} \frac{1}{k^3(k^2 - 1)}.$$

Here the nth term is approximately n^{-5} whereas before it was n^{-3}.

9. On the Telescoping Sums

9.4.4 Summation by computer — Gosper's algorithm

In 1977, in conjunction with his work on the development of one of the first symbolic algebra programs, Gosper in [2] discovered a beautiful way to sum $\{a_k\}$ whenever a_{k+1}/a_k is a rational function of k. As one of the landmarks in the history of computerization of the problem of closed form summation, Gosper's algorithm answers the following question affirmatively:

If a_{k+1}/a_k is rational in k, is the sum $\sum_{k=1}^{n} a_k$ telescoping?

Gosper's algorithm proceeds in two steps, each of which is fairly straightforward. Step 1 is to express the ratio in the special form

$$\frac{a_{k+1}}{a_k} = \frac{p(k+1)}{p(k)} \frac{q(k)}{r(k)},$$

where p, q, and r are polynomials satisfying

$$\gcd(q(k), r(k+h)) = 1, \text{ for all integers } h \geq 0.$$

Step 2 is to find a nonzero polynomial solution $c(k)$ of

$$q(k)c(k+1) - r(k-1)c(k) = p(k), \qquad (9.27)$$

whenever possible. If $c(k)$ exists, we find

$$b_k = \frac{r(k-1)c(k)}{p(k)} a_k, \quad a_k = b_{k+1} - b_k.$$

Let us use Gosper's algorithm to show that the sum

$$S_n = \sum_{k=0}^{n} \frac{(4k+1) \cdot k!}{(2k+1)!}$$

is telescoping. The term ratio

$$\frac{a_{k+1}}{a_k} = \frac{4k+5}{2(4k+1)(2k+3)}$$

is rational in k as expected. The choices

$$p(k) = 4k+1, \quad q(k) = 1, \quad r(k) = 2(2k+3)$$

clearly satisfy the conditions in Step 1. Thus, equation (9.27) becomes

$$c(k+1) - 2(2k+1)c(k) = 4k+1,$$

and the constant polynomial $c(k) = -1$ is a solution. Hence

$$b_k = \frac{-2(2k+1)}{4k+1} \frac{(4k+1) \cdot k!}{(2k+1)!} = -2 \frac{k!}{(2k)!}$$

satisfies $a_k = b_{k+1} - b_k$ and

$$S_n = b_{n+1} - b_0 = 2 - \frac{n!}{(2n+1)!}.$$

(9.26) is also a good application of Gosper's algorithm. Rewriting (9.26) in the equivalent form

$$\sum_{k=0}^{\infty} \left(\frac{1}{k!(k^4 + k^2 + 1)} - \frac{1}{2\,k!} \right) = 0,$$

Gosper's algorithm finds that $b_k = \frac{k^2}{2[k!(k^2-k+1)]}$. This yields the sum as $b_\infty - b_0 = 0$ as expected.

Here we need not rely so much on cleverness and luck, as calculations can be done on the computer. Quite often, when the summand involves binomial coefficients, factorials, products of rational functions, there are very good chances that such a problem can be solved by Gosper's algorithm or by its generalization — the WZ method. The interested reader can find more details in Petkovsek-Wilf-Zeilberger's elegant book [3] and additional examples in their excellent article [4]. The latter contains 27 Monthly problems over the period of 1978–1997 solved by a computer. The related programs are available at www.math.rutgers.edu/~zeilberg/programs.html

In summary, we have given a partial account of everything you always wanted to know about the telescoping sums! It is interesting to notice that there are problems we may not be able to solve by this method, even though they may appear to be candidates. A good example related to our discussion in 9.3.3 is to find $\sum_{n=1}^{\infty} \arctan(1/n^2)$ in closed form. The answer turns out to be a curious identity (5.34), which cannot be derived by telescoping methods.

Exercises

1. Find the following sums in closed form:
   (a) $\frac{1^2}{1 \cdot 3} + \frac{2^2}{3 \cdot 5} + \cdots + \frac{n^2}{(2n-1)(2n+1)}$.

   (b) $\sum_{k=1}^{n} k^2 \binom{n}{k}$.

2. Evaluate
$$\sum_{k=0}^{n} \binom{n}{k} \cos kx \sin(n-k)x.$$

3. Prove that
$$\sum_{k=1}^{n} \sec kx \sec(k-1)x = \frac{\tan nx}{\sin x}.$$

4. Prove that
$$1 + \sum_{n=1}^{\infty} \frac{1}{n+1} \left(\frac{1 \cdot 3 \cdots (2n-1)}{2 \cdot 4 \cdots 2n} \right)^2 = \frac{4}{\pi}.$$

9. On the Telescoping Sums

5. Evaluate
$$\sum_{k=0}^{2n} (-1)^k \binom{4n}{2k} \cdot \binom{2n}{k}^{-1}.$$

6. Let n be a positive integer. Prove that
$$\frac{1}{2\sqrt{1}} + \frac{1}{3\sqrt{2}} + \cdots + \frac{1}{(n+1)\sqrt{n}} < 2\left(1 - \frac{1}{\sqrt{n+1}}\right).$$

7. Find the sum of the series
$$\sum_{k=2}^{\infty} (\zeta(k) - 1).$$

8. Find the sum of
$$\sum_{i=1}^{\infty} \sum_{j=1}^{\infty} \frac{i! \, j!}{(i+j+1)!}.$$

9. Prove that

   (a) $\sum_{k=1}^{\infty} \arctan\left(\frac{1}{k^2+k+1}\right) = \frac{\pi}{4}$.

   (b) $\sum_{k=1}^{\infty} \arctan\left(\frac{2ak}{a^2k^4 - a^2k^2 + 1}\right) = \frac{\pi}{2}$.

   (c) $\sum_{k=1}^{\infty} (-1)^{k+1} \arctan\left(\frac{2}{k^2}\right) = \frac{\pi}{4}$.

10. Let F_n be the nth Fibonacci number. Show that
$$\sum_{n=1}^{\infty} \arctan\left(\frac{F_{n+2} - F_n}{1 + F_n F_{n+2}}\right) = \frac{3\pi}{4}.$$

11. **Putnam Problem, 1977-A4.** For $0 < x < 1$, evaluate
$$\sum_{n=0}^{\infty} \frac{x^{2^n}}{1 - x^{2^{n+1}}}.$$

12. **Putnam Problem, 1978-B2.** Express
$$\sum_{n=1}^{\infty} \sum_{m=1}^{\infty} \frac{1}{m^2 n + mn^2 + 2mn}$$
as a rational number.

13. **Putnam Problem, 1977-B1.** Evaluate the infinite product
$$\prod_{n=2}^{\infty} \frac{n^3 - 1}{n^3 + 1}.$$

Hint. Using a telescoping product. A telescoping product is the multiplicative analogue of a telescoping sum. Here each factor is written as a quotient in such a way that successive factors cancel out.

14. **Putnam Problem, 1981-B1.** Find

$$\lim_{n\to\infty}\left[\frac{1}{n^5}\sum_{h=1}^{n}\sum_{k=1}^{n}(5h^4 - 18h^2k^2 + 5k^4)\right].$$

15. **Open Problem.** Let $P(x)/Q(x)$ be a rational function with real coefficients in which $\deg Q \geq 2 + \deg P$. Evaluate

$$\sum_{k=1}^{\infty} \arctan \frac{P(k)}{Q(k)}$$

in closed form. When the leading coefficient of $Q(x)$ is 1, show that

$$\sum_{k=1}^{\infty} \arctan \frac{P(k)}{Q(k)} = \sum_{j=1}^{n} \arg(\Gamma(1-a_j)) \pmod{2\pi},$$

where a_j ($1 \leq j \leq n$) are the roots of $Q(x) - iP(x)$.

References

[1] R. Apéry, Irrationalité de $\zeta(2)$ et $\zeta(3)$, *Astérisque*, **61** (1979) 11–13.

[2] R. Gosper, Indefinite hypergeometric sums in MACSYMA, *Proc. MACSYMA Users Conference*, Berkeley, CA, (1977) 237–252

[3] M. Petkovsek, H. Wilf, and D. Zeilberger, $A = B$, A. K. Peters, Ltd., Wellesley, MA, 1996.

[4] ———, How to do Monthly problems with your computer, *Amer. Math. Monthly*, **104** (1997) 505–519.

10

Summation of Subseries in Closed Form

One of the problems in the Second Putnam Mathematical Competition (1939) was to prove that if

$$u(x) = \sum_{n=0}^{\infty} \frac{x^{3n}}{(3n)!}, \quad v(x) = \sum_{n=0}^{\infty} \frac{x^{3n+1}}{(3n+1)!}, \quad w(x) = \sum_{n=0}^{\infty} \frac{x^{3n+2}}{(3n+2)!},$$

then $u^3 + v^3 + w^3 - 3uvw = 1$. The suggested solution relied on

$$u + v + w = e^x,$$
$$u + \omega v + \omega^2 w = e^{\omega x},$$
$$u + \omega^2 v + \omega w = e^{\omega^2 x},$$

where $\omega = e^{2\pi i/3}$, a primitive cubic root of unity. Notice that all u, v and w are subseries of e^x. As a byproduct, solving the above system of equations yields

$$u(x) = \frac{1}{3}\left(e^x + 2e^{-x/2}\cos\left(\frac{\sqrt{3}x}{2}\right)\right),$$

$$v(x) = \frac{1}{3}\left(e^x - e^{-x/2}\cos\left(\frac{\sqrt{3}x}{2}\right) + \sqrt{3}e^{-x/2}\sin\left(\frac{\sqrt{3}x}{2}\right)\right),$$

$$w(x) = \frac{1}{3}\left(e^x - e^{-x/2}\cos\left(\frac{\sqrt{3}x}{2}\right) - \sqrt{3}e^{-x/2}\sin\left(\frac{\sqrt{3}x}{2}\right)\right).$$

Later, for any positive integer k and $m = 0, 1, \ldots, k-1$, Rubel and Stolarsky in [12] proved

$$\sum_{n=0}^{\infty} \frac{x^{kn+m}}{(kn+m)!} = \frac{1}{k}\left(e^x + \frac{e^{\omega x}}{\omega^m} + \frac{e^{\omega^2 x}}{\omega^{2m}} + \cdots + \frac{e^{\omega^{k-1}x}}{\omega^{(k-1)m}}\right), \tag{10.1}$$

where $\omega = e^{2\pi i/k}$ is a primitive kth root of unity.

These results motivate a question that does not seem to have received much attention: let $f(x) = \sum_{n=0}^{\infty} a_n x^n$. Consider a subseries of f:

$$\sum_{n \in \mathbb{N}_1} a_n x^n,$$

where $\mathbb{N}_1$ is a subset of $\mathbb{N}$ with $\mathbb{N}_1 \neq \emptyset$ and $\mathbb{N}_1 \neq \mathbb{N}$. It is interesting to see whether there is a closed-form expression for the function $f_1(x)$ such that

$$f_1(x) = \sum_{n \in \mathbb{N}_1} a_n x^n.$$

In this chapter, we extend (10.1) to general analytical functions, and prove the following theorem.

Theorem 10.1. *Let k be a positive integer and let $\omega = e^{2\pi i / k}$ be a primitive kth root of unity. If $f(x) = \sum_{n=0}^{\infty} a_n x^n$, then for any integer m, we have*

$$\sum_{n \equiv m \,(\mathrm{mod}\, k)} a_n x^n = a_m x^m + a_{m+k} x^{m+k} + \cdots = \frac{1}{k} \sum_{j=0}^{k-1} \omega^{-jm} f(\omega^j x), \qquad (10.2)$$

where $a_j = 0$ when $j < 0$.

Proof. Since $f(x) = \sum_{n=0}^{\infty} a_n x^n$, we have

$$\sum_{j=0}^{k-1} \omega^{-jm} f(\omega^j x) = \sum_{j=0}^{k-1} \omega^{-jm} \sum_{n=0}^{\infty} a_n (\omega^j x)^n$$

$$= \sum_{n=0}^{\infty} a_n \left(\sum_{j=0}^{k-1} \omega^{(n-m)j} \right) x^n. \qquad (10.3)$$

If $n - m = kl$ for some integer l, appealing to $\omega^k = 1$, we have $\omega^{(n-m)j} = \omega^{klj} = (\omega^k)^{lj} = 1$ and so

$$\sum_{j=0}^{k-1} \omega^{(n-m)j} = k.$$

On the other hand, if $n - m \neq kl$,

$$\sum_{j=0}^{k-1} \omega^{(n-m)j} = \frac{1 - \omega^{(n-m)k}}{1 - \omega^{n-m}} = 0.$$

Thus, the summands in the right-hand side of (10.3) are for only those values of n for which $n - m$ are multiples of k, i.e., $n \equiv m \,(\mathrm{mod}\, k)$. This proves (10.2) as desired.

In view of (10.2), the left-hand side is a subseries of f that for each $n \in \mathbb{N}$ has a block of k terms of the form $a_n x^n$, a process known as multisection [8]; while the right-hand side is an appropriate linear combination of the original series. DeMorgan [5] called (10.2) a *series multisection formula*. It is easy to see that (10.1) is an immediate consequence

10. Summation of Subseries in Closed Form

of (10.2). Formula (10.2) somehow seems to be not widely known and remains underappreciated. For example, a recent University Math Contest asked contestants to determine the closed form of $u(x)$. Instead of applying (10.1) or (10.2), the proposed solution requires solving the differential equation $u'''(x) = u(x)$. The derivation of (10.2) is within easy reach of undergraduate analysis students. To demonstrate its power and wide applicability, we illustrate it by a variety of examples. Our discussion will lead to the discovery of results some of which have been known for a long time, some of which were found only recently, as well as some which appear to be new. However, these results do not require the extensive background and advanced tools that the original discoveries required.

Example 1. In [7], Hardy and Williams proposed to determine the real function of x whose power series is
$$\frac{x^3}{3!} + \frac{x^9}{9!} + \frac{x^{15}}{15!} + \cdots.$$
Clearly, this is a subseries of e^x. Consider the primitive 6th root of unity,
$$\omega = e^{2\pi i/6} = \frac{1}{2} + \frac{\sqrt{3}}{2} i.$$
Then $\omega^{-3} = \omega^3 = -1$. Appealing to (10.2) with $k = 6, m = 3$, we have
$$\frac{x^3}{3!} + \frac{x^9}{9!} + \frac{x^{15}}{15!} + \cdots = \sum_{n=0}^{\infty} \frac{x^{6n+3}}{(6n+3)!}$$
$$= \frac{1}{6}(e^x - e^{\omega x} + e^{\omega^2 x} - e^{-x} + e^{-\omega x} - e^{-\omega^2 x})$$
$$= \frac{1}{3}(\sinh x - \sinh(\omega x) + \sinh(\omega^2 x))$$
$$= \frac{1}{3}\left[\sinh x - 2\sinh\left(\frac{x}{2}\right)\cos\left(\frac{\sqrt{3}x}{2}\right)\right].$$

Example 2. It is well known that
$$-\ln(1-x) = \sum_{n=1}^{\infty} \frac{x^n}{n} = x + \frac{x^2}{2} + \frac{x^3}{3} + \cdots, \qquad |x| < 1. \tag{10.4}$$
Therefore, (10.2) yields
$$\sum_{n=0}^{\infty} \frac{x^{kn+m}}{kn+m} = -\frac{1}{k} \sum_{j=0}^{k-1} \omega^{-jm} \ln(1 - \omega^j x), \qquad |x| < 1. \tag{10.5}$$
In particular, for $|x| < 1$,
$$\sum_{n=0}^{\infty} \frac{x^{2n+1}}{2n+1} = x + \frac{x^3}{3} + \frac{x^5}{5} + \cdots = \frac{1}{2} \ln\left(\frac{1+x}{1-x}\right),$$
$$\sum_{n=0}^{\infty} \frac{x^{4n+3}}{4n+3} = \frac{x^3}{3} + \frac{x^7}{7} + \frac{x^{11}}{11} + \cdots = \frac{1}{4} \ln\left(\frac{1+x}{1-x}\right) - \frac{1}{2} \tan^{-1} x,$$

where in the second identity we have used that

$$\tan^{-1} x = \frac{1}{2i} \ln\left(\frac{1+ix}{1-ix}\right).$$

As an application, (10.5) will provide uniform proofs of many formulas that appear in the classical table [6].

Example 3. In the following, we evaluate some combinatorial sums in closed form. When (10.2) is applied to

$$f(x) = (1+x)^n = \sum_{i=0}^{n} \binom{n}{i} x^i,$$

it gives

$$\sum_{j=0}^{\lfloor (n-m)/k \rfloor} \binom{n}{kj+m} x^{kj+m} = \frac{1}{k} \sum_{j=0}^{k-1} \omega^{-jm}(1+x\omega^j)^n,$$

where $\lfloor x \rfloor$ denotes the greatest integer of x. This enables us to derive various identities listed in [4] such as

$$\sum_{j=0}^{\lfloor (n-m)/k \rfloor} \binom{n}{kj+m} = \frac{1}{k} \sum_{j=0}^{k-1} \left(2\cos\frac{j\pi}{k}\right)^n \cos\frac{(n-2m)j\pi}{k},$$

$$\sum_{j=0}^{\lfloor n/k \rfloor} \binom{n}{kj} = \frac{1}{k} \sum_{j=0}^{k-1} \left(2\cos\frac{j\pi}{k}\right)^n \cos\frac{nj\pi}{k},$$

$$\sum_{j=0}^{\lfloor n/3 \rfloor} \binom{n}{3j} = \frac{1}{3}\left(2^n + 2\cos\frac{n\pi}{3}\right).$$

A more challenging problem is to evaluate the combinatorial expression

$$T_n(k) = \sum_{i=1}^{n} \left\{ \binom{kn+1}{ki-1} - \binom{kn+1}{ki-2} \right\},$$

where the case $k = 3$ occurs in an oceanographic problem modeling the Gulf stream [9]. When $k = 1$, we have the telescoping sum

$$T_n(1) = \sum_{i=1}^{n} \left\{ \binom{n+1}{i-1} - \binom{n+1}{i-2} \right\} = \binom{n+1}{n-1} = \frac{1}{2}n(n+1).$$

Similarly, by the binomial theorem,

$$T_n(2) = \sum_{i=1}^{n} \left\{ \binom{2n+1}{2i-1} - \binom{2n+1}{2i-2} \right\}$$

$$= \sum_{i=0}^{2n+1} (-1)^{i+1} \binom{2n+1}{i} + \binom{2n+1}{2n} - 1$$

$$= -(1-x)^{2n+1}|_{x=1} + (2n+1) - 1 = 2n.$$

10. Summation of Subseries in Closed Form

For $k \geq 3$, we rewrite $T_n(k)$ as

$$T_n(k) = \sum_{i \equiv -1 \,(\mathrm{mod}\, k)} \binom{kn+1}{i} - \sum_{i \equiv -2 \,(\mathrm{mod}\, k)} \binom{kn+1}{i},$$

where i ranges from $k-2$ to $kn-1$. Applying (10.2) yields

$$T_n(k) = \frac{1}{k} \sum_{j=0}^{k-1} \omega^j (1+\omega^j)^{kn+1} - \frac{1}{k} \sum_{j=0}^{k-1} \omega^{2j} (1+\omega^j)^{kn+1}$$

$$= \frac{1}{k} \sum_{j=1}^{k-1} \omega^j (1 - \omega^{2j})(1+\omega^j)^{kn}$$

$$= \frac{1}{k} \sum_{j=1}^{k-1} \omega^{2j} (\omega^{-j} - \omega^j) \omega^{knj/2} (\omega^{j/2} + \omega^{-j/2})^{kn}$$

$$= \frac{1}{k} \sum_{j=1}^{k-1} \omega^{2j} (-2i \sin 2j\pi/k)(-1)^{nj} (2\cos j\pi/k)^{kn}.$$

But $\omega^{2(k-j)} = \omega^{-2j}$, $\sin 2(k-j)\pi/k = -\sin 2j\pi/k$, and $\cos(k-j)\pi/k = -\cos j\pi/k$, so the last sum can be cut in half. Thus

$$T_n(k) = \frac{1}{k} \sum_{j=1}^{\lfloor (k-1)/2 \rfloor} (-1)^{nj} (-2i \sin 2j\pi/k)(2\cos j\pi/k)^{kn} (\omega^{2j} - \omega^{-2j})$$

$$= \frac{1}{k} \sum_{j=1}^{\lfloor (k-1)/2 \rfloor} (-1)^{nj} (2\sin 2j\pi/k)(2\sin 4j\pi/k)(2\cos j\pi/k)^{kn}.$$

Example 4. We now demonstrate how to use (10.2) to compute series involving the central binomial coefficients. By the binomial series, we have the generating function for the central binomial coefficients as

$$\frac{1}{\sqrt{1-4x}} = \sum_{n=0}^{\infty} \binom{2n}{n} x^n = 1 + 2x + 6x^2 + 20x^3 + \cdots, \qquad |x| < 1/4.$$

Applying (10.2) to this series gives

$$\sum_{n=0}^{\infty} \binom{2(kn+m)}{kn+m} x^{kn+m} = \frac{1}{k} \sum_{j=0}^{k-1} \frac{1}{\omega^{jm} \sqrt{1-4\omega^j x}}, \qquad |x| < 1/4.$$

If we take $k = 2, 4$ and $m = 0$, respectively, then

$$\sum_{n=0}^{\infty} \binom{4n}{2n} x^{2n} = \frac{1}{2} \left(\frac{1}{\sqrt{1+4x}} + \frac{1}{\sqrt{1-4x}} \right),$$

$$\sum_{n=0}^{\infty} \binom{8n}{4n} x^{4n} = \frac{1}{4} \left(\frac{1}{\sqrt{1+4x}} + \frac{1}{\sqrt{1-4x}} + \sqrt{\frac{2(1+\sqrt{1+16x^2})}{1+16x^2}} \right),$$

where we have used
$$\sqrt{a \pm bi} = \sqrt{\frac{a + \sqrt{a^2 + b^2}}{2}} \pm \sqrt{\frac{-a + \sqrt{a^2 + b^2}}{2}} i.$$

Setting $x = 1/8$, we recover Lehmer's two sums that appeared in [10]:
$$\sum_{n=0}^{\infty} \frac{\binom{4n}{2n}}{8^{2n}} = \frac{3\sqrt{2} + \sqrt{6}}{6},$$

$$\sum_{n=0}^{\infty} \frac{\binom{8n}{4n}}{8^{4n}} = \frac{15\sqrt{2} + 5\sqrt{6} + 6\sqrt{5} + \sqrt{10}}{60}.$$

Similarly, since (see (15.1))
$$2(\sin^{-1} x)^2 = \sum_{n=1}^{\infty} \frac{1}{n^2} \binom{2n}{n}^{-1} (2x)^{2n},$$

applying $x(d/dx)$ twice on both sides gives
$$\frac{x^2}{1 - x^2} + \frac{x \sin^{-1} x}{(1 - x^2)^{3/2}} = \sum_{n=1}^{\infty} \binom{2n}{n}^{-1} (2x)^{2n}.$$

By (10.2), letting $k = 4$ and $m = 0$, appealing to $\sin^{-1} z = \sinh^{-1}(iz)/i$, we find that
$$\sum_{n=1}^{\infty} \binom{4n}{2n}^{-1} (2x)^{4n} = \frac{x^4}{1 - x^4} + \frac{x \sin^{-1} x}{2(1 - x^2)^{3/2}} - \frac{x \sinh^{-1} x}{(1 + x^2)^{3/2}}.$$

Setting $x = 1/2$ yields
$$\sum_{n=1}^{\infty} \binom{4n}{2n}^{-1} = \frac{1}{15} + \frac{\sqrt{3}}{27} \pi - \frac{2\sqrt{5}}{25} \ln \phi,$$

where $\phi = (1 + \sqrt{5})/2$, the golden ratio as we have seen in Chapter 6.

A comparable method studying the series involving the central binomial coefficients proposed in [1, 2, 13] is mostly based on the identity
$$\binom{n}{k}^{-1} = (n + 1) \int_0^1 t^k (1 - t)^{n-k} \, dt,$$

a particular case of Euler's beta function. Here (10.2) provides us an alternate method and maintains the exposition at an elementary level.

Example 5. We begin with an infinite sum, which eventually leads to an elementary proof of the Gauss formula on the digamma function at rational values. For $0 < m < k$, we try to evaluate
$$S(k, m) = \sum_{n=0}^{\infty} \left(\frac{1}{n + 1} - \frac{k}{m + kn} \right).$$

10. Summation of Subseries in Closed Form

in closed form. To use (10.2), we consider the power series

$$f_1(x) = \sum_{n=0}^{\infty} \left(\frac{1}{n+1} - \frac{k}{m+kn}\right) x^{m+kn}, \quad |x| < 1.$$

The second series in $f_1(x)$ was already evaluated in (10.5). Using the previous result, we get

$$f_1(x) = -x^{m-k} \ln(1-x^k) + \sum_{j=0}^{k-1} \omega^{-jm} \ln(1-\omega^j x)$$

$$= -x^{m-k} \ln \frac{1-x^k}{1-x} - (x^{m-k} - 1) \ln(1-x) + \sum_{j=1}^{k-1} \omega^{-jm} \ln(1-\omega^j x).$$

To see the connection between $S(k, m)$ and $f_1(x)$, we turn to Abel's continuity theorem, which claims: If $g(x) = \sum_{n=0}^{\infty} c_n x^n$ converges for $|x| < 1$ and if $\sum_{n=0}^{\infty} c_n$ converges, then

$$\lim_{x \to 1^-} g(x) = \sum_{n=0}^{\infty} c_n.$$

Thus, by Abel's continuity theorem, we have

$$S(k, m) = \lim_{x \to 1^-} f_1(x) = -\ln k + \sum_{j=1}^{k-1} \omega^{-jm} \ln(1-\omega^j).$$

When the sum on the right-hand side is translated into a sum of real and imaginary terms, it is easy to show that the imaginary terms cancel out and that the real part of the jth summand is given by

$$\frac{1}{2} \cos \frac{2mj\pi}{k} \ln\left(2 - 2\cos \frac{2j\pi}{k}\right) - \sin \frac{2mj\pi}{k} \arctan\left(\frac{\sin(2j\pi/k)}{1-\cos(2j\pi/k)}\right).$$

Finally, since

$$\arctan\left(\frac{\sin(2j\pi/k)}{1-\cos(2j\pi/k)}\right) = \arctan\left(\cot \frac{j\pi}{k}\right) = \frac{\pi}{2} - \frac{j\pi}{k},$$

appealing to the well-known identities

$$\sum_{j=1}^{n} \sin jx = \sin((n+1)x/2) \sin(nx/2) / \sin(x/2)$$

and

$$\sum_{j=1}^{n-1} j \sin jx = \frac{\sin nx}{4\sin^2(x/2)} - \frac{n \cos((2n-1)x/2)}{2 \sin(x/2)},$$

we obtain

$$\sum_{j=1}^{k-1} \sin \frac{2mj\pi}{k} \arctan\left(\frac{\sin(2j\pi/k)}{1-\cos(2j\pi/k)}\right) = \frac{\pi}{2} \cot \frac{m\pi}{k}$$

and so

$$S(k,m) = -\ln k - \frac{\pi}{2}\cot\frac{m\pi}{k} + \frac{1}{2}\sum_{j=1}^{k-1}\cos\frac{2mj\pi}{k}\ln\left(2 - 2\cos\frac{2j\pi}{k}\right). \quad (10.6)$$

It is remarkable that $S(k,m)$ is linked to the evaluation of the reciprocal of polygonal numbers in closed form.

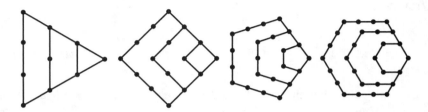

Figure 10.1. The first four polygons

A *polygonal number* is a number represented by dots arrayed in the shape of a polygon. Figure 10.1 graphically illustrates the process by which the polygonal numbers are built up. Starting with the nth triangular number T_n, we have

$$T_n = 1 + 2 + \cdots + n = \frac{n(n+1)}{2}.$$

Notice that

$$n + 2T_{n-1} = n^2 = S_n$$

gives the nth square number,

$$n + 3T_{n-1} = \frac{1}{2}n(3n-1) = P_n$$

gives the nth pentagonal number, and so on. In general, let $p_r(n)$ be the nth r-gon number. Then

$$p_r(n) = \frac{1}{2}n[(n-1)r - 2(n-2)] = \frac{1}{2}n[(r-2)n - (r-4)].$$

Recently, Problem 1147 [11] in the *Pi Mu Epsilon Journal* asks one to find the sum of the reciprocal of the pentagonal numbers. The published solution [3] to this problem used Euler's integral. In terms of $S(k,m)$, we obtain the following intriguing generalization: for $r \geq 5$,

$$\sum_{n=1}^{\infty}\frac{1}{p_r(n)} = -\frac{2}{r-4}\sum_{n=0}^{\infty}\left(\frac{1}{n+1} - \frac{r-2}{2+(r-2)n}\right) = -\frac{2}{r-4}S(r-2, 2).$$

10. Summation of Subseries in Closed Form

In particular, appealing to (10.6), we have

$$\sum_{n=1}^{\infty} \frac{1}{p_5(n)} = \frac{1}{3}(9\ln 3 - \sqrt{3}\pi),$$

$$\sum_{n=1}^{\infty} \frac{1}{p_8(n)} = \frac{1}{12}(9\ln 3 + \sqrt{3}\pi),$$

$$\sum_{n=1}^{\infty} \frac{1}{p_{10}(n)} = \frac{1}{6}(6\ln 2 + \pi),$$

$$\sum_{n=1}^{\infty} \frac{1}{p_{14}(n)} = \frac{1}{10}(4\ln 2 + 3\ln 3 + \sqrt{3}\pi).$$

For readers familiar with the *digamma function*, recall that

$$\psi(x) = \frac{d}{dx}[\ln \Gamma(x)] = \frac{\Gamma'(x)}{\Gamma(x)}.$$

Thus,

$$\psi(k/m) - \psi(1) = \sum_{n=0}^{\infty} \left(\frac{1}{n+1} - \frac{k}{m+kn} \right) = S(k,m).$$

The proceeding calculation based on (10.2) indeed recaptured the famous *Gauss formula*:

$$\psi(m/k) = -\gamma + S(k,m)$$
$$= -\gamma - \ln k - \frac{\pi}{2}\cot\frac{m\pi}{k} + \frac{1}{2}\sum_{j=1}^{k-1} \cos\frac{2mj\pi}{k} \ln\left(2 - 2\cos\frac{2j\pi}{k}\right),$$

where γ is Euler's constant. This shows that $\psi(x)$ can be evaluated by elementary functions when x is a rational number.

The above examples are exploratory, interesting and well within grasp of analysis students. Thus, (10.2) opens up a new world of results that would not otherwise have been obtained except by more specialized techniques. Combining with the differential and integral operators, we extend (10.2) to the formulas

$$\sum_{n \equiv m \pmod{k}} na_n x^n = \frac{1}{k} \sum_{j=0}^{k-1} \omega^{-jm} F(\omega^j x), \quad \text{where } F(x) = xf'(x),$$

$$\sum_{n \equiv m \pmod{k}} \frac{a_n}{n} x^n = \frac{1}{k} \sum_{j=0}^{k-1} \omega^{-jm} G(\omega^j x), \quad \text{where } G(x) = \int_0^x \frac{f(t)dt}{t}.$$

These will enable us to discover many more identities. The details are left to the reader.

Exercises

1. Show that
$$\sum_{j=0}^{\lfloor (n-1)/3 \rfloor} \binom{n}{3j+1} = \frac{1}{3}\left(2^n + 2\cos\frac{(n-2)\pi}{3}\right).$$

2. Show that
$$\sum_{j=0}^{\lfloor (n-1)/2 \rfloor} \binom{n}{2j+1} x^j = \frac{(1+\sqrt{x})^n - (1-\sqrt{x})^n}{2\sqrt{x}}.$$

3. Find a closed form of
$$\sum_{k=0}^{\lfloor n/2 \rfloor} \binom{n}{2k} \frac{x^{2k}}{2k+1}.$$

4. For any $a > 0$, prove that
$$\sum_{j=0}^{n} \binom{2n}{2j}(-a)^j = (1+a)^n \cos(2n \tan^{-1} \sqrt{a}).$$

5. Let the generating function of Fibonacci numbers be given by
$$\sum_{n=1}^{\infty} F_n x^n = \frac{x}{1-x-x^2}.$$
Find the generating functions for F_{2n} and F_{2n+1} respectively via (10.2).

6. Determine the real function of x whose power series is
$$\sum_{k=0}^{\infty} \frac{x^{4k+3}}{(4k+3)!} = \frac{x^3}{6} + \frac{x^7}{5040} + \frac{x^{11}}{39916800} + \cdots$$

7. Find
$$\sum_{n=0}^{\infty} \frac{1}{(4n+1)(4n+3)} x^{4n+1}$$
in closed form.

8. **Monthly Problem 2991** [1922, 356; 1924, 51–52]. Sum the infinite series
$$S_2(x) = 1 + \frac{3x^2}{2!} + \frac{4z^4}{4!} + \frac{6z^6}{6!} + \cdots,$$
where the numerators of the coefficients form a series of numbers whose third differences are all equal to 2.

9. It is well-known that e^x is bounded for $x < 0$. Determine all possible proper subseries of e^x which remain bounded for $x < 0$.

10. **Open Problem on Ramanujan q-series.** Define
$$A(q) = \prod_{n=1}^{\infty}(1-q^n) = 1 - q - q^2 + q^5 + q^7 - q^{12} - q^{15} + q^{22} + \cdots.$$
Now, we trisect the series into
$$A_0 = 1 - q^{12} - q^{15} + q^{51} + q^{57} - q^{117} - q^{126} + \cdots,$$
$$A_1 = -q + q^7 + q^{22} - q^{40} - q^{70} + q^{100} + q^{145} + \cdots,$$
$$A_2 = -q^2 + q^5 + q^{26} - q^{35} - q^{77} + q^{92} + q^{155} + \cdots.$$

10. Summation of Subseries in Closed Form

Look for algebraic relations among A_0, A_1, A_2 by using Mathematica or Maple. For example, Berndt and Hart recently found that $A_2 A_0^2 + A_0 A_1^2 + A_1 A_2^2 = 0$. For an introduction on this topic, see Michael Simos's "A multisection of q-series", which is available at cis.csuohio.edu/~somos/multiq.html

References

[1] J. Borwein, D. Bailey and R. Girgensohn, *Experimentation in Mathematics*, A. K. Peters, Massachusetts, 2004.

[2] A. Chen, Accelerated series for π and $\ln 2$, *Pi Mu Epsilon J.*, **13** (2008) 529–534.

[3] H. Chen, Solution to Problem 1147, *Pi Mu Epsilon J.*, **12** (2007) 433-434.

[4] A. DeMorgan, *The differential and integral calculus*, Robert Baldwin, London, 1942.

[5] H. W. Gould, *Combinatorial identities*, West Virginia University, Revised edition, 1972.

[6] E. R. Hansen, *A Tables of Series and Products*, Englewood Cliffs, Prentice-Hall, 1976.

[7] K. Hardy and K. Williams, *The Green Book — 100 practice problems for undergraduate mathematics competitions*, Integer Press, Ottawa, 1985.

[8] R. Honsberger, *Mathematical Gems* III, Washington, DC, MAA, 210–214, 1985.

[9] G. R. Ierley and O. G. Ruehr, Analytic and numerical solutions of a nonlinear boundary layer problem, *Stud. Appl. Math.*, **75** (1986) 1–36.

[10] D. H. Lehmer, Interesting series involving the central binomial coefficients, *Amer. Math. Monthly*, **92** (1985) 449-457.

[11] C. Roussean, Problem 1147, *Pi Mu Epsilon J.*, **12** (2007) 363.

[12] L. Rubel and K. Stolarsky, Subseries of the Power Series for e^x, *Amer. Math. Monthly*, **87** (1980) 371–376.

[13] T. Trif, Combinatorial sums and series involving inverses of binomial coefficients, *Fibonacci Quarterly*, **38** (2000) 79–84.

11

Generating Functions for Powers of Fibonacci Numbers

In this chapter, using Binet's formula and the power-reduction formulas, we derive explicit formulas of generating functions for powers of Fibonacci numbers. The corresponding results are extended to the famous Lucas and Pell numbers.

As we did in Chapter 6, by $F_{n+2} = F_{n+1} + F_n$, we found that the generating function for Fibonacci numbers is

$$\sum_{n=1}^{\infty} F_n x^n = \frac{x}{1-x-x^2}. \tag{11.1}$$

In general, for any positive integer k, let the generating function of $\{F_n^k\}$ be

$$G_k(x) = \sum_{n=1}^{\infty} F_n^k x^n.$$

It is natural to ask whether there is an explicit formula for $G_k(x)$. When $k = 2$, there is a recursive relation

$$F_{n+3}^2 = 2F_{n+2}^2 + 2F_{n+1}^2 - F_n^2. \tag{11.2}$$

To prove this, using Binet's formula

$$F_n = \frac{1}{\sqrt{5}}(\alpha^n - \beta^n), \tag{11.3}$$

where $\alpha = (1+\sqrt{5})/2$, $\beta = -1/\alpha = (1-\sqrt{5})/2$, appealing to $\alpha\beta = -1$, we square (11.3) to get

$$F_n^2 = \frac{1}{5}((\alpha^2)^n - 2(-1)^n + (\beta^2)^n).$$

This shows that F_n^2 is a linear combination of the nth power of $-1, \alpha^2$, and β^2. Thus, the corresponding characteristic equation becomes

$$(t+1)(t-\alpha^2)(t-\beta^2) = t^3 - 2t^2 - 2t + 1 = 0$$

from which (11.2) follows.

121

Another more accessible way to prove (11.2) relies on knowing in advance that such a formula as (11.2) exists. We search for it by setting

$$F_{n+3}^2 = aF_{n+2}^2 + bF_{n+1}^2 + cF_n^2,$$

where a, b and c are constants to be determined. Putting $n = 1, 2, 3$, respectively, yields the system of equations

$$3^2 = 2^2 a + 1^2 b + 1^2 c,$$
$$5^2 = 3^2 a + 3^2 b + 1^2 c,$$
$$8^2 = 5^2 a + 3^2 b + 2^2 c.$$

Solving for a, b, c leads to (11.2). Once (11.2) has been proved, we have

$$\begin{aligned}
G_2(x) &= F_1^2 x + F_2^2 x^2 + F_3^2 x^3 + F_4^2 x^4 + \cdots, \\
2x G_2(x) &= 2F_1^2 x^2 + 2F_2^2 x^3 + 2F_3^2 x^4 + 2F_4^2 x^5 + \cdots, \\
2x^2 G_2(x) &= 2F_1^2 x^3 + 2F_2^2 x^4 + 2F_3^2 x^5 + 2F_4^2 x^6 + \cdots, \\
x^3 G_2(x) &= F_1^2 x^4 + F_2^2 x^5 + F_3^2 x^6 + F_4^2 x^7 + \cdots.
\end{aligned}$$

Summing the terms vertically and then appealing to (11.2), we obtain

$$(1 - 2x - 2x^2 + x^3) G_2(x) = x - x^2,$$

and so

$$G_2(x) = \sum_{n=1}^{\infty} F_n^2 x^n = \frac{x(1-x)}{(1+x)(1-3x+x^2)}. \tag{11.4}$$

Similarly, we have

$$F_{n+4}^3 = 3F_{n+3}^3 + 6F_{n+2}^3 - 3F_{n+1}^3 - F_n^3 \tag{11.5}$$

which results in

$$G_3(x) = \sum_{n=1}^{\infty} F_n^3 x^n = \frac{x(1-2x-x^2)}{(1+x-x^2)(1-4x-x^2)}. \tag{11.6}$$

Theoretically, repeating this process, we can find recurrence relations for $F_n^4, F_n^5, \ldots$, and thereby determine the corresponding generating functions. However, this process becomes tedious and not feasible for larger k. As a matter of fact a proof of a general explicit formula in print has not been found. In 1962, Riordan [2], instead of giving an explicit form, found that $G_k(x)$ satisfies the recurrence relation

$$(1 - a_k x + (-1)^k x^2) G_k(x) = 1 + kx \sum_{i=1}^{\lfloor k/2 \rfloor} (-1)^i a_{ki} G_{k-2i}((-1)^i x)/i,$$

where the a_{ki} have a rather complicated structure. Recently, Mansour [1] gave an explicit formula in terms of determinants of two $k \times k$ matrices, which is difficult to calculate even for $k = 3, 4$.

Naturally, we ask whether there exists an elementary method for finding explicit formulas $G_k(x)$? In this chapter, we give an affirmative answer. First, we derive generating

11. Generating Functions for Powers of Fibonacci Numbers

functions for $\{F_{kn}\}$ and the Lucas numbers $\{L_{kn}\}$ based on Binet's formulas. Then, we introduce the power-reduction formula, which enables us to show that F_n^k is a linear combination of either $\{L_{kn}\}$ or $\{F_{kn}\}$, from which we establish an explicit formula for $G_k(x)$. Finally, we extend our results to more general sequences including Lucas and Pell numbers.

We start with establishing the generating functions for $\{F_{kn}\}$ and $\{L_{kn}\}$, where the *Lucas numbers* $\{L_n\}$ are defined by

$$L_1 = 1, \quad L_2 = 3,$$

$$L_n = L_{n-1} + L_{n-2} \quad \text{(for } n \geq 3\text{)}.$$

Similar to (11.3), we have *Binet's formula* for L_n, namely

$$L_n = \alpha^n + \beta^n. \tag{11.7}$$

We now prove the generating functions for $\{F_{kn}\}$ and $\{L_{kn}\}$ are

$$\sum_{n=0}^{\infty} F_{kn} x^n = \frac{F_k x}{1 - L_k x + (-1)^k x^2}, \tag{11.8}$$

$$\sum_{n=0}^{\infty} L_{kn} x^n = \frac{2 - L_k x}{1 - L_k x + (-1)^k x^2}. \tag{11.9}$$

To see this, appealing to $\alpha\beta = -1$, by (11.3) and (11.7), we have

$$\sum_{n=0}^{\infty} F_{kn} x^n = \sum_{n=0}^{\infty} \frac{\alpha^{kn} - \beta^{kn}}{\sqrt{5}} x^n$$

$$= \frac{1}{\sqrt{5}} \left(\sum_{n=0}^{\infty} \alpha^{kn} x^n - \sum_{n=0}^{\infty} \beta^{kn} x^n \right)$$

$$= \frac{1}{\sqrt{5}} \left(\frac{1}{1 - \alpha^k x} - \frac{1}{1 - \beta^k x} \right)$$

$$= \frac{1}{\sqrt{5}} \frac{(\alpha^k - \beta^k) x}{1 - L_k x + (-1)^k x^2}$$

$$= \frac{F_k x}{1 - L_k x + (-1)^k x^2},$$

as desired. (11.9) can be proved along the same lines.

Next, we establish an explicit formula of $G_k(x)$ for any positive integer k. In order to motivate the idea, we study two special cases, namely $k = 2$ and $k = 3$. This process not only reveals how to approach the general case but also leads to new proofs for $G_2(x)$ and $G_3(x)$.

For $k = 2$, by (11.9),

$$\sum_{n=0}^{\infty} L_{2n} x^n = \frac{2 - 3x}{1 - 3x + x^2}.$$

In terms of L_{2n}, we have

$$F_n^2 = \frac{1}{5}((\alpha^2)^n - 2(-1)^n + (\beta^2)^n) = \frac{1}{5}(L_{2n} - 2(-1)^n), \qquad (11.10)$$

and so

$$\sum_{n=1}^{\infty} F_n^2 x^n = \frac{1}{5}\left(\sum_{n=0}^{\infty} L_{2n} x^n - 2\sum_{n=0}^{\infty} (-1)^n x^n\right)$$

$$= \frac{1}{5}\left(\frac{2-3x}{1-3x+x^2} - \frac{2}{1+x}\right)$$

$$= \frac{x(1-x)}{(1+x)(1-3x+x^2)}.$$

For $k = 3$, by (11.8) and (11.1), we have

$$\sum_{n=0}^{\infty} F_{3n} x^n = \frac{2x}{1-4x-x^2}, \quad \sum_{n=0}^{\infty} F_n(-x)^n = \frac{3x}{1+x-x^2}.$$

Again, appealing to

$$F_n^3 = \frac{\alpha^{3n} - 3\alpha^{2n}\beta^n + 3\alpha^n\beta^{2n} - \beta^{3n}}{5\sqrt{5}} = \frac{1}{5}(F_{3n} - 3(-1)^n F_n), \qquad (11.11)$$

we find that

$$\sum_{n=1}^{\infty} F_n^3 x^n = \frac{1}{5}\left(\sum_{n=0}^{\infty} F_{3n} x^n - 3\sum_{n=0}^{\infty} F_n(-x)^n\right)$$

$$= \frac{1}{5}\left(\frac{2x}{1-4x-x^2} + \frac{3x}{1+x-x^2}\right)$$

$$= \frac{x(1-2x-x^2)}{(1+x-x^2)(1-4x-x^2)}.$$

These cases are not isolated examples. In fact, they essentially reframe the problem as follows: once expressing F_n^k as a linear combination of F_{nk} and L_{nk}, we can find an explicit formula of the underlying $G_k(x)$ by using the generating functions of F_{nk} and L_{nk}. To this end, we show that (11.10) and (11.11) are the base case of the following generalizations.

Power-Reduction Formula. For each positive integer k,

$$5^k F_n^{2k} = \sum_{i=0}^{k-1} \binom{2k}{i}(-1)^{i(n+1)} L_{2(k-i)n} + (-1)^{k(n+1)}\binom{2k}{k}, \qquad (11.12)$$

$$5^k F_n^{2k+1} = \sum_{i=0}^{k} \binom{2k+1}{i}(-1)^{i(n+1)} F_{(2(k-i)+1)n}. \qquad (11.13)$$

11. Generating Functions for Powers of Fibonacci Numbers

To prove (11.12), starting with the binomial theorem, we have

$$5^k F_n^{2k} = (\alpha^n - \beta^n)^{2k} = \sum_{i=0}^{2k} (-1)^i \binom{2k}{i} \alpha^{in} \beta^{(2k-i)n}$$

$$= \sum_{i=0}^{k-1} (-1)^i \binom{2k}{i} \alpha^{in} \beta^{(2k-i)n} + (-1)^k \binom{2k}{k} \alpha^{kn} \beta^{kn}$$

$$+ \sum_{i=k+1}^{2k} (-1)^i \binom{2k}{i} \alpha^{in} \beta^{(2k-i)n}.$$

Replacing i by $2k - i$ in the last sum and appealing to

$$\binom{2k}{i} = \binom{2k}{2k-i},$$

we obtain

$$\sum_{i=k+1}^{2k} (-1)^i \binom{2k}{i} \alpha^{in} \beta^{(2k-i)n} = \sum_{i=0}^{k-1} (-1)^i \binom{2k}{i} \alpha^{(2k-i)n} \beta^{in}.$$

Hence

$$5^k F_n^{2k} = \sum_{i=0}^{k-1} (-1)^i \binom{2k}{i} (\alpha)^{in} (\beta)^{in} (\alpha^{2(k-i)n} + \beta^{2(k-i)n}) + (-1)^k \binom{2k}{k} (\alpha\beta)^{kn}.$$

By $\alpha\beta = -1$ and (11.7), this proves (11.12). Since $L_1 = 1$, we may rewrite (11.12) as

$$F_n^{2k} = \frac{1}{5^k} \left(\sum_{i=0}^{k-1} \binom{2k}{i} (-1)^{i(n+1)} L_{2(k-i)n} + (-1)^{k(n+1)} \binom{2k}{k} L_1 \right),$$

which indicates that F_n^{2k} is a linear combination of Lucas numbers. Similarly, we can prove the formula (11.13).

Finally, we are ready to derive an explicit formula for $G_k(x)$. Using the power-reduction formulas (11.12) and (11.13) respectively, we obtain

$$5^k G_{2k}(x) = \sum_{i=0}^{k-1} (-1)^i \binom{2k}{i} \sum_{n=0}^{\infty} L_{2(k-i)n}((-1)^i x)^n + (-1)^k \binom{2k}{k} \sum_{n=0}^{\infty} ((-1)^k x)^n$$

$$= \sum_{i=0}^{k-1} (-1)^i \binom{2k}{i} \frac{2 - L_{2(k-i)}(-1)^i x}{1 - L_{2(k-i)}(-1)^i x + x^2} + (-1)^k \binom{2k}{k} \frac{1}{1 - (-1)^k x},$$

$$5^k G_{2k+1}(x) = \sum_{i=0}^{k} (-1)^i \binom{2k+1}{i} \sum_{n=0}^{\infty} F_{(2(k-i)+1)n}((-1)^i x)^n$$

$$= \sum_{i=0}^{k} \binom{2k+1}{i} \frac{F_{2(k-i)+1} x}{1 - L_{2(k-i)+1}(-1)^i x - x^2}.$$

In particular, we find that the generating functions for G_4, G_5, G_6 are as follows:

$$G_4(x) = \frac{1}{25}\left(\frac{2-7x}{1-7x+x^2} - \frac{4(2+3x)}{1+3x+x^2} + \frac{6}{1-x}\right)$$

$$= \frac{x(1-4x-4x^2+x^3)}{(1-x)(1-7x+x^2)(1+3x+x^2)},$$

$$G_5(x) = \frac{1}{25}\left(\frac{5x}{1-11x-x^2} + \frac{10x}{1-x-x^2} + \frac{10x}{1+4x-x^2}\right)$$

$$= \frac{x(1-7x-16x^2+7x^3+x^4)}{(1+4x-x^2)(1-x-x^2)(1-11x-x^2)},$$

$$G_6(x) = \frac{1}{125}\left(\frac{2-18x}{1-18x+x^2} + \frac{15(2-3x)}{1-3x+x^2} - \frac{6(2+7x)}{1+7x+x^2} - \frac{20}{1+x}\right)$$

$$= \frac{x(1-12x-53x^2+53x^3+12x^4-x^5)}{(1-x)(1-18x+x^2)(1-3x+x^2)(1+7x+x^2)}.$$

When $k = 2, 3$, it is interesting to see that the homogeneous recursive relations (11.2) and (11.5) are used to derive the generating functions G_2 and G_3. Conversely, $G_k(x)$ can be employed to find new homogeneous recursive relations. For example, G_4 can be used to prove that

$$F_{n+5}^4 = 5F_{n+4}^4 + 15F_{n+3}^4 - 15F_{n+2}^4 - 5F_{n+1}^4 + F_n^4.$$

The details will be studied in the next chapter.

Generally, the ideas used to extract an explicit formula for $G_k(x)$ actually work for any power of numbers that satisfy a second-order recursive relation. In the following, we sketch the derivations of closed forms for the generating functions for powers of Lucas and Pell numbers. We leave the details and the general case for the reader.

Similar to (11.12) and (11.13), we have the following power-reduction formulas:

$$L_n^{2k} = \sum_{i=0}^{k-1} \binom{2k}{i}(-1)^{in} L_{2(k-i)n} + (-1)^{kn}\binom{2k}{k},$$

$$L_n^{2k+1} = \sum_{i=0}^{k}\binom{2k+1}{i}(-1)^{in} L_{(2(k-i)+1)n}.$$

Let the generating function of $\{L_n^k\}$ be $H_k(x)$. Then,

$$H_{2k}(x) = \sum_{i=0}^{k-1}\binom{2k}{i}\frac{2-L_{2(k-i)}(-1)^i x}{1-L_{2(k-i)}(-1)^i x + x^2} + \binom{2k}{k}\frac{1}{1-(-1)^k x},$$

$$H_{2k+1}(x) = \sum_{i=0}^{k}\binom{2k+1}{i}\frac{2-L_{2(k-i)+1}(-1)^i x}{1-L_{2(k-i)+1}(-1)^i x - x^2}.$$

11. Generating Functions for Powers of Fibonacci Numbers

Explicitly,

$$H_1(x) = \sum_{n=1}^{\infty} L_n x^n = \frac{2-x}{1-x-x^2},$$

$$H_2(x) = \sum_{n=1}^{\infty} L_n^2 x^n = \frac{4-7x-x^2}{(1+x)(1-3x+x^2)},$$

$$H_3(x) = \sum_{n=1}^{\infty} L_n^3 x^n = \frac{8-13x-24x^2+x^3}{(1+x-x^2)(1-4x-x^2)},$$

$$H_4(x) = \sum_{n=1}^{\infty} L_n^4 x^n = \frac{16-79x-164x^2+76x^3+x^4}{(1-x)(1+3x+x^2)(1-7x+x^2)}.$$

Next, recall that the *Pell numbers* are defined by $P_n = 2P_{n-1} + P_{n-2}$ with $P_1 = 1$, $P_2 = 2$ and the *Pell-Lucas numbers* satisfy the same recursive relation but with $Q_1 = 2$, $Q_2 = 6$. Setting

$$s = 1 + \sqrt{2}, \quad t = 1 - \sqrt{2},$$

we are able to find:

Binet-type Formulas

$$P_n = \frac{1}{2\sqrt{2}}(s^n - t^n), \quad Q_n = s^n + t^n;$$

the generating functions of $\{P_{kn}\}$ and $\{Q_{kn}\}$:

$$\sum_{n=0}^{\infty} P_{kn} x^n = \frac{P_k \, x}{1 - Q_k \, x + (-1)^k x^2},$$

$$\sum_{n=0}^{\infty} Q_{kn} x^n = \frac{2 - Q_k \, x}{1 - Q_k \, x + (-1)^k x^2};$$

and the power-reduction formulas:

$$2^{3k} P_n^{2k} = \sum_{i=0}^{k-1} \binom{2k}{i} (-1)^{i(n+1)} Q_{2(k-i)n} + (-1)^{k(n+1)} \binom{2k}{k},$$

$$2^{3k} P_n^{2k+1} = \sum_{i=0}^{k} \binom{2k+1}{i} (-1)^{i(n+1)} P_{(2(k-i)+1)n}.$$

Finally, setting $\mathcal{G}_k(x) = \sum_{n=0}^{\infty} P_n^k x^n$, we have

$$2^{3k} \mathcal{G}_{2k}(x) = \sum_{i=0}^{k-1} (-1)^i \binom{2k}{i} \frac{2 - Q_{2(k-i)}(-1)^i x}{1 - Q_{2(k-i)}(-1)^i x + x^2}$$

$$+ (-1)^k \binom{2k}{k} \frac{1}{1 - (-1)^k x},$$

$$2^{3k} \mathcal{G}_{2k+1}(x) = \sum_{i=0}^{k} \binom{2k+1}{i} \frac{P_{2(k-i)+1} x}{1 - Q_{2(k-i)+1}(-1)^i x - x^2}.$$

In particular,

$$\mathcal{G}_1(x) = \sum_{n=1}^{\infty} P_n x^n = \frac{x}{1 - 2x - x^2},$$

$$\mathcal{G}_2(x) = \sum_{n=1}^{\infty} P_n^2 x^n = \frac{x(1-x)}{(1+x)(1-6x+x^2)},$$

$$\mathcal{G}_3(x) = \sum_{n=1}^{\infty} P_n^3 x^n = \frac{x(1 - 4x - x^2)}{(1 + 2x - x^2)(1 - 14x - x^2)},$$

$$\mathcal{G}_4(x) = \sum_{n=1}^{\infty} P_n^4 x^n = \frac{x(1+x)(1-14x+x^2)}{(1-x)(1+6x+x^2)(1-34x-x^2)}.$$

Exercises

1. Let ϕ be the golden ratio. For positive integer $n \geq 2$, prove that

$$\phi^n = F_n \phi + F_{n-1}.$$

2. Define

$$u_0 = a, u_1 = b$$

and

$$u_n = p u_{n-1} - q u_{n-2}$$

with $p^2 \neq 4q$. Find the generating function for $\{u_n\}$.

3. Let

$$G_k(x) = \sum_{n=0}^{\infty} u_n^k x^n.$$

Prove that

$$(G_m - q G_{m-1}) G_1(x) = [a + (b - pa)x] \sum_{n=0}^{\infty} G_{m+n} x^n.$$

4. **Monthly Problem 3801** [1936, 580; 1938, 636–637]. Show that

$$\cot^{-1} 1 = \cot^{-1} 2 + \cot^{-1} 5 + \cot^{-1} 13 + \cot^{-1} 34 + \cdots,$$

where these integers constitute every other term of the Fibonacci series and satisfy the recurrence $u_{n+1} = 3u_n - u_{n-1}$.

5. **A Problem of Euler/Monthly Problem 3674** [1934, 269; 1935, 518–521]. In how many different ways can a convex n-gon be divided into triangles by nonintersecting diagonals?

Comment. Drawing two nonintersecting diagonals in a convex pentagon divides the interior into three triangles. This can be done in five distinct ways. Let a_n be the

11. Generating Functions for Powers of Fibonacci Numbers 129

required number of ways for an $(n+2)$-gon. Thus $a_3 = 5$. In general, show that a_n satisfies a recurrence relation

$$a_{n+1} = \sum_{k=0}^{n} a_k a_{n-k}.$$

Use the generating function to show that $a_n = (2n)!/(n!(n+1)!)$.

6. Find an explicit formula for a_n if $a_n = na_{n-1} + (n+1)!$ for $n \geq 1$, and $a_0 = 0$.

7. Let $a_1 = a_2 = 1$, and let

$$a_n = 1 - \frac{1}{n}\sum_{k=1}^{n-2}(n-k-1)ka_k, \quad \text{if } n > 2.$$

Find an explicit formula for a_n.

8. For $n \geq 2$, show that

$$\sum_{k=1}^{n}\frac{F_n F_{n+1} - F_k^2}{F_{n+2} - F_k - 1} \geq F_{n+2} - 1.$$

9. Prove that

$$F_{kn} = \frac{F_{(k-1)n}L_n + L_{(k-1)n}F_n}{2},$$

$$L_{kn} = \frac{5F_{(k-1)n}L_n + L_{(k-1)n}F_n}{2}.$$

10. Prove that

$$F_{kn} = \sum_{i=0}^{k}\binom{k}{i}F_n^i F_{n-1}^{k-i} F_i,$$

$$L_{kn} = \sum_{i=0}^{k}\binom{k}{i}F_n^i F_{n-1}^{k-i} L_i.$$

11. **Monthly Problem 10765** [1999, 864; 2001, 978–979]. Fix positive integers k and n with $n > 2k+1$. Prove that

$$\lfloor \sqrt[k]{F_n} \rfloor - \lfloor \sqrt[k]{F_{n-k}} + \sqrt[k]{f_{n-2k}} \rfloor$$

is 0 unless F_n is a kth power, and then it is 1.

12. **Monthly Problem 10825** [2000, 752; 2002, 762–763]. Given real numbers x and y, define $S_k(x,y)$ for $k \in \mathbb{N}$ by $S_0(x,y) = x$, $S_1(x,y) = y$, and the recurrence $S_n(x,y) = S_{n-1}(x,y) + S_{n-2}(x,y)$ for all $n \in \mathbb{N}$. Show that

$$\inf_{n \in \mathbb{N}}|S_n(x,y)| \leq \sqrt{\frac{|x^2 + xy - y^2|}{5}},$$

and determine when equality holds.

13. **Putnam Problem 1999-A3.** Consider the power series expansion

$$\frac{1}{1 - 2x - x^3} = \sum_{n=0}^{\infty} a_n x^n.$$

Prove that, for each integer $n \geq 0$, there is an integer m such that

$$a_n^2 + a_{n+1}^2 = a_m.$$

14. **Open Problem.** Let f_n satisfy that

$$f_n = af_{n-1} + bf_{n-2} + cf_{n-3}$$

with arbitrary initial values f_0, f_1 and f_2. Use Mathematica or Maple to find an explicit formula for second powers. Is there a similar formula for third powers? Also, what about formulas for

$$f_n = af_{n-1} + bf_{n-2} + cf_{n-3} + df_{n-4}?$$

References

[1] T. Mansour, A formula for the generating functions of powers of Horsdam's sequence, *Australasian J. Combinatorics*, **30** (2004) 207–212.

[2] J. Riordan, Generating functions for powers of Fibonacci numbers, *Duke Math. J.*, **29** (1962) 5–12.

[3] E. Weisstein, Fibonacci Number, Mathworld — A Wolfram Web Resource, available at mathworld.wolfram.com/FibonacciNumber.html

[4] E. Weisstein, Pell Numbers, Mathworld — A Wolfram Web Resource, available at mathworld.wolfram.com/PellNumber.html

12
Identities for the Fibonacci Powers

In this chapter, using generating functions, we establish the following identities for any positive integers n and k,

$$F_{n+k+1}^k = \sum_{i=0}^{k} a_i(k) F_{n+k-i}^k,$$

where the $a_i(k)$ are given explicitly in terms of Fibonomial coefficients (see (12.7) below). Along the way, we will focus on how to derive identities, instead of merely focusing on verification.

Recall that Fibonacci numbers $\{F_n\}$ are defined by

$$F_1 = F_2 = 1,$$
$$F_{n+2} = F_{n+1} + F_n \quad (\text{for } n \geq 1). \tag{12.1}$$

In the previous chapter, we proved that for any $n \geq 1$,

$$F_{n+3}^2 = 2F_{n+2}^2 + 2F_{n+1}^2 - F_n^2, \tag{12.2}$$
$$F_{n+4}^3 = 3F_{n+3}^3 + 6F_{n+2}^3 - 3F_{n+1}^3 - F_n^3. \tag{12.3}$$

In these identities, a power of a Fibonacci number is expressed as a linear combination of the same power of successive Fibonacci numbers. Naturally, we ask whether there are any more such curious and pretty identities as (12.1)–(12.3). Specifically, for any positive integers k and n, we look for a class of identities in the form

$$F_{n+k+1}^k = \sum_{i=0}^{k} a_i(k) F_{n+k-i}^k, \tag{12.4}$$

where $a_i(k)$ are integers independent of n.

For the fourth degree we make the reasonable guess that the formula has the form

$$F_{n+5}^4 = aF_{n+4}^4 + bF_{n+3}^4 + cF_{n+2}^4 + dF_{n+1}^4 + eF_n^4. \tag{12.5}$$

131

Start with

$$\begin{pmatrix} F_5^4 & F_4^4 & F_3^4 & F_2^4 & F_1^4 \\ F_6^4 & F_5^4 & F_4^4 & F_3^4 & F_2^4 \\ F_7^4 & F_6^4 & F_5^4 & F_4^4 & F_3^4 \\ F_8^4 & F_7^4 & F_6^4 & F_5^4 & F_4^4 \\ F_9^4 & F_8^4 & F_7^4 & F_6^4 & F_5^4 \end{pmatrix} \begin{pmatrix} a \\ b \\ c \\ d \\ e \end{pmatrix} = \begin{pmatrix} F_6^4 \\ F_7^4 \\ F_8^4 \\ F_9^4 \\ F_{10}^4 \end{pmatrix}$$

which results from setting $n = 1, 2, 3, 4, 5$ in (12.5) respectively. Solving this system, for example, by using Mathematica,

```
a[n_] = Fibonacci[n] ^ 4;
b[n_] = a[n + 5] - a*a[n + 4]  - b*a[n + 3] - c*a[n + 2] -
   d*a[n + 1] - e*a[n];
Solve[{b[1] == 0, b[2] == 0, b[3] == 0, b[4] == 0, b[5] == 0}, {a, b, c, d, e}]

{{a -> 5, b -> 15, c -> -15, d -> -5, e -> 1}}
```

we obtain
$$F_{n+5}^4 = 5F_{n+4}^4 + 15F_{n+3}^4 - 15F_{n+2}^4 - 5F_{n+1}^4 + F_n^4.$$

Along the same lines, we get

$$F_{n+6}^5 = 8F_{n+5}^5 + 40F_{n+4}^5 - 60F_{n+3}^5 - 40F_{n+2}^5 + 8F_{n+1}^5 + F_n^5;$$

$$F_{n+7}^6 = 13F_{n+6}^6 + 104F_{n+5}^6 - 260F_{n+4}^6 - 260F_{n+3}^6 + 104F_{n+2}^6 + 13F_{n+1}^6 - F_n^6;$$

$$F_{n+8}^7 = 21F_{n+7}^7 + 273F_{n+6}^7 - 1092F_{n+5}^7 - 1820F_{n+4}^7 + 1092F_{n+3}^7 + 273F_{n+2}^7 \\ - 21F_{n+1}^6 - F_n^6;$$

$$F_{n+9}^8 = 34F_{n+8}^8 + 714F_{n+7}^8 - 4641F_{n+6}^8 - 12376F_{n+5}^8 + 12376F_{n+4}^8 + 4641F_{n+3}^8 \\ - 714F_{n+2}^8 - 34F_{n+1}^8 + F_n^8;$$

$$F_{n+10}^9 = 55F_{n+9}^9 + 1870F_{n+8}^9 - 19635F_{n+7}^9 - 85085F_{n+6}^9 + 136136F_{n+5}^9 \\ + 85085F_{n+4}^9 - 19635F_{n+3}^9 - 1870F_{n+2}^9 + 55F_{n+1}^9 + F_n^9;$$

$$F_{n+11}^{10} = 89F_{n+10}^{10} + 4895F_{n+9}^{10} - 83215F_{n+8}^{10} - 582505F_{n+7}^{10} + 1514513F_{n+6}^{10} \\ + 1514513F_{n+5}^{10} - 582505F_{n+4}^{10} - 83215F_{n+3}^{10} + 4895F_{n+2}^{10} + 89F_{n+1}^{10} - F_n^{10}.$$

The coefficients of the above identities are listed below:

$$\begin{pmatrix} 1 & 1 & & & & & & & & & \\ 2 & 2 & -1 & & & & & & & & \\ 3 & 6 & -3 & -1 & & & & & & & \\ 5 & 15 & -15 & -5 & 1 & & & & & & \\ 8 & 40 & -60 & -40 & 8 & 1 & & & & & \\ 13 & 104 & -260 & -260 & 104 & 13 & -1 & & & & \\ 21 & 273 & -1092 & -1820 & 1092 & 273 & -21 & -1 & & & \\ 34 & 714 & -4641 & -12376 & 12376 & 4641 & -714 & -34 & 1 & & \\ 55 & 1870 & -19635 & -85085 & 136136 & 85085 & -19635 & -1870 & 55 & 1 & \\ 89 & 4895 & -83125 & -582505 & 1514513 & 1514513 & -582505 & -83215 & 4895 & 89 & -1 \end{pmatrix}.$$

12. Identities for the Fibonacci Powers

This matrix reveals a wealth of patterns:

- The first column is the Fibonacci numbers.
- There is an alternating pattern of signs by double columns.
- There is a symmetry similar to Pascal's triangle up to sign.

A similar table was obtained by Brousseau [1] via a different approach. However, it seems that there are no analytical formulas for determining the general $a_i(k)$ in (12.4). To recognize the underlying pattern of the $a_i(k)$, we apply Sloane's On-Line Encyclopedia of Integer Sequences [5] to these results. This enables us to conjecture that the coefficients $a_i(k)$ in (12.4) are:

$$a_0(k) = F_{k+1} = \text{Fibonomial}(k+1, 1),$$
$$a_1(k) = F_{k+1} F_k = \text{Fibonomial}(k+1, 2),$$
$$\vdots$$
$$a_i(k) = (-1)^{i + i(i+1)/2} \text{Fibonomial}(k+1, i+1); \qquad (12.6)$$
$$\vdots$$
$$a_k(k) = (-1)^{k + k(k+1)/2},$$

where a *Fibonomial coefficient* is defined by

$$\text{Fibonomial}(n, k) = \frac{F_n F_{n-1} \cdots F_{n-k+1}}{F_k F_{k-1} \cdots F_1}. \qquad (12.7)$$

When we look back over the discussion above, we find that the whole story is built on a hunch that a formula in the form of (12.4) exists. In the previous chapter, we saw how to use the identity (12.2) to derive the generating functions for Fibonacci squares, namely

$$G_2(x) = \sum_{n=0}^{\infty} F_n^2 x^n = \frac{x(1-x)}{(1+x)(1-3x+x^2)}. \qquad (12.8)$$

It is natural to ask about the converse of this derivation:

Can we establish (12.2) based on the generating function (12.8)?

Indeed, we can. To see this, rewrite (12.8) as

$$(1 - 2x - 2x^2 + x^3)G_2(x) = x - x^2.$$

Now if $G_2(x)$ is the generating function of $\{F_n^2\}$ then $xG_2(x)$ is the generating function for the *shifted* sequence $\{F_{n-1}^2\}$, and consequently we have

$$(1 - 2x - 2x^2 + x^3)G_2(x) = x - x^2 + \sum_{n=3}^{\infty} (F_n^2 - 2F_{n-1}^2 - 2F_{n-2}^2 + F_{n-3}^2)x^n,$$

and so

$$F_n^2 - 2F_{n-1}^2 - 2F_{n-2}^2 + F_{n-3}^2 \equiv 0 \quad \text{for all } n \geq 3,$$

which is equivalent to (12.2).

It is no coincidence that the $G_2(x)$ is a rational function if and only if (12.2) holds. In fact, the relationship between $G_2(x)$ and (12.2) is the base case of the following generalization.

Theorem 12.1. *A sequence $\{u_n\}$ satisfies the recursive formula*

$$u_n = a_1 u_{n-1} + a_2 u_{n-2} + \cdots + a_k u_{n-k} \tag{12.9}$$

if and only if its generating function, $G(x)$, has the form

$$G(x) = \frac{g(x)}{1 - a_1 x - a_2 x^2 - \cdots - a_k x^k}$$

for some polynomial $g(x)$.

Proof. By the definition of a generating function, $G(x) = \sum_{n=0}^{\infty} u_n x^n$, we have

$$a_j x^j G(x) = a_j \sum_{n=j}^{\infty} u_{n-j} x^n$$

for each $j = 1, 2, \ldots, k$. Hence

$$G(x)[1 - a_1 x - a_2 x^2 - \cdots - a_k x^k]$$
$$= G(x) - a_1 x G(x) - a_2 x^2 G(x) - \cdots - a_k x^k G(x)$$
$$= g(x) + \sum_{n=k}^{\infty} (u_n - a_1 u_{n-1} - a_2 u_{n-2} \cdots - a_k u_{n-k}) x^n,$$

where $g(x)$ is a polynomial. Thus, $\{u_n\}$ satisfies (12.9) if and only if $G(x)$ has the asserted rational form.

Therefore a sequence defined by a homogeneous linear recursive formula always has a rational generating function and vice versa. Moreover, the numerator contains information about the first few terms of the sequence, while the denominator describes the linear recursive relation. Based on Theorem 12.1, we can show the existence of (12.4) and also confirm the formulas stated in (12.6).

Theorem 12.2. *For any positive integers k and n, there exists an identity in the form of (12.4) with the coefficients $a_i(k)$ determined by (12.6).*

To show the existence of (12.4), by Theorem 12.1, we need

Lemma 12.1. *Let $\alpha = (1 + \sqrt{5})/2$, $\beta = (1 - \sqrt{5})/2$. If the generating function of $\{F_n^k\}$ is in the form of*

$$G_k(x) = \sum_{n=0}^{\infty} F_n^k x^n = \frac{P_k(x)}{Q_k(x)}, \tag{12.10}$$

then

$$Q_k(x) = \prod_{i=0}^{k} (1 - \alpha^i \beta^{k-i} x) \tag{12.11}$$

and the degree of $P_k(x)$ is less than the degree of $Q_k(x)$.

12. Identities for the Fibonacci Powers

Proof. By Binet's formula and the binomial theorem, we have

$$\sum_{n=0}^{\infty} F_n^k x^n = \sum_{n=0}^{\infty} \frac{1}{5^{k/2}} \left(\sum_{i=0}^{k} \binom{k}{i} (-1)^{k-i} \alpha^{ni} \beta^{n(k-i)} \right) x^n$$

$$= \frac{1}{5^{k/2}} \sum_{i=0}^{k} \binom{k}{i} (-1)^{k-i} \sum_{n=0}^{\infty} \alpha^{ni} \beta^{n(k-i)} x^n$$

$$= \frac{1}{5^{k/2}} \sum_{i=0}^{k} \binom{k}{i} \frac{(-1)^{k-i}}{1 - \alpha^i \beta^{k-i} x}.$$

If k is odd, grouping the i-th term and the $(k-i)$-th term together, we get

$$\sum_{n=0}^{\infty} F_n^k x^n = \frac{1}{5^{k/2}} \sum_{i=0}^{(k-1)/2} \binom{k}{i} \left(\frac{(-1)^{k-i}}{1 - \alpha^i \beta^{k-i} x} + \frac{(-1)^i}{1 - \alpha^{k-i} \beta^i x} \right)$$

$$= \frac{1}{5^{k/2}} \sum_{i=0}^{(k-1)/2} \binom{k}{i} (-1)^i \left(\frac{1}{1 - \alpha^{k-i} \beta^i x} - \frac{1}{1 - \alpha^i \beta^{k-i} x} \right)$$

$$= \frac{1}{5^{k/2}} \sum_{i=0}^{(k-1)/2} \binom{k}{i} (-1)^i \frac{(\alpha^{k-i} \beta^i - \alpha^i \beta^{k-i}) x}{(1 - \alpha^{k-i} \beta^i x)(1 - \alpha^i \beta^{k-i} x)}.$$

If k is even, in view of the fact that $\alpha\beta = -1$, grouping the i-th term and the $(k-i)$-th term together, with the exception of the middle term, we get

$$\sum_{n=0}^{\infty} F_n^k x^n$$

$$= \frac{1}{5^{k/2}} \sum_{i=0}^{k/2-1} \binom{k}{i} \left(\frac{(-1)^{k-i}}{1 - \alpha^i \beta^{k-i} x} + \frac{(-1)^i}{1 - \alpha^{k-i} \beta^i x} \right) + \binom{k}{k/2} \frac{(-1)^{k/2}}{1 - (-1)^{k/2} x}$$

$$= \frac{1}{5^{k/2}} \sum_{i=0}^{k/2-1} \binom{k}{i} (-1)^i \left(\frac{1}{1 - \alpha^i \beta^{k-i} x} + \frac{1}{1 - \alpha^{k-i} \beta^i x} \right) + \binom{k}{k/2} \frac{(-1)^{k/2}}{1 - (-1)^{k/2} x}$$

$$= \frac{1}{5^{k/2}} \sum_{i=0}^{k/2-1} \binom{k}{i} (-1)^i \frac{2 - (\alpha^{k-i} \beta^i + \alpha^i \beta^{k-i}) x}{(1 - \alpha^{k-i} \beta^i x)(1 - \alpha^i \beta^{k-i} x)} + \binom{k}{k/2} \frac{(-1)^{k/2}}{1 - (-1)^{k/2} x}.$$

This proves (12.11) as desired and confirms that the degree of $P_k(x)$ is less than the degree of $Q_k(x)$.

Appealing to Theorem 12.1 again, we see the coefficients in (12.4) are determined by the expansion of $Q_k(x)$. To perform the expansion comfortably, we show

Lemma 12.2. *Let*

$$\prod_{i=0}^{k} (1 - r^i x) = 1 + \sum_{i=1}^{k+1} b_i x^i. \tag{12.12}$$

Then, for $i = 1, 2, \ldots, k, k+1$,

$$b_i = (-1)^i r^{i(i-1)/2} \frac{(r^{k+1} - 1)(r^k - 1) \cdots (r^{k-i+2} - 1)}{(r - 1)(r^2 - 1) \cdots (r^i - 1)}. \tag{12.13}$$

Proof. Set
$$f(x) = \prod_{i=0}^{k}(1-r^i x).$$

Then f satisfies the functional equation
$$(1-x)f(rx) = f(x)(1-r^{k+1}x),$$

which is equivalent to
$$(1-x)\left(1+\sum_{i=1}^{k+1} b_i r^i x^i\right) = \left(1+\sum_{i=1}^{k+1} b_i x^i\right)(1-r^{k+1}x).$$

Equating the coefficients of x^i ($i = 1, 2, \ldots, k, k+1$) on both sides, we have
$$r^i b_i - r^{i-1} b_{i-1} = b_i - r^{k+1} b_{i-1},$$

and so
$$b_i = \frac{r^{k+1} - r^{i-1}}{1-r^i} b_{i-1} = -\frac{r^{i-1}(r^{k-i+2}-1)}{r^i-1} b_{i-1}.$$

Iterating this relationship to write b_{i-1} in terms of b_{i-2} and continuing until b_i is expressed as a multiple of b_1, we have
$$b_i = (-1)^2 \frac{r^{i-1}r^{i-2}(r^{k-i+2}-1)(r^{k-i+3}-1)}{(r^i-1)(r^{i-1}-1)} b_{i-2}$$
$$\vdots$$
$$= (-1)^{i-1} \frac{r^{i-1}r^{i-2}\cdots 1(r^{k-i+2}-1)(r^{k-i+3}-1)\cdots(r^k-1)}{(r^i-1)(r^{i-1}-1)\cdots(r^2-1)} b_1.$$

Since
$$b_1 = -(1+r+r^2+\cdots+r^k) = -\frac{r^{k+1}-1}{r-1},$$

we obtain
$$b_i = (-1)^i \frac{r^{i-1}r^{i-2}\cdots 1(r^{k+1}-1)(r^k-1)\cdots(r^{k-i+2}-1)}{(r-1)(r^2-1)\cdots(r^i-1)},$$

which is equivalent to (12.13) and this ends the proof of Lemma 12.2.

Now, we are ready to prove Theorem 12.2. Set
$$Q_k(x) = 1 - \sum_{i=0}^{k} a_i(k) x^{i+1}. \tag{12.14}$$

Multiplying (12.10) by (12.14), we have
$$P_k(x) = \sum_{n=0}^{\infty} F_n^k x^n - \sum_{n=0}^{\infty}\sum_{i=0}^{k} a_i(k) F_n^k x^{n+i+1}.$$

12. Identities for the Fibonacci Powers

Equating the coefficients of x^{n+k+1} on both sides gives

$$0 = F^k_{n+k+1} - \sum_{i=0}^{k} a_i(k) F^k_{n+k-i}.$$

This proves (12.4) as desired. Thus, it remains to verify the formulas stated in (12.6). Replacing r by α/β and x by $\beta^k x$ in (12.12), in view of the exponents of β in (12.13), namely

$$k(i+1) - \frac{1}{2} i(i+1) - \frac{1}{2}(i+1)(i+2) - \frac{1}{2}(i+1)(2k-i+2) = \frac{1}{2} i(i+1)$$

and $\alpha\beta = -1$, we have

$$a_i(k) = -b_{i+1} \beta^{k(i+1)}$$
$$= (-1)^i (-1)^{i(i+1)/2} \frac{(\alpha^{k+1} - \beta^{k+1})(\alpha^k - \beta^k) \cdots (\alpha^{k-i+1} - \beta^{k-i+1})}{(\alpha - \beta)(\alpha^2 - \beta^2) \cdots (\alpha^{i+1} - \beta^{i+1})}.$$

By the definition of Fibonomial coefficients (12.7), we get

$$a_i(k) = (-1)^i (-1)^{i(i+1)/2} \text{Fibonomial}(k+1, i+1).$$

This proves (12.6) as required.

To demonstrate the power of the both theorems, we highlight two applications.

1. Let the sequence $\{U_n\}$ satisfy $U_{n+2} = aU_{n+1} + bU_n$ with $a^2 + 4b > 0$ and the generating function for the kth power be given by

$$\sum_{n=0}^{\infty} U_n^k x^n = \frac{p_k(x)}{q_k(x)}.$$

Based on the proof of Lemma 12.1, we find that

$$q_k(x) = \prod_{i=0}^{k} (1 - s^i t^{k-i} x),$$

where s and t are roots of $x^2 - ax - b = 0$. Furthermore, the degree of $p_k(x)$ is less than the degree of $q_k(x)$. Thus, similar to Theorem 12.2, we have

$$U_{n+k+1}^k = \sum_{i=0}^{k} c_i(k) U_{n+k-i}^k,$$

where the coefficients c_i, $i = 0, 1, 2, \ldots, k$, are given by

$$c_i = (-1)^i (-b)^{i(i+1)/2} \frac{(s^{k+1} - t^{k+1})(s^k - t^k) \cdots (s^{k-i+1} - t^{k-i+1})}{(s-t)(s^2 - t^2) \cdots (s^{i+1} - t^{i+1})}.$$

In particular, for Lucas numbers $\{L_n\}$, since

$$L_{n+2} = L_{n+1} + L_n \text{ and } L_n = \alpha^n + \beta^n,$$

we observe that the denominator of the generating function of $\{L_n^k\}$ is the same as that of $\{F_n^k\}$. Thus, Theorem 1 implies that identity (12.4) holds for Lucas numbers:

$$L_{n+k+1}^k = \sum_{i=0}^{k} a_i(k) L_{n+k-i}^k,$$

where $a_i(k)$ are given by (12.6). Explicitly, for any positive integer n,

$$L_{n+3}^2 = 2L_{n+2}^2 + 2L_{n+1}^2 - L_n^2,$$
$$L_{n+4}^3 = 3L_{n+3}^3 + 6L_{n+2}^3 - 3L_{n+1}^3 - L_n^3,$$
$$L_{n+5}^4 = 5L_{n+4}^4 + 15L_{n+3}^4 - 15L_{n+2}^4 - 5L_{n+1}^4 + L_n^4.$$

For the Pell numbers $\{P_n\}$, which are defined by

$$P_{n+2} = 2P_{n+1} + P_n,$$

we have $a = 2, b = 1, s = 1 + \sqrt{2}, t = 1 - \sqrt{2}$. Thus, Theorem 1 implies that

$$P_{n+k+1}^k = \sum_{i=0}^{k} (-1)^{i+i(i+1)/2} \frac{(s^{k+1} - t^{k+1})(s^k - t^k) \cdots (s^{k-i+1} - t^{k-i+1})}{(s-t)(s^2 - t^2) \cdots (s^{i+1} - t^{i+1})} P_{n+k-i}^k.$$

Explicitly, for any positive integer n,

$$P_{n+3}^2 = 5P_{n+2}^2 + 5P_{n+1}^2 - P_n^2,$$
$$P_{n+4}^3 = 12P_{n+3}^3 + 30P_{n+2}^3 - 12P_{n+1}^3 - P_n^3,$$
$$P_{n+5}^4 = 29P_{n+4}^4 + 174P_{n+3}^4 - 174P_{n+2}^4 - 29P_{n+1}^4 + P_n^4.$$

2. Define

$$D_k(F_n) = \begin{vmatrix} F_n^k & F_{n+1}^k & \cdots & F_{n+k}^k \\ F_{n+1}^k & F_{n+2}^k & \cdots & F_{n+k+1}^k \\ \vdots & \vdots & \ddots & \vdots \\ F_{n+k}^k & F_{n+k+1}^k & \cdots & F_{n+2k}^k \end{vmatrix}.$$

We show that Theorem 12.2 can be used to simplify the computation of $D_k(F_n)$ as

$$D_k(F_{n+1}) = (-1)^{nk(k+1)/2} D_k(F_1).$$

Indeed, since

$$D_k(F_{n+1}) = \begin{vmatrix} F_{n+1}^k & F_{n+2}^k & \cdots & F_{n+k}^k & F_{n+k+1}^k \\ F_{n+2}^k & F_{n+3}^k & \cdots & F_{n+k+1}^k & F_{n+k+2}^k \\ \vdots & \vdots & \ddots & \vdots & \vdots \\ F_{n+k+1}^k & F_{n+k+2}^k & \cdots & F_{n+2k}^k & F_{n+2k+1}^k \end{vmatrix},$$

12. Identities for the Fibonacci Powers

applying (12.4) on the last column and then repeatedly using the column operations, we have

$$D_k(F_{n+1}) = \begin{vmatrix} F_{n+1}^k & F_{n+2}^k & \cdots & F_{n+k}^k & a_k(k)F_n^k \\ F_{n+2}^k & F_{n+3}^k & \cdots & F_{n+k+1}^k & a_k(k)F_n^k \\ \vdots & \vdots & \ddots & \vdots & \vdots \\ F_{n+k+1}^k & F_{n+k+2}^k & \cdots & F_{n+2k}^k & a_k(k)F_n^k \end{vmatrix}$$

$$= (-1)^k a_k(k) \begin{vmatrix} F_n^k & F_{n+1}^k & \cdots & F_{n+k}^k \\ F_{n+1}^k & F_{n+2}^k & \cdots & F_{n+k+1}^k \\ \vdots & \vdots & \ddots & \vdots \\ F_{n+k}^k & F_{n+k+1}^k & \cdots & F_{n+2k}^k \end{vmatrix}$$

$$= (-1)^k a_k(k) D_k(F_n).$$

Iterating this relationship, we get

$$D_k(F_{n+1}) = (-1)^{2k} a_k^2(k) D_k(F_{n-1})$$
$$= \cdots = (-1)^{nk} a_k^{nk}(k) D_k(F_1)$$
$$= (-1)^{nk(k+1)/2} D_k(F_1).$$

In particular, for $k = 1$, we have

$$D_1(F_n) = \begin{vmatrix} F_n & F_{n+1} \\ F_{n+1} & F_{n+2} \end{vmatrix}$$
$$= (-1)^{n-1} D_1(F_1)$$
$$= (-1)^{n-1} \begin{vmatrix} F_1 & F_2 \\ F_2 & F_3 \end{vmatrix}$$
$$= (-1)^{n-1} \begin{vmatrix} 1 & 1 \\ 1 & 2 \end{vmatrix}$$
$$= (-1)^{n-1}.$$

That is, $F_n F_{n+2} - F_{n+1}^2 = (-1)^{n-1}$, which is the well-known *Cassini's identity*.

It is interesting to see that symbolic software provides a powerful tool for verifying identities. At the beginning of the chapter, we exhibited how to use Mathematica to verify the identity of Fibonacci fourth powers, but such a proof is not very attractive since it does not reveal why there is such an identity. A sophisticated application of a symbolic software should not only yield a deeper understanding but also offer new formulas. In [2], Dobbs presented an algorithmic method to manipulate trigonometric identities. This method was later simplified by Klamkin in [4]. Roughly speaking, they proved that all trigonometric identities can be derived from the basic identity $\sin^2 \theta + \cos^2 \theta = 1$. Thus, it is well worth knowing if the process of manipulating Fibonacci identities is also algorithmic. This should be a nice project for undergraduate research.

Exercises

1. For $n \geq 2$, prove that
$$5(F_{n-1}^2 + F_{n+1}^2) \equiv 2(-1)^{n+1} \pmod{(F_{n-1} + F_{n+1})}.$$

2. Show that
$$F_n^4 - F_{n-2}F_{n-1}F_{n+1}F_{n+2} = 1.$$

3. Use a computer to show that
$$256 F_{9n} = 625 F_n^9 + 4500 F_n^7 L_n^2 + 3150 F_n^5 L_n^4 + 420 F_n^3 L_n^6 + 6 F_n L_n^8.$$

   Can you find a similar formula for L_{9n}?

4. Let $\sqrt{-1} = i$. Prove that
$$F_n = \prod_{k=1}^{n-1}\left(1 - 2i \cos\frac{k\pi}{n}\right) = \prod_{k=1}^{\lfloor (n-1)/2 \rfloor}\left(3 + 2\cos\frac{2k\pi}{n}\right)$$
$$= \frac{2}{\sqrt{5}} i^{n+1} \sin\left(-\frac{ni}{2} \ln\frac{1+\sqrt{5}}{1-\sqrt{5}}\right) = \frac{2}{\sqrt{5}} i^n \sinh\left(\frac{n}{2} \ln\frac{1+\sqrt{5}}{1-\sqrt{5}}\right),$$
$$L_n = \prod_{k=1}^{n-1}\left(1 - 2i \cos\frac{(2k+1)\pi}{2n}\right) = \prod_{k=1}^{\lfloor (n-2)/2 \rfloor}\left(3 + 2\cos\frac{(2k+1)\pi}{n}\right)$$
$$= 2i^n\, i^{n+1} \cos\left(-\frac{ni}{2} \ln\frac{1+\sqrt{5}}{1-\sqrt{5}}\right) = 2i^n\, i^{n+1} \cosh\left(\frac{n}{2} \ln\frac{1+\sqrt{5}}{1-\sqrt{5}}\right).$$

   Comment: This should strengthen our conviction that manipulating Fibonacci identities is algorithmic.

5. Pascal's triangle exhibits some interesting patterns of divisibility by various numbers. Rewrite (12.4) as
$$\sum_{i=0}^{k+1} (-1)^{i(i+1)/2} \text{Fibonomial}\,(k+1, i+1)\, F_{n+k-i}^k = 0.$$

   The triangle of Fibonomial coefficients begins with (see Sloane A010048 [5])

   ```
                   1
                 1   1
               1   1   1
             1   2   2   1
           1   3   6   3   1
         1   5  15  15   5   1
   ```

   Show that this triangle also has some nearly identical patterns of divisibility by the same numbers.

12. Identities for the Fibonacci Powers

6. **Putnam Problem 1986-A6.** Let $a_1, a_2, \ldots, a_n$ be real numbers, and let $b_1, b_2, \ldots, b_n$ be distinct positive integers. Suppose there is a polynomial $f(x)$ satisfying the identity
$$(1-x)^n f(x) = 1 + \sum_{i=1}^{\infty} a_i x^{b_i}.$$
Find a simple expression (not involving any sums) for $f(1)$ in terms of $b_1, b_2, \ldots, b_n$ and n (but independent of $a_1, a_2, \ldots, a_n$.)

7. **Putnam Problem 1993-A2.** Let $\{x_n\}$ be a sequence of nonzero real numbers such that
$$x_n^2 - x_{n-1}x_{n+1} = 1 \quad \text{for } n = 1, 2, 3 \ldots.$$
Prove there exists a real number a such that $x_{n+1} = ax_n - x_{n-1}$ for all $n \geq 1$.

8. Define
$$u_0 = 0, u_1 = 1$$
and
$$u_n = au_{n-1} + bu_{n-2}.$$
Also, for any nonnegative integer m, define
$$\binom{m}{j}_u = \begin{cases} 1, & \text{if } j = 0, \\ \frac{u_m \cdots u_{m-j+1}}{u_j \cdots u_1}, & \text{if } j = 1, \ldots m \end{cases}.$$
Prove that
$$\sum_{i=0}^{k+1} (-1)^{i(i+1)/2} b^{i(i-1)/2} \binom{k+1}{i}_u u_{n-i}^k = 0.$$
In particular, for $k = 2$,
$$u_{n+3}^2 = (a^2+b)u_{n+2}^2 + b(a^2+b)u_{n+1}^2 - b^3 u_n^2.$$

9. Let
$$a_{ij} = \binom{i}{n-j} a^{i+j-n} b^{n-j}, \quad 0 \leq i, j \leq n$$
and $A_{n+1} = (a_{ij})$ be a matrix of order $n+1$. Find all the eigenvalues of A_{n+1} and prove that
$$\operatorname{tr}(A_{n+1}^k) = \frac{u_{k(n+1)}}{u_k}.$$

10. Define the q–Fibonacci polynomials as
$$f(n, a, b) = af(n-1, a, b) + q^{n-2} bf(n-2, a, b)$$
with initial values $f(0, a, b) = 0$, $f(1, a, b) = 1$. Prove that
$$f(n, a, b) = \sum_{j=0}^{n-1} \begin{bmatrix} n-1-j \\ j \end{bmatrix} q^{j^2} a^{n-1-2j} b^j,$$
where
$$\begin{bmatrix} n \\ k \end{bmatrix} = \frac{(1-q^n)(1-q^{n-1}) \cdots (1-q^{n-k-1})}{(1-q)(1-q^2) \cdots (1-q^k)}$$
denotes a q–binomial coefficient.

References

[1] A. Brousseau, A sequence of power formulas, *Fibonacci Quarterly*, **6** (1968) 81–83.

[2] D. E. Dobbs, Proving trigonometric identities to fresh persons, *MATYC J.*, **14** (1980) 39–42.

[3] A. F. Horadam, Generating functions for powers of a certain generalized sequence of numbers, *Duke Math. J.*, **32** (1965) 437–446.

[4] M. S. Klamkin, On proving trigonometric identities, *Math. Mag.*, **56** (1983) 215–220.

[5] N. J. A. Sloane, The On-line Encyclopedia of Integer Sequences, available at `www.research.att.com/~njas/sequences/`

13
Bernoulli Numbers via Determinants

In this chapter, we represent Bernoulli numbers via determinants based on their recursive relations and Cramer's rule. These enable us to evaluate a class of determinants involving factorials where the evaluation of these determinants by row and column manipulation is either quite challenging or almost impossible [3].

Recall the recursive relation for the Bernoulli numbers B_k (see (6.12)):

$$B_0 = 1, \quad \sum_{k=0}^{n-1} \binom{n}{k} B_k = 0. \tag{13.1}$$

Let $b_k = B_k/k!$. Rewriting (13.1) explicitly yields the following system of linear equations:

$$\begin{cases} b_1 + \frac{1}{2!} = 0, \\ \frac{b_1}{2!} + b_2 + \frac{1}{3!} = 0, \\ \frac{b_1}{3!} + \frac{b_2}{2!} + b_3 + \frac{1}{4!} = 0, \\ \cdots \cdots \\ \frac{b_1}{(n-1)!} + \frac{b_2}{(n-2)!} + \cdots + b_{n-1} + \frac{1}{n!} = 0, \\ \frac{b_1}{n!} + \frac{b_2}{(n-1)!} + \cdots + b_n + \frac{1}{(n+1)!} = 0, \end{cases}$$

or equivalently

$$\begin{cases} b_1 = -\frac{1}{2!}, \\ \frac{b_1}{2!} + b_2 = -\frac{1}{3!}, \\ \frac{b_1}{3!} + \frac{b_2}{2!} + b_3 = -\frac{1}{4!}, \\ \cdots \cdots \\ \frac{b_1}{(n-1)!} + \frac{b_2}{(n-2)!} + \cdots + b_{n-1} = -\frac{1}{n!}, \\ \frac{b_1}{n!} + \frac{b_2}{(n-1)!} + \cdots + b_n = -\frac{1}{(n+1)!}. \end{cases}$$

Solving for b_n by means of Cramer's rule, we find

$$b_n = \begin{vmatrix} 1 & 0 & 0 & \cdots & -\frac{1}{2!} \\ \frac{1}{2!} & 1 & 0 & \cdots & -\frac{1}{3!} \\ \frac{1}{3!} & \frac{1}{2!} & 1 & \cdots & -\frac{1}{4!} \\ \vdots & \vdots & \vdots & \ddots & \vdots \\ \frac{1}{(n-1)!} & \frac{1}{(n-2)!} & \frac{1}{(n-3)!} & \cdots & -\frac{1}{n!} \\ \frac{1}{n!} & \frac{1}{(n-1)!} & \frac{1}{(n-2)!} & \cdots & -\frac{1}{(n+1)!} \end{vmatrix}.$$

Using the basic properties of the determinants, factoring out the coefficient -1 and then permutating the columns, we obtain

$$B_n = n! b_n = (-1)^n n! \begin{vmatrix} \frac{1}{2!} & 1 & 0 & \cdots & 0 \\ \frac{1}{3!} & \frac{1}{2!} & 1 & \cdots & 0 \\ \frac{1}{4!} & \frac{1}{3!} & \frac{1}{2!} & \cdots & 0 \\ \vdots & \vdots & \vdots & \ddots & \vdots \\ \frac{1}{n!} & \frac{1}{(n-1)!} & \frac{1}{(n-2)!} & \cdots & 1 \\ \frac{1}{(n+1)!} & \frac{1}{n!} & \frac{1}{(n-1)!} & \cdots & \frac{1}{2!} \end{vmatrix}, \quad \text{for all } n \in \mathbb{N}. \quad (13.2)$$

In particular, appealing to the well-known fact (see (6.10))

$$B_{2n+1} = 0, \quad n = 1, 2, 3, \ldots.$$

we get an added bonus from (13.2), namely

$$\begin{vmatrix} \frac{1}{2!} & 1 & 0 & \cdots & 0 \\ \frac{1}{3!} & \frac{1}{2!} & 1 & \cdots & 0 \\ \frac{1}{4!} & \frac{1}{3!} & \frac{1}{2!} & \cdots & 0 \\ \vdots & \vdots & \vdots & \ddots & \vdots \\ \frac{1}{n!} & \frac{1}{(n-1)!} & \frac{1}{(n-2)!} & \cdots & 1 \\ \frac{1}{(n+1)!} & \frac{1}{n!} & \frac{1}{(n-1)!} & \cdots & \frac{1}{2!} \end{vmatrix} = 0, \quad \text{for } n = 3, 5, 7, \ldots.$$

This was the proposed problem 3784 which appeared in the *American Mathematical Monthly*, 1936.

Furthermore, using $B_0 = 1, B_1 = -1/2$ and $B_{2n+1} = 0$ for $n \geq 1$, we may rewrite

$$\frac{x}{e^x - 1} = 1 - \frac{1}{2}x + \sum_{n=1}^{\infty} \frac{B_{2n}}{(2n)!} x^{2n}$$

as

$$x = \left(x + \frac{x^2}{2!} + \frac{x^3}{3!} + \cdots\right)\left(1 - \frac{x}{2} + b_2 x^2 + b_4 x^4 + b_6 x^6 \cdots\right). \quad (13.3)$$

13. Bernoulli Numbers via Determinants

Applying the Cauchy product of power series on the right hand, and then equating the coefficients of x^{2k+1} for $k = 1, 2, \ldots n$ in (13.3), we have

$$\begin{cases} b_2 + \frac{1}{3!} - \frac{1}{2(2!)} = 0, \\ \frac{b_2}{3!} + b_4 + \frac{1}{5!} - \frac{1}{2(4!)} = 0, \\ \frac{b_2}{5!} + \frac{b_4}{3!} + b_6 + \frac{1}{7!} - \frac{1}{2(6!)} = 0, \\ \cdots \cdots \\ \frac{b_2}{(2n-3)!} + \frac{b_4}{(2n-5)!} + \cdots + b_{2(n-1)} + \frac{1}{(2n-1)!} - \frac{1}{2[2(n-1)]!} = 0, \\ \frac{b_2}{(2n-1)!} + \frac{b_4}{(2n-3)!} + \cdots + b_{2n} + \frac{1}{(2n+1)!} - \frac{1}{2[(2n)!]} = 0, \end{cases}$$

or equivalently

$$\begin{cases} b_2 = \frac{1}{2\,(3!)}, \\ \frac{b_2}{3!} + b_4 = \frac{3}{2\,(5!)}, \\ \frac{b_2}{5!} + \frac{b_4}{3!} + b_6 = \frac{5}{2\,(7!)}, \\ \cdots \cdots \\ \frac{b_2}{(2n-3)!} + \frac{b_4}{(2n-5)!} + \cdots + b_{2(n-1)} = \frac{2n-3}{2[(2n-1)!]}, \\ \frac{b_2}{(2n-1)!} + \frac{b_4}{(2n-3)!} + \cdots + b_{2n} = \frac{2n-1}{2[(2n+1)!]}, \end{cases} \qquad (13.4)$$

where we have used the identity

$$\frac{1}{2\,[(2k)]!} - \frac{1}{(2k+1)!} = \frac{2k-1}{2\,[(2k+1)!]}.$$

Therefore, using Cramer's rule for the system of linear equations (13.4) and recalling $B_{2n} = (2n)!\, b_{2n}$, we find

$$B_{2n} = (-1)^{n-1} \frac{(2n)!}{2} \begin{vmatrix} \frac{1}{3!} & 1 & 0 & \cdots & 0 \\ \frac{3}{5!} & \frac{1}{3!} & 1 & \cdots & 0 \\ \frac{5}{7!} & \frac{1}{5!} & \frac{1}{3!} & \cdots & 0 \\ \vdots & \vdots & \vdots & \ddots & \vdots \\ \frac{2n-3}{(2n-1)!} & \frac{1}{(2n-3)!} & \frac{1}{(2n-5)!} & \cdots & 1 \\ \frac{2n-1}{(2n+1)!} & \frac{1}{(2n-1)!} & \frac{1}{(2n-3)!} & \cdots & \frac{1}{3!} \end{vmatrix}.$$

On the other hand, equating the coefficients of x^{2k} for $k = 2, 3, \ldots, n+1$ in (13.3), we have

$$\begin{cases} \frac{b_2}{2!} + \frac{1}{4!} - \frac{1}{2(3!)} = 0, \\ \frac{b_2}{4!} + \frac{b_4}{2!} + \frac{1}{6!} - \frac{1}{2(5!)} = 0, \\ \frac{b_2}{6!} + \frac{b_4}{4!} + \frac{b_6}{2!} + \frac{1}{8!} - \frac{1}{2(7!)} = 0, \\ \cdots \cdots \\ \frac{b_2}{[2(n-1)]!} + \frac{b_4}{[2(n-2)]!} + \cdots + \frac{b_{2(n-1)}}{2!} + \frac{1}{[2n]!} - \frac{1}{2[(2n-1)!]} = 0, \\ \frac{b_2}{(2n)!} + \frac{b_4}{[2(n-1)]!} + \cdots + \frac{b_{2n}}{2!} + \frac{1}{[2(n+1)]!} - \frac{1}{2[(2n+1)!]} = 0, \end{cases}$$

or equivalently

$$\begin{cases} \frac{b_2}{2!} = \frac{1}{4!}, \\ \frac{b_2}{4!} + \frac{b_4}{2!} = \frac{2}{6!}, \\ \frac{b_2}{6!} + \frac{b_4}{4!} + \frac{b_6}{2!} = \frac{3}{8!}, \\ \cdots \cdots \\ \frac{b_2}{[2(n-1)]!} + \frac{b_4}{[2(n-2)]!} + \cdots + \frac{b_{2(n-1)}}{2!} = \frac{n-1}{2[(2n)!]}, \\ \frac{b_2}{(2n)!} + \frac{b_4}{2(n-1)!} + \cdots + \frac{b_{2n}}{2!} = \frac{n}{[2(n+1)]!}, \end{cases} \qquad (13.5)$$

where we have used the identity

$$\frac{1}{2[(2k+1)]!} - \frac{1}{[2(k+1)]!} = \frac{k}{[2(k+1)]!}.$$

Therefore, applying Cramer's rule to (13.5) and recognizing $B_{2n} = (2n)! \, b_{2n}$, we find another determinant expression for non-vanishing Bernoulli numbers, namely

$$B_{2n} = (-1)^{n-1} 2^n (2n)! \begin{vmatrix} \frac{1}{4!} & \frac{1}{2!} & 0 & \cdots & 0 \\ \frac{2}{6!} & \frac{1}{4!} & \frac{1}{2!} & \cdots & 0 \\ \frac{3}{8!} & \frac{1}{6!} & \frac{1}{4!} & \cdots & 0 \\ \vdots & \vdots & \vdots & \ddots & \vdots \\ \frac{n-1}{(2n)!} & \frac{1}{[2(n-1)]!} & \frac{1}{[2(n-2)]!} & \cdots & \frac{1}{2!} \\ \frac{n}{[2(n+1)]!} & \frac{1}{(2n)!} & \frac{1}{[2(n-1)]!} & \cdots & \frac{1}{4!} \end{vmatrix}.$$

Now, we conclude this chapter with three observations.

1. The determinant expressions for Bernoulli numbers provide us an alternate method to explore the nature of some determinants. For example, recall that B_{2n} alternate in

13. Bernoulli Numbers via Determinants

sign. Thus,

$$\begin{vmatrix} \frac{1}{4!} & \frac{1}{2!} & 0 & \cdots & 0 \\ \frac{2}{6!} & \frac{1}{4!} & \frac{1}{2!} & \cdots & 0 \\ \frac{3}{8!} & \frac{1}{6!} & \frac{1}{4!} & \cdots & 0 \\ \vdots & \vdots & \vdots & \ddots & \vdots \\ \frac{n-1}{(2n)!} & \frac{1}{[2(n-1)]!} & \frac{1}{[2(n-2)]!} & \cdots & \frac{1}{2!} \\ \frac{n}{[2(n+1)]!} & \frac{1}{(2n)!} & \frac{1}{[2(n-1)]!} & \cdots & \frac{1}{4!} \end{vmatrix} > 0, \quad \text{for all } n \geq 1.$$

The positivity of determinants has caught our attention since it can not be confirmed by the well-known Hadamard Theorem [4].

2. The method used in this note can be extended to some other important numbers. For instance, let

$$S_k(n) = 1^k + 2^k + \cdots + n^k.$$

For positive integer m, the binomial theorem gives

$$(x+1)^m - x^m = \sum_{i=0}^{m-1} \binom{m}{i} x^i.$$

Adding up these identities for $x = 1, 2, \ldots, n$, we get

$$(n+1)^m - 1 = \sum_{i=0}^{m-1} \binom{m}{i} S_i(n). \tag{13.6}$$

Taking $m = 1, 2, \ldots, k$ in (13.6) respectively, we obtain a system of linear equations for $S_i(n), 0 \leq i \leq k-1$. Cramer's rule gives, after some simplifications,

$$S_{k-1}(n) = \frac{1}{k!} \begin{vmatrix} 1 & 0 & 0 & \cdots & 0 & n+1 \\ 1 & 2 & 0 & \cdots & 0 & (n+1)^2 \\ 1 & 3 & 3 & \cdots & 0 & (n+1)^3 \\ \vdots & \vdots & \vdots & \ddots & \vdots & \vdots \\ 1 & \binom{k-1}{1} & \binom{k-1}{2} & \cdots & \binom{k-1}{k-3} & (n+1)^{k-1} \\ 1 & \binom{k}{1} & \binom{k}{2} & \cdots & \binom{k}{k-2} & (n+1)^k \end{vmatrix}.$$

Appealing to the definition of determinant, we see that $S_{k-1}(n)$ is a polynomial in n with degree k. In particular, we have

$$S_2(n) = \sum_{i=1}^{n} i^2 = \frac{1}{3!} \begin{vmatrix} 1 & 0 & n+1 \\ 1 & 2 & (n+1)^2 \\ 1 & 3 & (n+1)^3 \end{vmatrix} = \frac{n(n+1)(2n+1)}{6},$$

$$S_3(n) = \sum_{i=1}^{n} i^3 = \frac{1}{4!} \begin{vmatrix} 1 & 0 & 0 & n+1 \\ 1 & 2 & 0 & (n+1)^2 \\ 1 & 3 & 3 & (n+1)^3 \\ 1 & 4 & 6 & (n+1)^4 \end{vmatrix} = \frac{n^2(n+1)^2}{4}.$$

The interested reader is encouraged to find the determinant expressions for Fibonacci and other special numbers.

3. Our discussion above pivoted on the application of Cramer's rule on recursive relations. There are many other uses of linear algebra in analysis. For example, we can use linear algebra to study sequences defined by a linear recursive relation that is associated with a matrix. Let

$$A = \begin{pmatrix} a & b \\ c & d \end{pmatrix}$$

be an arbitrary 2×2 matrix and let $T = a + d = \mathrm{tr}(A)$, $D = ad - bc$. Define

$$x_{n+1} = T x_n - D x_{n-1}.$$

It is not hard to show that

$$x_n = \sum_{i=0}^{\lfloor n/2 \rfloor} \binom{n-i}{i} T^{n-2i} (-D)^i, \qquad (13.7)$$

and

$$A^n = \begin{pmatrix} a x_n - D x_{n-1} & b x_n \\ c x_n & d x_n - D x_{n-1} \end{pmatrix}. \qquad (13.8)$$

As an illustration, applying (13.7) and (13.8) to the Fibonacci numbers F_n, where $T = -D = 1$, we recover the famous identities

$$F_n = \sum_{i=0}^{\lfloor n/2 \rfloor} \binom{n-i}{i}$$

and

$$\begin{pmatrix} 1 & 1 \\ 1 & 0 \end{pmatrix}^n = \begin{pmatrix} F_{n+1} & F_n \\ F_n & F_{n-1} \end{pmatrix}.$$

Finally, it is interesting to notice that to solve difficult enumeration problems on plane partitions, George Andrews in [1] brought to the attention of the mathematical community the study of determinants involving binomial numbers. In [2], Bressoud gives an interesting brief account on the historical facts surrounding the discovery of these determinants. As might be expected, there are still a large number of open problems. For a list of such determinants and the contexts in which they have appeared, you might enjoy the excellent survey paper by Krattenthaler [5].

Exercises

1. Let
$$a_{ij} = \int_a^b x^{i+j}\,dx.$$
Show that
$$\det(a_{ij}) = \frac{2^n n!}{(2n+1)!}(b-a)^{(n+1)^2} \prod_{k=1}^n \binom{2k}{k}^{-2}.$$

2. Define
$$A_n(b_1, b_2, \ldots, b_n) = \begin{vmatrix} b_1 & -1 & 0 & \cdots & 0 \\ b_2 & b_1 & -2 & \cdots & 0 \\ \vdots & \vdots & \vdots & \ddots & \vdots \\ b_{n-1} & b_{n-2} & b_{n-3} & \cdots & 1-n \\ b_n & b_{n-1} & b_{n-2} & \cdots & b_1 \end{vmatrix}.$$
Evaluate A_n when $b_n = n^n/n!$.

3. Define
$$D_n(a_1, a_2, \ldots, a_n) = \begin{vmatrix} a_1 & a_2 & a_3 & \cdots & a_n \\ 1 & a_1 & a_2 & \cdots & a_{n-1} \\ 0 & 1 & a_1 & \cdots & a_{n-2} \\ \vdots & \vdots & \vdots & \ddots & \vdots \\ 0 & 0 & 0 & \cdots & 1 \end{vmatrix}.$$
Prove the following:

   (a) If
$$A_n = D_n\left(\frac{1}{2!}, \frac{1}{4!}, \ldots, \frac{1}{(2n)!}\right),$$
then
$$\lim_{n \to \infty} \left|\frac{A_{n-1}}{A_n}\right| = \frac{\pi^2}{4}.$$

   (b) Let B_n be the Bernoulli numbers. Then
$$D_{2n}\left(\frac{1}{2!}, \frac{1}{3!}, \ldots, \frac{1}{(2n+1)!}\right) = \frac{B_n}{(2n)!}.$$

4. **Reciprocal of a power series.** Let
$$f(x) = \sum_{n=0}^\infty a_n x^n.$$
If $a_0 \neq 0$, then the reciprocal of the power series exists and is determined by
$$[f(x)]^{-1} = \sum_{n=0}^\infty b_n x^n,$$

where
$$b_n = \frac{(-1)^n}{a_0^{n+1}} \begin{vmatrix} a_1 & a_0 & 0 & \cdots & 0 \\ a_2 & a_1 & a_0 & \cdots & 0 \\ \vdots & \vdots & \vdots & \ddots & \vdots \\ a_n & a_{n-1} & a_{n-2} & \cdots & a_1 \end{vmatrix}.$$

5. Let
$$\frac{\sinh x}{\sin x} = \sum_{n=0}^{\infty} \frac{a_{2n}}{(2n)!} x^{2n}.$$

Show that
$$a_{2n} = \frac{(-1)^n}{(2n+1)!} \begin{vmatrix} 1 & 1 & 0 & \cdots & 0 \\ 1 & -\binom{3}{3} & \binom{3}{1} & \cdots & 0 \\ \vdots & \vdots & \vdots & \ddots & \vdots \\ 1 & (-1)^n \binom{2n+1}{2n-1} & (-1)^{n-1}\binom{2n+1}{2n-3} & \cdots & -\binom{2n+1}{3} \end{vmatrix}.$$

6. Let
$$A_{ij} = \begin{cases} 1/(2i - 2j + 3)!, & j \leq i+1, \\ 0, & j > i+1 \end{cases}.$$

Prove that
$$\det(A_{ij}) = \frac{(-1)^{n+1}(2^{2n} - 2)B_{2n}}{(2n)!}.$$

7. Prove the *Vandermonde determinant evaluation*
$$\det_{1 \leq i,j \leq n} (x_i^{j-1}) = \prod_{1 \leq i < j \leq n} (x_j - x_i).$$

Extend the above result to evaluate
$$\det_{1 \leq i,j \leq n} (p_j(x_i)),$$

where $p_j = a_j x^{j-1} +$ lower terms.

8. Let $S_k(n) = \sum_{i=1}^{n} i^k$ and
$$A_k(n) = \sum_{i_1,i_2,\ldots,i_k=1, i_1 < i_2 < \cdots < i_k}^{n} i_1 i_2 \cdots i_k.$$

Find a closed form for $A_k(n)$ in terms of $S_k(n)$. For example,
$$A_2(n) = \sum_{i,j=1, i<j}^{n} ij = \sum_{j=1}^{n} \frac{(j-1)j^2}{2} = \frac{1}{2}(S_1^2(n) - S_2(n)).$$

13. Bernoulli Numbers via Determinants

9. **Putnam Problem 1978-A2.** Let $a, b, p_1, p_2, \ldots, p_n$ be real numbers with $a \neq b$. Define $f(x) = (p_1 - x)(p_2 - x) \cdots (p_n - x)$. Show that

$$\det \begin{vmatrix} p_1 & a & a & \cdots & a & a \\ b & p_2 & a & \cdots & a & a \\ b & b & p_3 & \cdots & a & a \\ \vdots & \vdots & \vdots & \ddots & \vdots & \vdots \\ b & b & b & \cdots & p_{n-1} & a \\ b & b & b & \cdots & b & p_1 \end{vmatrix} = \frac{bf(a) - af(b)}{b - a}.$$

10. **Putnam Problem 1992-B5.** Let D_n denote the value of the $(n-1) \times (n-1)$ determinant

$$\begin{vmatrix} 3 & 1 & 1 & 1 & \cdots & 1 \\ 1 & 4 & 1 & 1 & \cdots & 1 \\ 1 & 1 & 5 & 1 & \cdots & 1 \\ \vdots & \vdots & \vdots & \vdots & \ddots & \vdots \\ 1 & 1 & 1 & 1 & \cdots & n+1 \end{vmatrix}.$$

Is the set $\{D_n/n!\}$ bounded?

11. **More well-known numbers via determinants**. Given a sequence $\{a_n\}_0^\infty$, define

$$D_n(a_0, a_1, a_2, \ldots, a_n) = \begin{vmatrix} a_1 & a_0 & 0 & \cdots & 0 \\ a_2 & a_1 & a_0 & \cdots & 0 \\ \vdots & \vdots & \vdots & \ddots & \vdots \\ a_{n-1} & a_{n-2} & a_{n-3} & \cdots & a_0 \\ a_n & a_{n-1} & a_{n-2} & \cdots & a_1 \end{vmatrix}.$$

Problem 3 indicates

$$B_{2n} = \frac{(-1)^{n+1}(2n)!}{2^{2n} - 2} D_n(1, 1/3!, \ldots, 1/(2n+1)!).$$

Motivated by this result, we have the following interesting results:

(a) Let $f(x) = \sum_{n=0}^\infty a_n x^n$, $g(x) = \sum_{n=0}^\infty b_n x^n$. If $f(x)g(x) = 1$, then $a_0 \neq 0$ and

$$b_n = (-1)^n \frac{D_n(a_0, a_1, \ldots, a_n)}{a_0^{n+1}}.$$

(b) Let $D(x) = \sum_{n=0}^\infty D_n(a_0, a_1, a_2, \ldots, a_n) x^n$. If $a_n = 1/(2n+1)!$, show that

$$D(x^2) = \frac{x}{\sin x}, \quad D(-x^2) = \frac{2xe^x}{e^{2x} - 1}.$$

(c) Let H_n be the harmonic numbers. Prove that
$$H_n = \frac{(-1)^{n-1}}{(n-1)!} B_{n-1}^{n+1}.$$

(d) Let the *Stirling numbers of the second kind* $\left\{ \begin{array}{c} n \\ m \end{array} \right\}$ be defined by
$$\frac{(e^x-1)^m}{m!} = \sum_{n=m}^{\infty} \left\{ \begin{array}{c} n \\ m \end{array} \right\} \frac{x^n}{n!}.$$

Define
$$S_n^{(m)} = \left\{ \begin{array}{c} m+n \\ m \end{array} \right\} \bigg/ \binom{m+n}{m},$$

and prove that
$$H_n = D_{n-1}(S_0^{(n+1)}, S_1^{(n+1)}/1!, \ldots, S_{n-1}^{(n+1)}/(n-1)!).$$

12. **Bernoulli polynomials via determinants**. Let $B_n(x)$ be the Bernoulli polynomials. Prove that

$$B_n(x) = \frac{(-1)^n}{(n-1)!} \begin{vmatrix} 1 & x & x^2 & x^3 & \cdots & x^{n-1} & x^n \\ 1 & 1/2 & 1/3 & 1/4 & \cdots & 1/n & 1/(n+1) \\ 0 & 1 & 1 & 1 & \cdots & 1 & 1 \\ 0 & 0 & 2 & 3 & \cdots & n-1 & n \\ 0 & 0 & 0 & \binom{3}{2} & \cdots & \binom{n-1}{2} & \binom{n}{2} \\ \vdots & \vdots & \vdots & \vdots & \ddots & \vdots & \vdots \\ 0 & 0 & 0 & 0 & \cdots & \binom{n-1}{n-2} & \binom{n}{n-2} \end{vmatrix}.$$

References

[1] G. Andrews and D. Stanton, Determinants in plane partitions enumeration, *Europe J. Combinatorics*, **19** (1998) 273–282.

[2] D. M. Bressoud, *Proofs and Confirmations: The Story of the Alternating Sign Matrix Conjecture*, The Mathematical Association of America, 1999.

[3] C. Dubbs and D. Siegel, Computing determinants, *College Math. J.*, **18** (1987) 48–50.

[4] R. A. Horn and C. R. Johnson, *Matrix Analysis*, Cambridge University Press, 1985.

[5] C. Krattenthaler, Advanced determinant Calculus, available at
www.mat.univie.ac.at/~slc/wpapers/s42kratt.pdf

14

On Some Finite Trigonometric Power Sums

In Chapter 5, we established the trigonometric identity (see (5.22))

$$\sum_{k=0}^{n-1} \frac{1}{1 - 2x\cos\left(\frac{2k\pi}{n}\right) + x^2} = \frac{n(1+x^n)}{(1-x^n)(1-x^2)},$$

and reproduced the following elegant formula (see (5.2))

$$\sum_{k=1}^{n-1} \csc^2\left(\frac{k\pi}{n}\right) = \frac{(n-1)(n+1)}{3}$$

along with

$$\sum_{k=1}^{n-1} \tan^2\left(\frac{k\pi}{n}\right) = n(n-1), \quad \text{if } n \text{ is odd.}$$

In contrast to Fourier series, these finite power sums are over angles equally dividing the upper half plane. Moreover, these beautiful and somewhat surprising sums often arise in both analysis and number theory. In this chapter, by using generating functions, we extend the above results to the power sums as shown in identities (14.9)–(14.20) and in the Appendix.

We begin with establishing two auxiliary trigonometric identities. To this end, recall (5.15)

$$\prod_{k=1}^{n-1} (\cos\theta - \cos(k\pi/n)) = 2^{1-n} \frac{\sin n\theta}{\sin\theta}. \tag{14.1}$$

Taking the logarithm of both sides gives

$$\sum_{k=1}^{n-1} \ln(\cos\theta - \cos(k\pi/n)) = (1-n)\ln 2 + \ln\sin n\theta - \ln\sin\theta.$$

Differentiating with respect to θ yields

$$\sum_{k=1}^{n-1} \frac{\sin\theta}{\cos\theta - \cos(\frac{k\pi}{n})} = -n\cot n\theta + \cot\theta. \tag{14.2}$$

153

Replacing θ by $\pi - \theta$ leads to

$$\sum_{k=1}^{n-1} \frac{\sin\theta}{\cos\theta + \cos(\frac{k\pi}{n})} = -n\cot n\theta + \cot\theta. \qquad (14.3)$$

Adding (14.2) and (14.3) and dividing both sides by 2 we obtain

$$\sum_{k=1}^{n-1} \frac{\sin\theta \cos\theta}{\cos^2\theta - \cos^2(\frac{k\pi}{n})} = -n\cot n\theta + \cot\theta.$$

Now, dividing by $\sin\theta$ and consolidating yields

$$\sum_{k=0}^{n-1} \frac{\cos\theta}{\cos^2\theta - \cos^2(\frac{k\pi}{n})} = -n\cot n\theta \csc\theta. \qquad (14.4)$$

Next, substituting $\cos^2 x = 1 - \sin^2 x$ into (14.4), we have

$$\sum_{k=0}^{n-1} \frac{\cos\theta}{\sin^2(\frac{k\pi}{n}) - \sin^2\theta} = -n\cot n\theta \csc\theta. \qquad (14.5)$$

In the following, let n and p be positive integers. We show how to use the identities (14.4) and (14.5) to establish explicit formulas for various finite trigonometric power sums.

14.1 Sums involving $\sec^{2p}(k\pi/n)$

Let

$$S_p(n) = \sum_{k=0}^{n-1} \sec^{2p}\left(\frac{k\pi}{n}\right).$$

There are two cases, depending on whether n is even or odd.

Case 1: n odd. To construct a generating function for $\{S_p(n)\}$ of the form

$$G(n,t) = \sum_{p=1}^{\infty} S_p(n) t^{2p}, \qquad (14.6)$$

we rewrite (14.4) as

$$n \cot n\theta \cot\theta = \sum_{k=0}^{n-1} \frac{-\cos^2\theta}{\cos^2\theta - \cos^2(\frac{k\pi}{n})}. \qquad (14.7)$$

By expanding the summand as a geometric series we have

$$n \cot n\theta \cot\theta = \sum_{k=0}^{n-1} \frac{\left[\frac{\cos\theta}{\cos(\frac{k\pi}{n})}\right]^2}{1 - \left[\frac{\cos\theta}{\cos(\frac{k\pi}{n})}\right]^2} = \sum_{p=1}^{\infty} \left(\sum_{k=0}^{n-1} \sec^{2p}(\frac{k\pi}{n})\right) \cos^{2p}\theta, \qquad (14.8)$$

14. On Some Finite Trigonometric Power Sums

where the last summation order has been changed. Thus, comparing (14.8) and (14.6) with $t = \cos\theta$, we find the generating function for $\{S_p(n)\}$, namely

$$G(n,t) = n\cot n\theta \cot\theta$$
$$= \frac{nt}{\sqrt{1-t^2}} \cot(n \arccos t) = \frac{nt}{\sqrt{1-t^2}} \tan(n \arcsin t),$$

where we have used the fact that n is odd and the relation $\arccos\theta = \pi/2 - \arcsin\theta$. Therefore,

$$S_p(n) = \sum_{k=1}^{n-1} \sec^{2p}\left(\frac{k\pi}{n}\right) = G^{(2p)}(n,0)/(2p)!,$$

where $G^{(2p)}$ indicates the $2p$-th derivative of G with respect to t.

For $p = 1$, computing the indicated derivative directly, we find

$$\sum_{k=0}^{n-1} \sec^2\left(\frac{k\pi}{n}\right) = \frac{G''(n,0)}{2!} = n^2. \tag{14.9}$$

However, for $p \geq 2$, the expression of $G^{(2p)}(n,t)$ becomes very complicated and inconvenient. So, we now demonstrate a derivation that is carried out with the help of Mathematica rather than by hand. For example, in Mathematica, the identity (14.9) is reproduced with the following commands:

G[t_]:=n*t*Tan[n*ArcSin[t]]/Sqrt[1-t^2];
D[G[t],{t,2}]/2!/.t->0
n^2

Similarly, we have

D[G[t],{t,4}]/4!/.t->0;
Simplify[%]
1/3 n^2(2 + n^2)

Thus,

$$\sum_{k=0}^{n-1} \sec^4\left(\frac{k\pi}{n}\right) = \frac{n^2}{3}(n^2 + 2). \tag{14.10}$$

Further results obtained by Mathematica are listed at the end of the chapter.

Case 2: n even. Deleting the term for $k = n/2$ from $S_p(n)$, we reset

$$S_p(n) = \sum_{k=0, k \neq n/2}^{n-1} \sec^{2p}\left(\frac{k\pi}{n}\right).$$

Similarly, separating the term corresponding to $k = n/2$ from (14.7), we have

$$1 + n\cot n\theta \cot\theta = \sum_{k=0, k \neq n/2}^{n-1} \frac{-\cos^2\theta}{\cos^2\theta - \cos^2\left(\frac{k\pi}{n}\right)}.$$

As in (14.8), we find the generating function of $\{\mathcal{S}_p(n)\}$ has the form

$$\mathcal{G}(n,t) = 1 + n \cot n\theta \cot\theta,$$

where $t = \cos\theta$. Thus, explicitly, we have

$$\mathcal{G}(n,t) = 1 - \frac{nt}{\sqrt{1-t^2}} \cot(n \arcsin t),$$

where that n is even and the identity $\arccos\theta = \pi/2 - \arcsin\theta$ have been used. Therefore,

$$\mathcal{S}_p(n) = \sum_{k=0, k\neq n/2}^{n-1} \sec^{2p}\left(\frac{k\pi}{n}\right) = \lim_{t\to 0} \frac{\mathcal{G}^{(2p)}(n,t)}{(2p)!}.$$

As in Case 1, we may use Mathematica.

```
g[t_]:=1 - n*t*Cot[n*ArcSin[t]]/Sqrt[1-t^2];
D[g[t],{t,2}]/2!;
Limit[%,t->0]
1/3 (n^2 - 1)
```

Thus,

$$\sum_{k=0, k\neq n/2}^{n-1} \sec^2\left(\frac{k\pi}{n}\right) = \frac{1}{3}(n^2 - 1) \qquad (14.11)$$

and

```
g[t_]:=1 - n*t*Cot[n*ArcSin[t]]/Sqrt[1-t^2];
D[g[t],{t,4}]/4!;
Limit[%,t->0]
1/45 (-11 + 10 n^2 + n^4)
```

giving

$$\sum_{k=0, k\neq n/2}^{n-1} \sec^4\left(\frac{k\pi}{n}\right) = \frac{1}{45}(n^2 - 1)(n^2 + 11). \qquad (14.12)$$

14.2 Sums involving $\csc^{2p}(k\pi/n)$

Let

$$\mathcal{C}_p(n) = \sum_{k=1}^{n-1} \csc^{2p}\left(\frac{k\pi}{n}\right)$$

and its generating function be

$$\mathcal{F}(n,t) = \sum_{p=1}^{\infty} \mathcal{C}_p(n) t^{2p}.$$

14. On Some Finite Trigonometric Power Sums

On separating off the term for $k = 0$ from (14.5), and multiplying the remaining sums by $\tan\theta \sin\theta$, we find

$$1 - n\cot n\theta \tan\theta = \sum_{k=1}^{n-1} \frac{\sin^2\theta}{\sin^2(\frac{k\pi}{n}) - \sin^2\theta}$$

$$= \sum_{k=1}^{n-1} \frac{\left[\frac{\sin\theta}{\sin(\frac{k\pi}{n})}\right]^2}{1 - \left[\frac{\sin\theta}{\sin(\frac{k\pi}{n})}\right]^2} = \sum_{p=1}^{\infty} \left(\sum_{k=1}^{n-1} \csc^{2p}(\frac{k\pi}{n})\right) \sin^{2p}\theta.$$

This implies

$$\mathcal{F}(n,t) = 1 - n\cot n\theta \tan\theta$$

with $t = \sin\theta$. Explicitly, we have

$$\mathcal{F}(n,t) = 1 - \frac{nt}{\sqrt{1-t^2}} \cot(n \arcsin t).$$

Since this coincides with the generating function of $\mathcal{S}_p$,

$$C_p(n) = \mathcal{F}^{(2p)}(n,0)/(2p)! = \mathcal{S}_p(n).$$

Therefore, in view of (14.11) and (14.12), we find

$$\sum_{k=1}^{n-1} \csc^2\left(\frac{k\pi}{n}\right) = \sum_{k=0, k\neq n/2}^{n-1} \sec^2\left(\frac{k\pi}{n}\right) = \frac{1}{3}(n^2 - 1), \qquad (14.13)$$

$$\sum_{k=1}^{n-1} \csc^4\left(\frac{k\pi}{n}\right) = \sum_{k=0, k\neq n/2}^{n-1} \sec^4\left(\frac{k\pi}{n}\right) = \frac{1}{45}(n^2 - 1)(n^2 + 11). \qquad (14.14)$$

14.3 Sums involving $\tan^{2p}(k\pi/n)$

Let

$$T_p(n) = \sum_{k=0}^{n-1} \tan^{2p}\left(\frac{k\pi}{n}\right)$$

and its generating function be

$$\mathcal{T}(n,t) = \sum_{p=0}^{\infty} T_p(n) t^{2p}.$$

Then

$$T(n,t) = \sum_{p=0}^{\infty} \left(\sum_{k=0}^{n-1} \tan^{2p}\left(\frac{k\pi}{n}\right) \right) t^{2p}$$

$$= \sum_{k=0}^{n-1} \frac{1}{1 - t^2 \tan^2(k\pi/n)}$$

$$= \sum_{k=0}^{n-1} \frac{\cos^2(k\pi/n)}{(1+t^2)\cos^2(k\pi/n) - t^2}$$

$$= \frac{n}{1+t^2} + \frac{1}{1+t^2} \sum_{k=0}^{n-1} \frac{t^2/(1+t^2)}{\cos^2(k\pi/n) - t^2/(1+t^2)}.$$

Substituting $t = \cot\theta$ yields

$$\sum_{k=0}^{n-1} \frac{t^2/(1+t^2)}{\cos^2(k\pi/n) - t^2/(1+t^2)} = \sum_{k=0}^{n-1} \frac{\cos^2\theta}{\cos^2(k\pi/n) - \cos^2\theta}. \tag{14.15}$$

If n is odd, then $\cot(n\theta) = \tan(n \arctan t)$. Thus, in view of (14.4), we arrive at

$$T(n,t) = \frac{n}{1+t^2}(1 + t\tan(n\arctan t)). \tag{14.16}$$

If n is even, after deleting the term for $k = n/2$ from $T_p(n)$, we define

$$T_p^*(n) = \sum_{k=0, k \neq n/2}^{n-1} \tan^{2p}\left(\frac{k\pi}{n}\right)$$

and its generating function by

$$T^*(n,t) = \sum_{p=0}^{\infty} T_p^*(n) t^{2p}.$$

Then

$$T^*(n,t) = \sum_{p=0}^{\infty} \left(\sum_{k=0, k \neq n/2}^{n-1} \tan^{2p}\left(\frac{k\pi}{n}\right) \right) t^{2p}$$

$$= \sum_{k=0, k \neq n/2}^{n-1} \frac{1}{1 - t^2 \tan^2(k\pi/n)}$$

$$= \sum_{k=0, k \neq n/2}^{n-1} \frac{\cos^2(k\pi/n)}{(1+t^2)\cos^2(k\pi/n) - t^2}$$

$$= \frac{n}{1+t^2} + \frac{1}{1+t^2} \sum_{k=0}^{n-1} \frac{t^2/(1+t^2)}{\cos^2(k\pi/n) - t^2/(1+t^2)}.$$

14. On Some Finite Trigonometric Power Sums

Setting $t = \cot\theta$ and in view of (14.15), (14.4) and noticing that $\cot(n\theta) = -\cot(n\arctan t)$, we arrive at

$$T^*(n,t) = \frac{n}{1+t^2}(1 - t\cot(n\arctan t)). \tag{14.17}$$

Implementing (14.16) and (14.17) on Mathematica yields

```
T[t_]:=n*(1 + t*Tan[n*ArcTan[t]])/(1 + t^2);
D[T[t],{t,2}]/2!/.t->0; Factor[%]
(-1+n) n
```

$$\sum_{k=1}^{n-1} \tan^2\left(\frac{k\pi}{n}\right) = n(n-1); \tag{14.18}$$

```
D[T[t],{t,4}]/4!/.t->0; Factor[%]
1/3 (-1 +n)n(-3+ n+n^2)
```

Thus

$$\sum_{k=1}^{n-1} \tan^4\left(\frac{k\pi}{n}\right) = \frac{n(n-1)}{3}(n^2 + n - 3); \tag{14.19}$$

```
Tc[t_]:=n*(1 - t*Cot[n*ArcTan[t]])/(1 + t^2); D[Tc[t],{t,2}]/2!;
Limit[%, t->0]
1/3 (-2 + n)(-1 + n)
```

Thus

$$\sum_{\substack{k=1,k\neq n/2}}^{n-1} \tan^2\left(\frac{k\pi}{n}\right) = \frac{1}{3}(n-1)(n-2); \tag{14.20}$$

```
D[Tc[t],{t,4}]/4!; Factor[ Limit[%, t->0]]
1/45 (-2 + n)(-1 + n)(-13 + 3n + n^2)
```

Thus

$$\sum_{\substack{k=1,k\neq n/2}}^{n-1} \tan^4\left(\frac{k\pi}{n}\right) = \frac{(n-1)(n-2)}{45}(n^2 + 3n - 13). \tag{14.21}$$

14.4 Sums involving $\cot^{2p}(k\pi/n)$

Let

$$C_p(n) = \sum_{k=1}^{n-1} \cot^{2p}\left(\frac{k\pi}{n}\right)$$

and its generating function be

$$\mathcal{A}(n,t) = \sum_{p=0}^{\infty} C_p(n)t^{2p}.$$

Then
$$\mathcal{A}(n,t) = \sum_{p=0}^{\infty}\left(\sum_{k=1}^{n-1}\cot^{2p}\left(\frac{k\pi}{n}\right)\right)t^{2p}$$
$$= \sum_{k=1}^{n-1}\frac{1}{1-t^2\cot^2(k\pi/n)}$$
$$= \sum_{k=1}^{n-1}\frac{1-\cos^2(k\pi/n)}{1-(1+t^2)\cos^2(k\pi/n)}$$
$$= \frac{n}{1+t^2} - \frac{t^2}{1+t^2}\sum_{k=0}^{n-1}\frac{1/(1+t^2)}{\cos^2(k\pi/n)-1/(1+t^2)}.$$

Substituting $t = \tan\theta$ yields
$$\sum_{k=0}^{n-1}\frac{1/(1+t^2)}{\cos^2(k\pi/n)-1/(1+t^2)} = \sum_{k=0}^{n-1}\frac{\cos^2\theta}{\cos^2(k\pi/n)-\cos^2\theta}.$$

Therefore,
$$\mathcal{A}(n,t) = \frac{n}{1+t^2}(1-t\cot(n\arctan t)) = \mathcal{T}^*(n,t).$$

As an immediate consequence, (14.20) and (14.21) deduce that
$$\sum_{k=1}^{n-1}\cot^2\left(\frac{k\pi}{n}\right) = \frac{(n-1)(n-2)}{3}, \tag{14.22}$$

$$\sum_{k=1}^{n-1}\cot^4\left(\frac{k\pi}{n}\right) = \frac{(n-1)(n-2)}{45}(n^2+3n-13). \tag{14.23}$$

We now have established a variety of finite trigonometric power sum identities. The reader should be aware that different approaches to the explicit evaluations of finite trigonometric sums often yield distinctly different types of closed form evaluations. Notably, our preceding method yields evaluation in terms of binomial coefficients (see Exercise 9), while the comprehensive paper [1] offers a different approach and yields sums in terms of Bernoulli numbers. Clearly, as the powers of the trigonometric functions increase, explicit computations become increasingly tedious. Consequently, it is reasonable to use Mathematica to perform such calculations. For completeness, we provide additional power sum identities below.

A table of more power sums

1. Power sums of secant
When n is odd:
$$\sum_{k=0}^{n-1}\sec^6\left(\frac{k\pi}{n}\right) = \frac{n^2}{15}(2n^4+5n^2+8),$$

$$\sum_{k=0}^{n-1}\sec^8\left(\frac{k\pi}{n}\right) = \frac{n^2}{315}(17n^6+56n^4+98n^2+144),$$

14. On Some Finite Trigonometric Power Sums

$$\sum_{k=0}^{n-1} \sec^{10}\left(\frac{k\pi}{n}\right) = \frac{n^2}{2835}(62n^8 + 255n^6 + 546n^4 + 820n^2 + 1152).$$

When n is even:

$$\sum_{k=0, k \neq n/2}^{n-1} \sec^6\left(\frac{k\pi}{n}\right) = \frac{(n^2-1)}{945}(2n^4 + 23n^2 + 191),$$

$$\sum_{k=0, k \neq n/2}^{n-1} \sec^8\left(\frac{k\pi}{n}\right) = \frac{(n^2-1)}{14175}(3n^6 + 43n^4 + 337n^2 + 2497),$$

$$\sum_{k=0, k \neq n/2}^{n-1} \sec^{10}\left(\frac{k\pi}{n}\right) = \frac{(n^2-1)}{93555}(2n^8 + 35n^6 + 321n^4 + 2125n^2 + 14797).$$

2. Power sums of cosecant

$$\sum_{k=1}^{n-1} \csc^6\left(\frac{k\pi}{n}\right) = \frac{(n^2-1)}{945}(2n^4 + 23n^2 + 191),$$

$$\sum_{k=1}^{n-1} \csc^8\left(\frac{k\pi}{n}\right) = \frac{(n^2-1)}{14175}(3n^6 + 43n^4 + 337n^2 + 2497),$$

$$\sum_{k=1}^{n-1} \csc^{10}\left(\frac{k\pi}{n}\right) = \frac{(n^2-1)}{93555}(2n^8 + 35n^6 + 321n^4 + 2125n^2 + 14797).$$

3. Power sums of tangent
When n is odd:

$$\sum_{k=1}^{n-1} \tan^6\left(\frac{k\pi}{n}\right) = \frac{n(n-1)}{15}(2n^4 + 2n^3 - 8n^2 - 8n + 15),$$

$$\sum_{k=1}^{n-1} \tan^8\left(\frac{k\pi}{n}\right) = \frac{n(n-1)}{315}(17n^6 + 17n^5 - 95n^4 - 95n^3 + 213n^2 + 213n - 315).$$

$$\sum_{k=1}^{n-1} \tan^{10}\left(\frac{k\pi}{n}\right) = \frac{n(n-1)}{2385}(62n^8 + 62n^7 - 448n^6 - 48n^5 + 1358n^4 + 1358n^3 \\ - 2232n^2 - 2232n + 2835).$$

When n is even:

$$\sum_{k=1, k \neq n/2}^{n-1} \tan^6\left(\frac{k\pi}{n}\right) = \frac{(n-1)(n-2)}{945}(2n^4 + 6n^3 - 28n^2 - 96n + 251),$$

$$\sum_{k=1, k\neq n/2}^{n-1} \tan^8\left(\frac{k\pi}{n}\right) = \frac{(n-1)(n-2)}{14175}(3n^6 + 9n^5 - 59n^4 - 195n^3 + 457n^2 + 1761n - 3551),$$

$$\sum_{k=1, k\neq n/2}^{n-1} \tan^{10}\left(\frac{k\pi}{n}\right) = \frac{(n-1)(n-2)}{93555}(2n^8 + 6n^7 - 52n^6 - 168n^5 + 546n^4 + 1974n^3 - 3068n^2 - 13152n + 22417).$$

4. Power sums of cotangent

$$\sum_{k=1}^{n-1} \cot^6\left(\frac{k\pi}{n}\right) = \frac{(n-1)(n-2)}{945}(2n^4 + 6n^3 - 28n^2 - 96n + 251),$$

$$\sum_{k=1}^{n-1} \cot^8\left(\frac{k\pi}{n}\right) = \frac{(n-1)(n-2)}{14175}(3n^6 + 9n^5 - 59n^4 - 195n^3 + 475n^2 + 1761n - 3551),$$

$$\sum_{k=1}^{n-1} \cot^{10}\left(\frac{k\pi}{n}\right) = \frac{(n-1)(n-2)}{93555}(2n^8 + 6n^7 - 52n^6 - 168n^5 + 546n^4 + 1974n^3 - 3068n^2 - 13152n + 22417).$$

Exercises

1. Prove that
$$\sum_{k=0}^{n-1} \frac{\sin x}{\cos x - \cos(\theta + 2k\pi/n)} = \frac{n \sin nx}{\cos nx - \cos n\theta}.$$

2. Let ω be the nth primitive root of unity and $0 \leq j \leq n-1$. Prove that
$$\sum_{i \neq j} |\omega^i - \omega^j|^{-2} = \frac{1}{12}n(n^2 - 1).$$

3. Let ω be the nth primitive root of unity and let
$$g_N(n) = \sum_{k=1}^{n-1} (\omega^k - 1)^{-N}.$$

   Show that $g_N(n)$ is a polynomial with rational coefficients and degree at most N, and determine $g_N(n)$ for $N = 1, 2, 3$.

4. Evaluate
$$\sum_{k=1}^{n} \frac{\cos N(2k-1)\pi/n}{\sin^2(2k-1)\pi/(2n)},$$
in terms of the integers N and n, $0 \leq N \leq n$.

14. On Some Finite Trigonometric Power Sums

5. Let a and b be positive integers such that $0 < ab < n$. Then for any $x, y \in \mathbb{R}$,

$$\sum_{k=0}^{n-1} \sin^{2a}\left(\frac{bk\pi}{n} + x\right) = \sum_{k=0}^{n-1} \cos^{2a}\left(\frac{bk\pi}{n} + y\right) = \frac{n}{2^{2a}}\binom{2a}{a}.$$

6. **Monthly Problem 4152** [1945, 163; 1946, 344–346]. For $-N < x < N$, show the following trigonometric expansion for the binomial coefficient

$$\frac{2^N}{N}\sum_{m=1}^{N}\left(\cos\frac{m\pi}{N}\right)^N \cos\frac{m\pi x}{N} = \frac{N!}{\left(\frac{N+x}{2}\right)!\left(\frac{N-x}{2}\right)!}.$$

7. Prove that

    (a) If n is even, then

    $$\sum_{k=0}^{n-1} \frac{(-1)^k \cos\theta}{\cos^2\theta - \cos^2\left(\frac{k\pi}{n}\right)} = -n\csc n\theta \csc\theta.$$

    (b) If n is odd, then

    $$\sum_{k=0}^{n-1} \frac{(-1)^k \cos(k\pi/n)}{\cos^2\theta - \cos^2\left(\frac{k\pi}{n}\right)} = -n\csc n\theta \csc\theta.$$

8. For odd integer n, find the generating function for the alternating sums of $\sec^{2p+1}(k\pi/n)$.

9. If n is odd, prove that

$$S_p(n) = n\sum_{k=1}^{2p-1}(-1)^{p+k}\binom{p-1+kn}{2p-1}\sum_{j=k}^{2p-1}\binom{2p}{j+1}.$$

10. Let n, p be positive integers and let

$$S_{n,p} = \left(\frac{\pi}{2n}\right)^{2p}\sum_{k=1}^{n-1}\csc^{2p}\left(\frac{k\pi}{2n}\right).$$

    Find a closed form expression for $S_{n,p}$.

11. Evaluate

$$S = \sum_{i=1}^{N-1}\sum_{j=1}^{N-1}\sum_{k=1}^{N-1}\frac{\sin(kj\pi/N)\sin(ij\pi/N)}{1-\cos(j\pi/N)}.$$

12. Let

$$a_{ij} = \sin\frac{i\pi}{n}\sin\frac{j\pi}{n}\cos\frac{(i-j)\pi}{n}, \quad i,j = 1, 2, \ldots, n-1.$$

    Find the eigenvalues of the matrix $A = (a_{ij})$.

References

[1] B. Berndt and B. Yeap, Explicit evaluations and reciprocity theorems for finite trigonometric sums, *Advances in Appl. Math.*, **29** (2002) 358–385.

[2] H. Chen, On some trigonometric power sums, *Int. J. Math. Math. Anal.*, **30** (2002) 185–191.

[3] W. Chu, Summations on trigonometric functions, *Appl. Math. Comp.*, **141** (2003) 161–176.

[4] K. R. Stromberg, *Introduction to Classical Real Analysis*, Wadsworth, Belmont, 1981.

15
Power Series of $(\arcsin x)^2$

The power series expansion

$$(\arcsin x)^2 = \frac{1}{2} \sum_{n=1}^{\infty} \frac{(2x)^{2n}}{n^2 \binom{2n}{n}}, \quad |x| \leq 1 \tag{15.1}$$

plays an important role in the evaluation of series involving the central binomial coefficient. Some classical examples [1] are

$$\sum_{n=1}^{\infty} \frac{1}{n^2 \binom{2n}{n}} = \frac{\pi^2}{18} = \frac{1}{3}\zeta(2),$$

$$\sum_{n=1}^{\infty} \frac{(-1)^{n+1}}{n^3 \binom{2n}{n}} = \frac{2}{5}\zeta(3),$$

$$\sum_{n=1}^{\infty} \frac{1}{n^4 \binom{2n}{n}} = \frac{17}{3456}\pi^4 = \frac{17}{36}\zeta(4),$$

where $\zeta(x)$ is the Riemann zeta function. Since Euler, the series (15.1) has been established in a variety of ways, for example, by using Euler's series transformation [2] or by using the Gregory series for $\arctan x$ [1]. There are also a number of elegant and clever elementary proofs. In this chapter, we present two such proofs and display some applications. The first proof is based on the power series solution to a differential equation while the second is based on the evaluation of a parametric integral.

15.1 First Proof of the Series (15.1)

Let $y(x) = (\arcsin x)^2$. We begin with deriving a differential equation for $y(x)$. Differentiating $y(x)$ gives

$$\sqrt{1-x^2}\, y'(x) = 2\arcsin x. \tag{15.2}$$

Differentiating (15.2) and then multiplying by $\sqrt{1-x^2}$ shows that $y(x)$ satisfies the ordinary differential equation

$$(1-x^2)y''(x) - xy'(x) = 2. \tag{15.3}$$

Since this is a differential equation with variable coefficients, it is natural to seek a solution via the power series representation

$$y(x) = \sum_{n=0}^{\infty} a_n x^n. \tag{15.4}$$

Termwise differentiation of (15.4) yields

$$y'(x) = a_1 + 2a_2 x + 3a_3 x^2 + \cdots = \sum_{n=1}^{\infty} n a_n x^{n-1} = \sum_{n=0}^{\infty} (n+1) a_{n+1} x^n$$

and

$$y''(x) = 2a_2 + 6a_3 x + \cdots = \sum_{n=1}^{\infty} n(n+1) a_{n+1} x^{n-1} = \sum_{n=0}^{\infty} (n+2)(n+1) a_{n+2} x^n.$$

Since $a_0 = y(0) = 0, a_1 = y'(0) = 0$, substituting the series of $y'(x)$ and $y''(x)$ into (15.3) yields

$$2a_2 + 6a_3 x + \sum_{n=2}^{\infty} \left[(n+2)(n+1) a_{n+2} - n^2 a_n \right] x^n = 2.$$

Equating coefficients of x^n successively, we obtain $a_2 = 1, a_3 = 0$ and, in general,

$$a_{n+2} = \frac{n^2}{(n+2)(n+1)} a_n, \quad \text{for } n \geq 2.$$

For $n \geq 2$, this determines

$$a_{2n+1} = 0,$$

$$a_{2n} = \frac{[2^{n-1}(n-1)!]^2}{(2n)(2n-1)\cdots 4 \cdot 3} = \frac{1}{2} \frac{2^{2n}}{n^2 \binom{2n}{n}},$$

where we have used the fact that

$$\binom{2n}{n} = \frac{(2n)!}{(n!)^2}.$$

Thus, we find that

$$y(x) = \frac{1}{2} \sum_{n=1}^{\infty} \frac{(2x)^{2n}}{n^2 \binom{2n}{n}},$$

which converges for $|x| \leq 1$ by the Ratio Test. This proves (15.1) as desired.

15.2 Second Proof of the Series (15.1)

The second proof uses two different evaluations of the parametric integral

$$f(x) = \int_0^{\pi/2} \frac{d\theta}{1 - x \sin \theta}, \quad \text{for } |x| < 1. \tag{15.5}$$

For the first evaluation, setting $t = \tan(\theta/2)$, we have

$$\int_0^{\pi/2} \frac{d\theta}{1 - x \sin \theta} = 2 \int_0^1 \frac{dt}{(1 - x^2) + (t - x)^2}$$

$$= \frac{2}{\sqrt{1 - x^2}} \arctan\left(\frac{t - x}{\sqrt{1 - x^2}}\right) \Big|_0^1$$

$$= \frac{2}{\sqrt{1 - x^2}} \arctan\left(\sqrt{\frac{1 + x}{1 - x}}\right),$$

where we have used the identity

$$\arctan \alpha + \arctan \beta = \arctan\left(\frac{\alpha + \beta}{1 - \alpha\beta}\right).$$

Next, appealing to the identity

$$2 \arctan \sqrt{s} = \arcsin\left(\frac{s - 1}{s + 1}\right) + \frac{\pi}{2},$$

which can be verified by showing that both sides have the same derivatives, and setting $s = (1 + x)/(1 - x)$, our original function becomes

$$f(x) = \frac{\pi}{2\sqrt{1 - x^2}} + \frac{\arcsin x}{\sqrt{1 - x^2}}. \tag{15.6}$$

The second evaluation of (15.5) comes from expanding the integrand of (15.5) into a geometric series and then integrating term by term. This gives

$$f(x) = \sum_{n=0}^{\infty} \left(\int_0^{\pi/2} \sin^n \theta \, d\theta\right) x^n.$$

To proceed, for $n \geq 2$, we integrate by parts to get

$$\int \sin^n \theta \, d\theta = -\frac{1}{n} \sin^{n-1} \theta \cos \theta + \frac{n-1}{n} \int_0^{\pi/2} \sin^{n-2} \theta \, d\theta.$$

Evaluating at 0 and $\pi/2$ yields

$$\int_0^{\pi/2} \sin^n \theta \, d\theta = \frac{n-1}{n} \int_0^{\pi/2} \sin^{n-2} \theta \, d\theta.$$

Since

$$\int_0^{\pi/2} \sin^0 \theta \, d\theta = \frac{\pi}{2}, \quad \int_0^{\pi/2} \sin \theta \, d\theta = 1,$$

by induction, we obtain the following famous Wallis's formulas:

$$\int_0^{\pi/2} \sin^{2k}\theta \, d\theta = \frac{1 \cdot 3 \cdots (2k-1)}{2 \cdot 4 \cdots (2k)} \frac{\pi}{2} = \frac{\pi}{2^{2k+1}} \binom{2k}{k} \quad (15.7)$$

and

$$\int_0^{\pi/2} \sin^{2k+1}\theta \, d\theta = \frac{2 \cdot 4 \cdots (2k)}{1 \cdot 3 \cdots (2k+1)} = \frac{2^{2k}}{(2k+1)\binom{2k}{k}}.$$

Therefore,

$$f(x) = \frac{\pi}{2} \sum_{n=0}^{\infty} \frac{1}{2^{2n}} \binom{2n}{n} x^{2n} + \sum_{n=0}^{\infty} \frac{2^{2n}}{(2n+1)\binom{2n}{n}} x^{2n+1}. \quad (15.8)$$

Recalling the binomial series expansion

$$\frac{1}{\sqrt{1-x^2}} = \sum_{n=0}^{\infty} \frac{1}{2^{2n}} \binom{2n}{n} x^{2n}, \quad (15.9)$$

and comparing (15.6) and (15.8), we obtain

$$\frac{\arcsin x}{\sqrt{1-x^2}} = \sum_{n=0}^{\infty} \frac{2^{2n}}{(2n+1)\binom{2n}{n}} x^{2n+1}. \quad (15.10)$$

Finally, integrating (15.10) yields

$$\frac{1}{2}(\arcsin x)^2 = \sum_{n=0}^{\infty} \frac{2^{2n}}{(2n+1)(2n+2)\binom{2n}{n}} x^{2n+2},$$

that is, after replacing $n+1$ by n,

$$(\arcsin x)^2 = \sum_{n=1}^{\infty} \frac{2^{2n-1}}{(2n-1)(2n)\binom{2(n-1)}{n-1}} x^{2n},$$

which is equivalent to (15.1). Here, in addition to the required series, we get (15.10) as a bonus.

To demonstrate the power of (15.1), let us give some applications within the realm of elementary analysis.

1. The series expansion (15.1) yields the $(2n)$th derivative

$$(\arcsin^2(0))^{(2n)} = 2^{2n-1}[(n-1)!]^2, \quad \text{for all } n \geq 1.$$

The direct calculation is quite tedious.

15. Power Series of $(\arcsin x)^2$

2. Integrating (15.9) gives

$$\arcsin x = \int_0^x \frac{dt}{\sqrt{1-t^2}} = \sum_{n=0}^{\infty} \frac{1}{2^{2n}(2n+1)} \binom{2n}{n} x^{2n+1}.$$

In view of the Cauchy product of $\arcsin x$ and itself, equating the coefficients of x^{2n}, we obtain the combinatorial identity

$$\sum_{k=0}^{n} \frac{1}{(2k+1)(2n-2k-1)} \binom{2k}{k} \binom{2(n-k-1)}{n-k-1} = \frac{2^{4n-3}}{n^2 \binom{2n}{n}}.$$

Furthermore, the Cauchy product of $\arcsin x$ and $(\arcsin x)^2$ leads to

$$(\arcsin x)^3 = \sum_{n=1}^{\infty} \left(\frac{3!}{(2n+1)2^{2n}} \binom{2n}{n} \sum_{k=0}^{n-1} \frac{1}{(2k+1)^2} \right) x^{2n+1}.$$

In particular, for $x = 1$, we get

$$\sum_{n=1}^{\infty} \left(\frac{3!}{(2n+1)2^{2n}} \binom{2n}{n} \sum_{k=0}^{n-1} \frac{1}{(2k+1)^2} \right) = \frac{\pi^3}{8}.$$

3. Letting $x = \sin t$ in (15.1), we have

$$t^2 = \sum_{n=1}^{\infty} \frac{2^{2n-1}}{n^2 \binom{2n}{n}} \sin^{2n} t, \quad \text{for } |t| \leq \pi/2.$$

Integrating with respect to t from 0 to $\pi/2$ gives

$$\frac{\pi^3}{24} = \sum_{n=1}^{\infty} \frac{2^{2n-1}}{n^2 \binom{2n}{n}} \int_0^{\pi/2} \sin^{2n} t \, dt.$$

Applying Wallis's formula (15.7), we get the well-known Euler formula

$$\frac{\pi^2}{6} = \sum_{n=1}^{\infty} \frac{1}{n^2} = \zeta(2).$$

4. Dividing (15.1) by x and integrating from 0 to x, we have

$$\int_0^x \frac{(\arcsin t)^2}{t} dt = \frac{1}{4} \sum_{n=1}^{\infty} \frac{(2x)^{2n}}{n^3 \binom{2n}{n}}. \tag{15.11}$$

This function is viewed as a "higher transcendent" by Lehmer [1]. A symbolic evaluation by Mathematica gives

```
Sum[(2*x)^(2*n)/(n^3*Binomial[2 n, n]), {n, 1, Infinity}]

2 x^2 HypergeometricPFQ[{1, 1, 1, 1}, {3/2, 2, 2}, x^2]
```

That is,

$$\sum_{n=1}^{\infty} \frac{(2x)^{2n}}{n^3 \binom{2n}{n}} = 2x^2 \text{Hypergeometric}_4 F_3[\{1, 1, 1, 1\}, \{\frac{3}{2}, 2, 2\}, x^2].$$

Mathematica also deduces

$$\sum_{n=1}^{\infty} \frac{2^{2n}}{n^3 \binom{2n}{n}} = \pi^2 \ln 2 - \frac{7}{2} \zeta(3). \tag{15.12}$$

Moreover, replacing x by $\sin x$ and then setting $t = \sin \theta$ in (15.11), we have

$$\int_0^x \theta^2 \cot \theta \, d\theta = \sum_{n=1}^{\infty} \frac{2^{2n-2}}{n^3 \binom{2n}{n}} \sin^{2n} x. \tag{15.13}$$

Setting $x = \pi/2$ and manipulating (15.12) and (15.13) yields an integral representation for $\zeta(3)$:

$$\zeta(3) = \frac{2\pi^2}{7} \ln 2 - \frac{8}{7} \int_0^{\pi/2} \theta^2 \cot \theta \, d\theta.$$

Integrating by parts converts this into

$$\zeta(3) = \frac{2\pi^2}{7} \ln 2 + \frac{16}{7} \int_0^{\pi/2} \theta \ln \sin \theta \, d\theta.$$

5. Using the power series expansion of $\arcsin x$, Ewell in [3] rediscovered the Euler formula

$$\zeta(3) = \frac{\pi^2}{7} \left\{ 1 - 4 \sum_{n=1}^{\infty} \frac{\zeta(2n)}{(2n+1)(2n+2)2^{2n}} \right\}. \tag{15.14}$$

In the following, we give a different proof of (15.14). Indeed, recall (see (6.23) and (6.26))

$$\theta \cot \theta = 1 - 2 \sum_{n=1}^{\infty} \frac{\zeta(2n) \theta^{2n}}{\pi^{2n}}; \tag{15.15}$$

by (15.13), we get

$$\frac{1}{2} x^2 - \sum_{n=0}^{\infty} \frac{\zeta(2n)}{(n+1)\pi^{2n}} x^{2(n+1)} = \sum_{n=1}^{\infty} \frac{2^{2n-2}}{n^3 \binom{2n}{n}} \sin^{2n} x. \tag{15.16}$$

For $x = \pi/2$, appealing to (15.12), we obtain

$$\frac{1}{8} \pi^2 - \frac{\pi^2}{4} \sum_{n=1}^{\infty} \frac{\zeta(2n)}{(n+1)2^{2n}} = \frac{1}{4} (\pi^2 \ln 2 - \frac{7}{2} \zeta(3)).$$

15. Power Series of $(\arcsin x)^2$

Thus,
$$\zeta(3) = \frac{2\pi^2}{7}\left(\ln 2 - \frac{1}{2} + \sum_{n=1}^{\infty}\frac{\zeta(2n)}{(n+1)2^{2n}}\right). \tag{15.17}$$

By the identity
$$\frac{1}{n+1} = 2\left(\frac{1}{2n+1} - \frac{1}{(2n+1)(2n+2)}\right),$$
and appealing to the existing sum (see Exercise 10)
$$\sum_{n=1}^{\infty}\frac{\zeta(2n)}{(2n+1)2^{2n}} = \frac{1}{2}(1 - \ln 2) \tag{15.18}$$
we see that Euler's formula (15.14) follows from (15.17) immediately.

Furthermore, integrating (15.16) we obtain
$$\frac{1}{6}x^3 - \sum_{n=0}^{\infty}\frac{\zeta(2n)}{(n+1)(2n+3)\pi^{2n}}x^{2n+3} = \sum_{n=1}^{\infty}\frac{2^{2n-2}}{n^3\binom{2n}{n}}\int_0^x \sin^{2n} t\, dt.$$

For $x = \pi/2$, this gives another series representation for $\zeta(3)$:
$$\zeta(3) = \pi^2\left\{1 - 2\sum_{n=1}^{\infty}\frac{\zeta(2n)}{(2n+2)(2n+3)2^{2n}}\right\}. \tag{15.19}$$

We have seen a variety of consequences of (15.1). With the aid of these expressions (15.1) and (15.9)–(15.19), the interested reader is encouraged to find additional applications.

Exercises

1. Show that
$$\int_a^b f(x)\cot x\, dx = 2\sum_{n=1}^{\infty}\int_a^b f(x)\sin 2nx\, dx.$$
Using this result verifies (5.12) directly.

2. Setting $x = \sqrt{2}/2$ in (15.10) yields
$$\pi = \sum_{n=0}^{\infty}\frac{(n!)^2\, 2^{n+1}}{(2n+1)!}.$$
Can you give another proof?

3. Using the power series expansion of (15.9) and
$$\frac{1}{2}(\arcsin x)^2 = \int_0^x \frac{\arcsin t}{\sqrt{1-t^2}}\, dt,$$
prove that
$$\frac{\pi^2}{8} = 1 + \frac{1}{3^2} + \frac{1}{5^2} + \cdots + \frac{1}{(2k+1)^2} + \cdots.$$

4. Prove that

$$(\arcsin x)^4 = \frac{3}{2} \sum_{n=1}^{\infty} \left\{ \sum_{i=1}^{n-1} \frac{1}{i^2} \right\} \frac{(2x)^{2n}}{n^2 \binom{2n}{n}},$$

$$(\arcsin x)^6 = \frac{45}{4} \sum_{n=1}^{\infty} \left\{ \sum_{i=1}^{n-1} \frac{1}{i^2} \sum_{j=1}^{i-1} \frac{1}{j^2} \right\} \frac{(2x)^{2n}}{n^2 \binom{2n}{n}},$$

$$(\arcsin x)^8 = \frac{315}{2} \sum_{n=1}^{\infty} \left\{ \sum_{i=1}^{n-1} \frac{1}{i^2} \sum_{j=1}^{i-1} \frac{1}{j^2} \sum_{k=1}^{j-1} \frac{1}{k^2} \right\} \frac{(2x)^{2n}}{n^2 \binom{2n}{n}}.$$

5. Prove that, via a differential equation argument,

$$e^{\arcsin x} = \sum_{n=0}^{\infty} \frac{\prod_{i=1}^{n-1}(1+(2i)^2)}{(2n)!} x^{2n} + \sum_{n=0}^{\infty} \frac{\prod_{i=1}^{n}(1+(2i-1)^2)}{(2n+1)!} x^{2n+1}.$$

6. Setting $x = 1/2$ in (15.11) yields

$$S := \sum_{n=1}^{\infty} \frac{1}{n^3 \binom{2n}{n}} = 4 \int_0^{1/2} \frac{(\arcsin x)^2}{x} dx.$$

Show that

$$S = 2 \sum_{n=1}^{\infty} \frac{1}{n} \int_0^{\pi/3} x \cos nx \, dx$$

and

$$\zeta(3) = \frac{\pi}{2} \sum_{n=1}^{\infty} \frac{\sin \frac{n\pi}{3}}{n^2} - \frac{3}{4} S.$$

7. Show that

$$1 + xe^{x^2/4} \int_0^{x/2} e^{-t^2} dt = \sum_{n=0}^{\infty} \frac{n!}{(2n)!} x^{2n}.$$

8. **Monthly Problem 4293** [1948, 254; 1950, 46–47]. Evaluate

$$\sum_{n=1}^{\infty} \frac{(3-\sqrt{5})^n}{2^n n^3}$$

in closed form.

9. **Monthly Problem 11400** [2008, 948]. Evaluate

$$\sum_{n=1}^{\infty} \frac{\zeta(2n)}{n(n+1)}$$

in closed form.

15. Power Series of $(\arcsin x)^2$

10. Prove that

$$\sum_{n=1}^{\infty} \frac{\zeta(2n)}{(2n)2^{2n}} = \frac{1}{2}(\ln \pi - \ln 2) \quad \text{and} \quad \sum_{n=1}^{\infty} \frac{\zeta(2n)}{(2n+1)2^{2n}} = \frac{1}{2}(1 - \ln 2).$$

11. Prove that

$$\zeta(2) = \frac{10}{3}\left\{1 + 12\sum_{n=2}^{\infty} \frac{(n-1)\zeta(2n)}{(n+1)(n+2)(n+3)2^n}\right\},$$

$$\zeta(3) = \frac{4\pi^2}{27}\left\{1 - 6\sum_{n=1}^{\infty} \frac{\zeta(2n)}{(2n+1)(2n+3)2^{2n}}\right\}.$$

12. **Ramanujan Identity.** Show that

$$\zeta(3) = \frac{7\pi^3}{180} - 2\sum_{k=1}^{\infty} \frac{1}{k^3(e^{2k\pi} - 1)}.$$

This is a *hyperbolic series* approximation in which the "error" is

$$\left|\zeta(3) - \frac{7\pi^3}{180}\right| \approx 0.003743.$$

References

[1] D. H. Lehmer, Interesting series involving the central binomial coefficient, *Amer. Math. Monthly*, **92** (1985) 449–457.

[2] Z. A. Melzak, *Companion to Concrete Mathematics*, John Wiley & Sons, New York, 1973.

[3] J. A. Ewell, A new series representation for $\zeta(3)$, *Amer. Math. Monthly*, **97** (1990) 219–220.

16

Six Ways to Sum $\zeta(2)$

> *Read Euler, read Euler, he is the master of us all.* — P. Laplace

The history of mathematics is valuable because it shows the motivations underlying the developments of many branches of mathematics. Euler's recorded work on infinite series provides a particularly instructive example. In the eighteenth century infinite series were, as they are today, considered an essential part of analysis. In comparison, at the dawn of the seventeenth century infinite series were little understood and infrequently encountered. For example, Jakob Bernoulli handled the divergence of the harmonic series perfectly, but he could not find the exact value of the convergent series

$$\zeta(2) := 1 + \frac{1}{2^2} + \frac{1}{3^2} + \cdots + \frac{1}{n^2} + \cdots.$$

Indeed, he included in his *Tractatus* [3] a plea for help in summing this series:

> *If anyone finds and communicates to us that which thus far has eluded our efforts, great will be our gratitude.*

Before Euler, it appeared that all efforts at seeking an exact evaluation had proven futile, and even an accurate numerical evaluation was extremely difficult because of the slow decay of the terms. Indeed, since

$$\frac{1}{n} - \frac{1}{n+1} = \frac{1}{n(n+1)} < \frac{1}{n^2} < \frac{1}{n(n-1)} = \frac{1}{n-1} - \frac{1}{n}$$

we have

$$\frac{1}{N+1} < \sum_{n=N+1}^{\infty} \frac{1}{n^2} < \frac{1}{N},$$

so that to compute directly the sum with an accuracy of six decimal places would require at least a million terms.

In 1731, at the age of 24, Euler worked on this problem by calculating a numerical approximation. To this end, he transformed the series into the following rapidly convergent one:

$$\zeta(2) = (\ln 2)^2 + 2 \sum_{n=1}^{\infty} \frac{1}{n^2 \cdot 2^n}$$

175

with
$$\ln 2 = \sum_{n=1}^{\infty} \frac{1}{n \cdot 2^n}.$$

Using this he calculated $\zeta(2)$ accurately to six places and obtained the value
$$\zeta(2) = 1.644944\ldots.$$

Four years later, he finally arrived at the following brilliant result:
$$\zeta(2) = 1 + \frac{1}{2^2} + \frac{1}{3^2} + \cdots + \frac{1}{n^2} + \cdots = \frac{\pi^2}{6}. \tag{16.1}$$

This truly remarkable formula yielded $\zeta(2)$, the first nontrivial value of the Riemann zeta function. Since the finite sums of the series in (16.1) are always rational, no one expected the irrational number π, the ratio of the circumference of a circle to the diameter, to appear in the value of the sum. One of the most memorable gems of 18th century mathematics, (16.1) has a prominent place in mathematical history. A number of extremely elegant and clever proofs have been discovered (and rediscovered). In this chapter, we present six of these proofs.

16.1 Euler's Proof

We begin with the heuristic reasoning that led Euler to his wonderful discovery. Euler's idea was based on the extension of Newton's symmetric functions of the polynomial roots to the case where the polynomial is replaced by a power series. Let
$$\left(1 - \frac{x^2}{r_1^2}\right)\left(1 - \frac{x^2}{r_2^2}\right)\cdots\left(1 - \frac{x^2}{r_n^2}\right) = 1 - \alpha x^2 + \beta x^4 - \cdots + (-1)^n \gamma x^{2n}.$$

We have
$$\alpha = \frac{1}{r_1^2} + \frac{1}{r_2^2} + \cdots + \frac{1}{r_n^2}, \quad \beta = \frac{1}{r_1^2 r_2^2} + \frac{1}{r_1^2 r_3^2} + \cdots + \frac{1}{r_{n-1}^2 r_n^2}$$

and so on. In particular
$$\frac{1}{r_1^4} + \frac{1}{r_2^4} + \cdots + \frac{1}{r_n^4} = \alpha^2 - 2\beta.$$

Euler's idea was to apply these relations to the case when the polynomial is replaced by a power series
$$\frac{\sin x}{x} = 1 - \frac{x^2}{3!} + \frac{x^4}{5!} - \frac{x^6}{7!} + \cdots.$$

The function $\sin x / x$ has the roots $\pm \pi, \pm 2\pi, \pm 3\pi, \ldots$, and so the above formulas give the following. First
$$\frac{1}{\pi^2} + \frac{1}{4\pi^2} + \frac{1}{9\pi^2} + \cdots = \frac{1}{3!},$$

which leads at once to
$$\zeta(2) = 1 + \frac{1}{2^2} + \frac{1}{3^2} + \cdots + \frac{1}{n^2} + \cdots = \frac{\pi^2}{6}.$$

16. Six Ways to Sum $\zeta(2)$

One can go on and on, which is what Euler did, calculating $\zeta(2k)$ up to $k = 6$. In particular,

$$\zeta(4) = 1 + \frac{1}{2^4} + \frac{1}{3^4} + \cdots + \frac{1}{n^4} + \cdots = \frac{\pi^4}{90}.$$

Euler's proof is not considered valid today since the power series is not polynomial and does not share all properties of polynomials. Indeed, when Euler communicated (16.1) to his friends, the following objection was brought up: if $f(x)$ is any function to which this method is applied, $f(x)$ and $e^x f(x)$ both have the same roots and yet they should lead to different formulas. It took him about ten years to finally succeed in obtaining *Euler's infinite product for the sine* (see (5.26))

$$\frac{\sin x}{x} = \prod_{n=1}^{\infty} \left(1 - \frac{x^2}{n^2 \pi^2}\right). \tag{16.2}$$

Once this formula is established, all the objections disappear. Moreover, based on (16.2), Euler then succeeded in getting a closed formula for all the $\zeta(2k)$, as we showed in Chapter 6.

Euler returned to the evaluation of $\zeta(n)$ many times. In 1737, he discovered the fabulous product formula

$$\zeta(s) = \prod_p (1 - 1/p^s), \quad \text{where } p \text{ runs over all primes.}$$

Investigations on $\zeta(n)$ for odd n led him in 1749 to establish the *functional equation of the zeta function*

$$\zeta(1-s) = 2(2\pi)^{-s} \cos\frac{s\pi}{2} \Gamma(s) \zeta(s),$$

subsequently forgotten for over a century until resurrected by Riemann in 1859. Euler was one of the greatest masters of the theory of infinite series, one whose height very few can hope to reach. Yet we can all learn from his impact on modern mathematics and inspire students to look at Euler's work for new insights and focus.

16.2 Proof by Double Integrals

The proof presented first appeared as an exercise in William LeVeque's number theory textbook from 1956, and was later rediscovered in the *Mathematical Intelligencer* by Apostol in 1983.

The proof consists of two different evaluations of the double integral

$$I := \int_0^1 \int_0^1 \frac{1}{1-xy} \, dx \, dy.$$

First, we expand the integrand as a geometric series, apply the monotone convergence theorem, and then integrate term by term. This results in

$$I = \int_0^1 \int_0^1 \sum_{n=0}^{\infty} x^n y^n \, dx \, dy$$
$$= \sum_{n=0}^{\infty} \left(\int_0^1 x^n \, dx \right) \left(\int_0^1 y^n \, dy \right)$$
$$= \sum_{n=0}^{\infty} \frac{1}{(n+1)^2} = \zeta(2).$$

Next, we evaluate I by substitution

$$u := \frac{x+y}{2}, \quad v := \frac{y-x}{2}.$$

This yields

$$I = 2 \iint_S \frac{du \, dv}{1 - u^2 + v^2}$$

where S is the square with vertices $(0, 0), (1/2, -1/2), (1, 0)$ and $(1/2, 1/2)$ as shown in Figure 16.1.

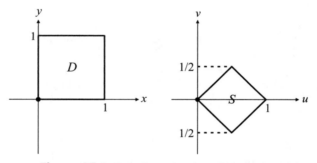

Figure 16.1. Transformed region of integration

In view of the symmetry of the square we get

$$I = 4 \int_0^{1/2} \left(\int_0^u \frac{dv}{1 - u^2 + v^2} \right) du + 4 \int_{1/2}^1 \left(\int_0^{1-u} \frac{dv}{1 - u^2 + v^2} \right) du.$$

Using

$$\int \frac{dx}{a^2 + x^2} = \frac{1}{a} \arctan(x/a) + C$$

yields

$$I = 4 \int_0^{1/2} \frac{1}{\sqrt{1-u^2}} \arctan\left(\frac{u}{\sqrt{1-u^2}}\right) du + 4 \int_{1/2}^1 \frac{1}{\sqrt{1-u^2}} \arctan\left(\frac{1-u}{\sqrt{1-u^2}}\right) du.$$

Set

$$g(u) := \arctan\left(\frac{u}{\sqrt{1-u^2}}\right), \quad h(u) := \arctan\left(\frac{1-u}{\sqrt{1-u^2}}\right).$$

16. Six Ways to Sum $\zeta(2)$

We have
$$g'(u) = \frac{1}{\sqrt{1-u^2}}, \quad h'(u) = -\frac{1}{2}\frac{1}{\sqrt{1-u^2}}.$$

Thus
$$I = 4\int_0^{1/2} g'(u)g(u)du + 4\int_{1/2}^1 -2h'(u)h(u)du$$
$$= 2\left[g^2(u)\right]_0^{1/2} - 4\left[h^2(u)\right]_{1/2}^1$$
$$= 2(\pi/6)^2 - 0 - 0 + 4(\pi/6)^2 = \frac{\pi^2}{6}$$

as expected.

Note that
$$\frac{1}{2^2} + \frac{1}{4^2} + \frac{1}{6^2} + \cdots = \frac{1}{2^2}\left(1 + \frac{1}{2^2} + \frac{1}{3^2} + \cdots\right) = \frac{1}{4}\zeta(2),$$

so the sum of odd terms
$$\frac{1}{1^2} + \frac{1}{3^2} + \frac{1}{5^2} + \cdots = \sum_{n=0}^\infty \frac{1}{(2n+1)^2} = \frac{3}{4}\zeta(2).$$

Thus Euler's formula (16.1) is equivalent to
$$\sum_{n=0}^\infty \frac{1}{(2n+1)^2} = \frac{\pi^2}{8}. \tag{16.3}$$

In a similar fashion, we may express the series in (16.3) as a double integral, namely
$$J := \sum_{n=0}^\infty \frac{1}{(2n+1)^2} = \int_0^1 \int_0^1 \frac{1}{1-x^2y^2}\, dxdy.$$

To compute the double integral, we ignore the boundary of the domain and consider x, y in the interior of the unit square
$$D = \{(x, y) : 0 < x, y < 1\}.$$

By making the substitution
$$(u, v) = \left(\tan^{-1} x\sqrt{\frac{1-y^2}{1-x^2}},\ \tan^{-1} y\sqrt{\frac{1-x^2}{1-y^2}}\right)$$

we have
$$(x, y) = \left(\frac{\sin u}{\cos v},\ \frac{\sin v}{\cos u}\right) \tag{16.4}$$

and u, v lie in the triangle
$$S = \{(u, v) : u > 0, v > 0, u + v < \pi/2\}$$

whenever $(x, y) \in D$. See Figure 16.2.

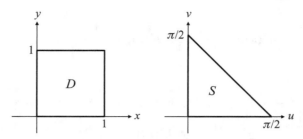

Figure 16.2. Transformed region of integration

The Jacobian determinant is

$$\det\begin{pmatrix} \dfrac{\cos u}{\cos v} & \dfrac{\sin u \sin v}{\cos^2 v} \\ \dfrac{\sin u \sin v}{\cos^2 u} & \dfrac{\cos v}{\cos u} \end{pmatrix} = 1 - \dfrac{\sin^2 u \sin^2 v}{\cos^2 u \cos^2 v} = 1 - x^2 y^2.$$

Thus (16.4) is a bijective transformation between D and S. Hence

$$J = \iint_S 1\, du\, dv = \text{Area of } S = \frac{1}{2}\left(\frac{\pi}{2}\right)^2 = \frac{\pi^2}{8}$$

as required.

Recently, Noam Elkies in [5] generalized (16.4) to the n-variable transformation

$$x_1 = \frac{\sin u_1}{\cos u_2}, \quad x_2 = \frac{\sin u_2}{\cos u_3}, \quad \ldots, \quad x_{n-1} = \frac{\sin u_{n-1}}{\cos u_n}, \quad x_n = \frac{\sin u_n}{\cos u_1} \quad (16.5)$$

and proved the following two properties:

1. The Jacobian determinant of the transformation (16.5) is

$$\frac{\partial(x_1, x_2, \ldots, x_n)}{\partial(u_1, u_2, \ldots, u_n)} = 1 \pm (x_1 x_2 \cdots x_n)^2,$$

the sign $-$ and $+$ chosen according to whether n is even or odd.

2. The transformation (16.5) maps the polytope

$$\Pi_n := \{(u_1, u_2, \ldots, u_n) \ :\ u_i > 0, u_i + u_{i+1} < \pi/2 \ (1 \le i \le n)\}$$

one-to-one into the open unit cube

$$S_n := \{(x_1, x_2, \ldots, x_n) \ :\ 0 < x_i < 1 \ (1 \le i \le n)\}.$$

In particular, he found that the volume of the polytope Π_n is

$$S(n) = \sum_{k=0}^{\infty} \frac{(-1)^{nk}}{(2k+1)^n} = \sum_{k=-\infty}^{\infty} \frac{1}{(4k+1)^n}.$$

16.3 Proof by Trigonometric Identities

After these two proofs via surprising and unexpected transformations, we can not resist the temptation to present two completely elementary proofs.

The first one begins with the identity

$$\frac{1}{\sin^2 x} = \frac{1}{4\sin^2(x/2)\cos^2(x/2)}$$
$$= \frac{1}{4}\left(\frac{1}{\sin^2(x/2)} + \frac{1}{\cos^2(x/2)}\right)$$
$$= \frac{1}{4}\left(\frac{1}{\sin^2(x/2)} + \frac{1}{\sin^2((\pi+x)/2)}\right).$$

Letting $x = \pi/2$ and repeatedly applying this identity yields

$$1 = \frac{1}{\sin^2(\pi/2)} = \frac{1}{4}\left(\frac{1}{\sin^2(\pi/4)} + \frac{1}{\sin^2(3\pi)/4}\right)$$
$$= \frac{1}{16}\left(\frac{1}{\sin^2(\pi/8)} + \frac{1}{\sin^2(3\pi)/8} + \frac{1}{\sin^2(5\pi/8)} + \frac{1}{\sin^2(7\pi)/8}\right)$$
$$= \frac{1}{4^n}\sum_{k=0}^{2^n-1}\frac{1}{\sin^2((2k+1)\pi/2^{n+1})}$$
$$= \frac{2}{4^n}\sum_{k=0}^{2^{n-1}-1}\frac{1}{\sin^2((2k+1)\pi/2^{n+1})}.$$

In the range of $0 < t < \pi/2$, we have $\sin t < t < \tan t$ and so

$$\frac{1}{\sin^2 t} > \frac{1}{t^2} > \cot^2 t = \frac{1}{\sin^2 t} - 1.$$

Thus, for $N = 2^n$ and $t = (2k+1)\pi/(2N)$, we have

$$1 > \frac{8}{\pi^2}\sum_{k=0}^{N/2-1}\frac{1}{(2k+1)^2} > 1 - \frac{1}{N}.$$

Taking the limit $N \to \infty$ yields

$$1 = \frac{8}{\pi^2}\sum_{k=0}^{\infty}\frac{1}{(2k+1)^2},$$

which is equivalent to (16.3).

The second one is based on the trigonometric identity

$$\sum_{k=1}^{N}\cot^2\left(\frac{k\pi}{2N+1}\right) = \frac{N(2N-1)}{3}. \qquad (16.6)$$

This identity initially appeared in a sequence of exercises in the problem book by the twin brothers Akiva and Isaak Yaglom, whose Russian original edition was published in 1954. Versions of closely related proofs were rediscovered and presented by F. Holme (1970), T. Apostol (1973) and by Ransford (1982). Indeed, Identity (16.6) is also given by (14.22). In view of

$$\cot^2 t < \frac{1}{t^2} < \csc^2 t,$$

using (16.6) yields

$$\frac{N(2N-1)}{3} < \sum_{k=1}^{N}\left(\frac{2N+1}{k\pi}\right)^2 < \frac{2N(N+1)}{3},$$

that is

$$\frac{\pi^2}{6}\frac{2N}{2N+1}\frac{2N-1}{2N+1} < \sum_{k=1}^{N}\frac{1}{k^2} < \frac{\pi^2}{6}\frac{2N}{2N+1}\frac{2N+2}{2N+1}.$$

Both the left-hand and the right-hand side converge to $\pi^2/6$ for $N \to \infty$, which ends the proof of (16.1).

16.4 Proof by Power Series

As we have seen in Chapter 15, the power series $(\arcsin x)^2$ together with Wallis's formula is used to derive (16.1). In the following, we present Boo Rim Choe's proof (*Monthly*, 1987), which uses only the power series of $\arcsin x$.

Note that

$$\arcsin x = \int_0^x \frac{dt}{\sqrt{1-t^2}} = \sum_{n=0}^{\infty}\frac{1}{2^{2n}(2n+1)}\binom{2n}{n}x^{2n+1}, \qquad (16.7)$$

which is valid for $|x| \leq 1$. Putting $x = \sin t$ we get

$$t = \sum_{n=0}^{\infty}\frac{1}{2^{2n}(2n+1)}\binom{2n}{n}\sin^{2n+1} t$$

for $|t| \leq \pi/2$. Integrating from 0 to $\pi/2$ and using Wallis's formula

$$\int_0^{\pi/2}\sin^{2n+1} t\, dt = \frac{2\cdot 4\cdots(2n)}{3\cdot 5\cdots(2n+1)}$$

yields (16.3) again.

Indeed, Choe's proof can be traced back to the following original proof of Euler. Let

$$I_k := \int_0^1 \frac{t^k}{\sqrt{1-t^2}}\, dt.$$

An integration by parts shows that

$$I_{k+2} = \frac{k+1}{k+2} I_k$$

from which I_k can be calculated for all k. In particular, we have

$$I_{2n+1} = \frac{2 \cdot 4 \cdots (2n)}{3 \cdot 5 \cdots (2n+1)}. \tag{16.8}$$

In view of

$$\frac{1}{2}(\arcsin x)^2 = \int_0^x \frac{\arcsin t}{\sqrt{1-t^2}} \, dt,$$

setting $x = 1$, then replacing $\arcsin t$ in the integrand by its power series expansion (16.7) and making use of (16.8), yields

$$\frac{\pi^2}{8} = 1 + \frac{1}{3^2} + \frac{1}{5^2} + \cdots + \frac{1}{(2n+1)^2} + \cdots$$

which is again equivalent to (16.1).

Euler tried to push this method for $\zeta(2k)$ for $k \geq 2$ but did not succeed. However, in this attempt, he found the following rapidly convergent formula

$$\zeta(2) = 3 \sum_{n=1}^{\infty} \frac{[(n-1)!]^2}{(2n)!} = 3 \sum_{n=1}^{\infty} \frac{1}{n^2 \binom{2n}{n}}. \tag{16.9}$$

16.5 Proof by Fourier Series

Let $f(x) = x(1-x)$. Consider the Fourier cosine series of $f(x)$ on $[0, 1]$. Notice that

$$a_0 = 2 \int_0^1 f(x) \, dx = \frac{1}{3},$$

$$a_n = 2 \int_0^1 f(x) \cos n\pi x \, dx = -2\frac{1+(-1)^n}{n^2 \pi^2}.$$

Therefore

$$f(x) = \frac{a_0}{2} + \sum_{n=1}^{\infty} a_n \cos n\pi x$$

$$= \frac{1}{6} - \sum_{n=1}^{\infty} \frac{\cos 2\pi n x}{\pi^2 n^2}.$$

Since f is continuous, of bounded variation on $[0, 1]$ and $f(0) = f(1)$, the Fourier series of f converges to f pointwise. Setting $x = 0$ we again get $\zeta(2) = \pi^2/6$.

16.6 Proof by Complex Variables

The final proof by Russell in [7] uses only Euler's formula

$$e^{ix} = \cos x + i \sin x$$

and the well-known logarithmic series. It begins with the definite integral

$$I := \int_0^{\pi/2} \ln(2\cos x)\, dx.$$

Now $2\cos x = e^{ix} + e^{-ix} = e^{ix}(1 + e^{-2ix})$ implies

$$\ln(2\cos x) = ix + \ln(1 + e^{-2ix}).$$

In terms of the logarithmic series expansion, we have

$$I = \int_0^{\pi/2} \left(ix - \sum_{n=1}^{\infty} \frac{(-1)^n}{n} e^{-2nix}\right) dx$$

$$= i\left(\frac{\pi^2}{8} - \sum_{k=0}^{\infty} \frac{1}{(2k+1)^2}\right).$$

But I is real, forcing

$$\frac{\pi^2}{8} = \sum_{k=0}^{\infty} \frac{1}{(2k+1)^2}$$

and thereby yielding an added bonus

$$\int_0^{\pi/2} \ln(\cos x)\, dx = -\frac{\pi}{2} \ln 2.$$

In general, the calculus of residues from complex analysis provides another tool to evaluate sums. A full account of this technique can be found in any introductory text on complex analysis. For example, see Marsden and Hoffman's *Basic Complex Analysis*.

We have seen a variety of proofs of Euler's famous formula for $\zeta(2)$. Some proofs appear in historical contexts. Some proofs have provided interesting connections to other topics. It is interesting to see how wide a range of mathematical subjects appeared in these proofs. Although one proof is enough to establish the truth of the formula, many generations of mathematicians have amused themselves by coming up with alternative proofs. It should come as no surprise if there are still more new proofs to come. The interested reader is encouraged to consult Robin Chapman's survey article, which compiles 14 different proofs and is available at
www.secamlocal.ex.ac.uk/people/staff/rjchapma/etc/zeta2.pdf.

Exercises

1. Prove that

$$\zeta(2) = (\ln 2)^2 + 2\sum_{n=1}^{\infty} \frac{1}{n^2 \cdot 2^n},$$

$$\zeta(3) = \frac{8}{7}\left(\frac{(\ln 2)^3}{3} + \sum_{n=1}^{\infty} \frac{1}{2^n}\left(\frac{\ln 2}{n^2} + \frac{1}{n^3}\right)\right).$$

16. Six Ways to Sum $\zeta(2)$

2. **Another proof of Euler's formula by double integral.** Prove that

   (a) $\dfrac{2}{2k+1} = \displaystyle\int_{-1}^{1} t^{2k}\,dt = \int_{-\infty}^{\infty} \dfrac{\tanh^{2k} x}{\cosh^2 x}\,dx,$

   (b) $\displaystyle\sum_{k=0}^{\infty} \dfrac{\tanh^{2k} x \tanh^{2k} y}{\cosh^2 x \cosh^2 y} = \dfrac{1}{\cosh(x+y)\cosh(x-y)},$

   (c) $\displaystyle\sum_{k=0}^{\infty} \dfrac{1}{(2k+1)^2} = \dfrac{1}{4}\int_{-\infty}^{\infty}\int_{-\infty}^{\infty} \dfrac{1}{\cosh(x+y)\cosh(x-y)}\,dx\,dy = \dfrac{\pi^2}{8}.$

3. Recall that (for example, see (5.28))

$$\sin x = (2n+1)\sin\left(\dfrac{x}{2n+1}\right)\prod_{k=1}^{n}\left(1 - \dfrac{\sin^2(x/(2n+1))}{\sin^2(k\pi/(2n+1))}\right).$$

   Let

$$u_k = \dfrac{\sin^2(x/(2n+1))}{\sin^2(k\pi/(2n+1))}.$$

   Show that

$$\dfrac{\sum_{k=1}^{n} u_k}{1 + \sum_{k=1}^{n} u_k} \le 1 - \dfrac{\sin x}{(2n+1)\sin(x/(2n+1))} \le \sum_{k=1}^{n} \dfrac{u_k}{1 - u_k}.$$

   Based on this inequality, give a proof of Euler's formula.

4. **Monthly Problem 3996** [1941, 341; 1942, 483–485]. Sum the series

$$\sum_{n=1}^{\infty} \dfrac{[(n-1)k]!}{(nk)!},$$

   where k is any integer greater than unity.

5. **Monthly Problem 4318** [1948, 586; 1950, 345–346]. Show that

$$\sum_{n=1}^{\infty} \dfrac{1}{(n\sinh n\pi)^2} = \dfrac{2}{3}G - \dfrac{11}{30}\zeta(2),$$

   where G is *Catalan's constant*, which is defined by

$$G = \sum_{n=0}^{\infty} \dfrac{(-1)^n}{(2n+1)^2}.$$

6. Show that

$$\sum_{n=1}^{\infty} \dfrac{1}{2^n}\sum_{k=1}^{n}\binom{n}{k}\dfrac{(-1)^{k+1}}{k} = 2\ln 2 \quad \text{and} \quad \sum_{n=1}^{\infty} \dfrac{1}{2^n}\sum_{k=1}^{n}\binom{n}{k}\dfrac{(-1)^{k+1}}{k^2} = \zeta(2).$$

Comment. Recall **Euler's Series Transformation**, which states that if $f(t) = \sum_{n=0}^{\infty} a_n t^n$ and $\sum_{n=0}^{\infty} a_n$ converges, then

$$\frac{1}{1-t} f\left(\frac{t}{1-t}\right) = \sum_{n=0}^{\infty} t^n \left(\sum_{k=0}^{n} \binom{n}{k} a_k\right).$$

One may establish that

$$\sum_{n=1}^{\infty} \frac{1}{2^n} \sum_{k=1}^{n} \binom{n}{k} \frac{x^k}{k^p} = 2 \sum_{n=1}^{\infty} \frac{x^n}{n^p}.$$

7. (H. Wilf) For every integer $n \geq 0$, show that

$$3 \sum_{i=1}^{n} \frac{1}{i^2 \binom{2i}{i}} + \frac{n!^4}{(2n)!} \sum_{k=0}^{\infty} \frac{k^2}{[(k+n+1)!]^2} = \frac{\pi^2}{6}.$$

Remark. This gives us a family of series which all converge to $\pi^2/6$. When $n = 0$ we have the usual series for $\zeta(2)$, and when $n \to \infty$ we have the faster convergent series (16.9).

8. Show that

$$\zeta(3) = \int_0^1 \int_0^1 \int_0^1 \frac{1}{1-xyz}\, dxdydz = \int_0^1 \int_0^1 \frac{-\ln(xy)}{1-xy}\, dxdy$$

$$= \int_0^1 \frac{\ln(1-x)\ln x}{x}\, dx = 4 \int_0^{\pi/2} \frac{\ln \sin x \ln \cos x}{\sin x \cos x}\, dx.$$

9. Show that

$$\zeta(3) = \frac{1}{2} \int_0^1 \frac{\ln(1-x) \ln x}{x(1-x)}\, dx.$$

Use this identity to show that

$$\zeta(3) = \frac{1}{2} \sum_{i=1}^{\infty} \sum_{j=1}^{\infty} \frac{1}{ij^2 \binom{i+j-1}{j}}.$$

10. Define

$$A_n = \sum_{i_1=1}^{\infty} \cdots \sum_{i_n=1}^{\infty} \frac{1}{i_1 i_2 \cdots i_n (i_1 + i_2 + \cdots + i_n)}.$$

Show that

$$A_n = (-1)^n \int_0^1 \frac{\ln^n x}{1-x}\, dx = n! \zeta(n+1).$$

11. Define

$$B_n = \sum_{i_1=1}^{\infty} \cdots \sum_{i_n=1}^{\infty} \frac{(-1)^{i_1+i_2+\cdots+i_n}}{i_1 i_2 \cdots i_n (i_1 + i_2 + \cdots + i_n)}.$$

16. Six Ways to Sum $\zeta(2)$

Show that

$$B_n = (-1)^n \int_0^1 \frac{\ln^n(1+x)}{x} dx$$
$$= (-1)^n \left(n!\zeta(n+1) - \frac{n}{n+1}\ln^{n+1} 2 - \sum_{k=1}^{\infty}\sum_{i=1}^{n} \frac{n(n-1)\cdots(n-i+1)}{2^k k^{i+1}} \ln^{n-i} 2 \right).$$

12. **Open Problem.** Motivated by

$$\sum_{i=1}^{\infty}\sum_{j=1}^{\infty} \frac{(-1)^i}{ij(i+j)} = -\int_0^1 \frac{\ln(1-x)\ln(1+x)}{x} dx = \frac{5}{8}\zeta(3),$$

define

$$C_n = \sum_{i_1=1}^{\infty} \cdots \sum_{i_n=1}^{\infty} \frac{(-1)^{i_1+i_2+\cdots+i_m}}{i_1 i_2 \cdots i_n (i_1 + i_2 + \cdots + i_n)}.$$

For $0 < m < n$, determine C_n in the perspective of Exercises 10 and 11 above. This is equivalent to evaluating the integrals

$$I(n,m) = \int_0^1 \frac{\ln^m(1-x)\ln^{n-m}(1+x)}{x} dx.$$

13. Let $0 < \alpha < 1/2$. Prove that

$$\zeta(p+1) = \sum_{n=0}^{\infty} \frac{1}{(1-\alpha)^{n+1}} \sum_{k=0}^{n} \binom{n}{k} \frac{(-\alpha)^{n-k}}{(k+1)^{p+1}}.$$

In particular,

$$\zeta(p+1) = 2\sum_{n=0}^{\infty}\sum_{k=0}^{n} \binom{n}{k} \frac{(-1)^{n-k} 2^k}{(k+1)^{p+1}}.$$

References

[1] T. M. Apostol, A proof that Euler missed: Evaluating $\zeta(2)$ the easy way, *Math. Intelligence*, **5** (1983) 59–60.

[2] A. Ayoub, Euler and the Zeta Function, *Amer. Math. Monthly*, **81** (1974) 1067–1086.

[3] J. Bernoulli, *Tractatus de seriebus infinitis*, 1689.
Available at www.kubkou.se/pdf/mh/jacobB.pdf

[4] B. R. Choe, An elementary proof of $\sum_{k=1}^{\infty} 1/n^2 = \pi^2/6$, *Amer. Math. Monthly*, **94** (1987) 662–663.

[5] N. D. Elkies, On the sums $\sum_{k=-\infty}^{\infty} (4k+1)^{-n}$, *Amer. Math. Monthly*, **110** (2003) 561–573.

[6] D. Kalman, Six ways to sum a series, *College Math. J.*, **24** (1993) 402–421.

[7] D. C. Russell, Another Eulerian-type proof, *Math. Magazine*, **64** (1991) 349.

[8] E. Weisstein, Riemann Zeta Function zeta(2), From MathWorld—A Wolfram Web Resource.
Available at mathworld.wolfram.com/RiemannZetaFunctionZeta2.html

17

Evaluations of Some Variant Euler Sums

In the mathematical world Euler is akin to the likes of Shakespeare and Mozart — universal, richly detailed, and inexhaustible. His work, dating back to the early eighteenth century, is still very much alive and generating intense interest.

In response to a letter from Goldbach, for integers $m \geq 1$ and $n \geq 2$, Euler considered sums of the form

$$S(m,n) := \sum_{k=1}^{\infty} \frac{1}{(k+1)^n} \left(1 + \frac{1}{2} + \cdots + \frac{1}{k}\right)^m = \sum_{k=1}^{\infty} \frac{1}{(k+1)^n} H_k^m, \qquad (17.1)$$

and was successful in obtaining several explicit values of these sums in terms of the Riemann zeta function $\zeta(k)$. Indeed, he established the beautiful formula

$$S(1,2) = \sum_{k=1}^{\infty} \frac{1}{(1+k)^2} H_k = \zeta(3)$$

as well as the more general relations: for all $n > 2$,

$$S(1,n) = \frac{n}{2} \zeta(n+1) - \frac{1}{2} \sum_{k=1}^{n-2} \zeta(k+1)\zeta(n-k). \qquad (17.2)$$

Two slight variant sums of Euler's definition, which are known nowadays as *Euler sums*, are defined as

$$\mathcal{S}(m,n) = \sum_{k=1}^{\infty} \frac{1}{k^n} H_k^{(m)}, \text{ with } H_k^{(m)} = 1 + 1/2^m + \cdots + 1/k^m, \qquad (17.3)$$

and

$$\mathcal{T}(m,n) = \sum_{k=1}^{\infty} \frac{1}{k^n} \left(1 + \frac{1}{2} + \cdots + \frac{1}{k}\right)^m. \qquad (17.4)$$

It is easy to see that

$$\mathcal{T}(2,n) = \mathcal{S}(2,n) + 2\mathcal{S}(1,n+1) + \zeta(n+2).$$

Over the centuries progress on evaluating $S(m,n)$ and $T(m,n)$ for $m \geq 2$ has been minimal. In 1948, as Monthly Problem 4305, Sandham proposed to prove that

$$T(2,2) = \sum_{k=1}^{\infty} \frac{1}{k^2} H_k^2 = \frac{17}{4}\zeta(4), \tag{17.5}$$

which apparently remained unnoticed until 1993 when Enrico Au-Yeung, an undergraduate at the University of Waterloo, numerically rediscovered (17.5). Shortly thereafter it was rigorously proven true by Borwein and Borwein in [2]. This empirical result launched a fruitful search for Euler sums through a profusion of methods: combinatorial, analytic, and algebraic. Based on extensive experimentation with computer algebra systems, a large class of Euler and related sums have been explicitly evaluated in terms of the Riemann zeta values. Some typical evaluations include

$$\sum_{k=1}^{\infty} \frac{1}{2^k k^2} H_k = \zeta(3) - \frac{\pi^2}{12}\ln 2,$$

$$\sum_{k=1}^{\infty} \frac{1}{k^2} H_k^{(2)} = \frac{7}{4}\zeta(4),$$

$$\sum_{k=1}^{\infty} \frac{1}{k^3} H_k^2 = \frac{7}{2}\zeta(5) - \zeta(2)\zeta(3),$$

and

$$\sum_{k=1}^{\infty} \frac{(-1)^{k-1}}{k^2} H_k = \frac{5}{8}\zeta(3).$$

In particular, Borwein and Bradley in [3] recently collected thirty-two beautiful proofs of $S(1,2) = \zeta(3)$.

Motivated by these results, replacing H_k by

$$h_k = 1 + \frac{1}{3} + \cdots + \frac{1}{2k-1}, \tag{17.6}$$

we study the following new variant Euler sums

$$\sum_{k=1}^{\infty} a_k h_k$$

where the a_k are relatively simple functions of k.

We begin by deriving some series involving h_k. Since

$$-\ln(1-x) = \int_0^x \frac{dt}{1-t} = \sum_{k=1}^{\infty} \frac{x^k}{k},$$

replacing x by $-x$ gives

$$\ln(1+x) = \sum_{k=1}^{\infty} \frac{(-1)^{k-1} x^k}{k}.$$

17. Evaluations of Some Variant Euler Sums

Averaging these two series yields

$$\frac{1}{2}\ln\left(\frac{1+x}{1-x}\right) = \sum_{k=1}^{\infty} \frac{1}{2k-1} x^{2k-1}. \tag{17.7}$$

In terms of the Cauchy product and partial fractions, we have

$$\frac{1}{4}\ln^2\left(\frac{1+x}{1-x}\right) = \sum_{k=1}^{\infty} \left(\frac{1}{(2k-1)\cdot 1} + \frac{1}{(2k-3)\cdot 3} + \cdots + \frac{1}{1\cdot(2k-1)}\right) x^{2k}$$

$$= \sum_{k=1}^{\infty} \frac{1}{2k}\left[\left(\frac{1}{2k-1}+\frac{1}{1}\right) + \left(\frac{1}{2k-3}+\frac{1}{3}\right) + \cdots + \left(\frac{1}{1}+\frac{1}{2k-1}\right)\right] x^{2k}$$

$$= \sum_{k=1}^{\infty} \left(1 + \frac{1}{3} + \cdots + \frac{1}{2k-1}\right) \frac{x^{2k}}{k}.$$

Appealing to (17.6), we have

$$\sum_{k=1}^{\infty} \frac{h_k}{k} x^{2k} = \frac{1}{4}\ln^2\left(\frac{1+x}{1-x}\right). \tag{17.8}$$

This enables us to evaluate a wide variety of interesting series via specialization, differentiation and integration.

First, setting $x = 1/2$, we find

$$\sum_{k=1}^{\infty} \frac{h_k}{2^{2k} k} = \frac{1}{4}\ln^2 3. \tag{17.9}$$

For $x = \sqrt{2}/2$,

$$\sum_{k=1}^{\infty} \frac{h_k}{2^k k} = \frac{1}{4}\ln^2(3 + 2\sqrt{2}). \tag{17.10}$$

Putting $x = (\sqrt{5}-1)/2 = \bar{\phi}$, the reciprocal of the golden ratio, we get

$$\sum_{k=1}^{\infty} \frac{h_k}{k} \bar{\phi}^{2k} = \frac{1}{4}\ln^2(2 + \sqrt{5}). \tag{17.11}$$

Furthermore, for any $\alpha \geq 2$, putting $x = (\sqrt{5}+1)/2\alpha$ and $x = (\sqrt{5}-1)/2\alpha$ in (17.8) respectively, we get

$$\sum_{k=1}^{\infty} \frac{h_k}{\alpha^{2k} k} \left(\frac{\sqrt{5}+1}{2}\right)^{2k} = \frac{1}{4}\ln^2\left(\frac{(2\alpha+1)+\sqrt{5}}{(2\alpha-1)-\sqrt{5}}\right) \tag{17.12}$$

and

$$\sum_{k=1}^{\infty} \frac{h_k}{\alpha^{2k} k} \left(\frac{\sqrt{5}-1}{2}\right)^{2k} = \frac{1}{4}\ln^2\left(\frac{(2\alpha-1)+\sqrt{5}}{(2\alpha+1)-\sqrt{5}}\right). \tag{17.13}$$

Recalling the Fibonacci numbers, which are defined by

$$F_1 = 1, \ F_2 = 1, \ F_k = F_{k-1} + F_{k-2} \quad \text{for } k \geq 2$$

and Binet's formula

$$F_k = \frac{1}{\sqrt{5}} \left(\left(\frac{\sqrt{5}+1}{2} \right)^k - \left(\frac{1-\sqrt{5}}{2} \right)^k \right),$$

and combining (17.12) and (17.13), we find

$$\sum_{k=1}^{\infty} \frac{h_k}{\alpha^{2k} k} F_{2k} = \frac{\sqrt{5}}{20} \ln \left(\frac{\alpha^2 + \alpha - 1}{\alpha^2 - \alpha - 1} \right) \ln \left(\frac{\alpha^2 + \alpha\sqrt{5} + 1}{\alpha^2 - \alpha\sqrt{5} + 1} \right). \tag{17.14}$$

In particular, for $\alpha = 2$,

$$\sum_{k=1}^{\infty} \frac{h_k}{2^{2k} k} F_{2k} = \frac{\sqrt{5}}{4} \ln 5 \ \ln(9 + 4\sqrt{5}). \tag{17.15}$$

Another step along this path is changing variables. Setting $x = \cos \theta$ in (17.8) leads to

$$\sum_{k=1}^{\infty} \frac{h_k}{k} \cos^{2k} \theta = \ln^2 (\cot(x/2)). \tag{17.16}$$

Integrating both sides from 0 to π, and using

$$\int_0^{\pi} \cos^{2k} \theta \, d\theta = \frac{\pi}{2^{2k}} \binom{2k}{k}$$

and

$$\int_0^{\pi} \ln^2 (\cot(x/2)) \, d\theta = \frac{\pi^3}{4},$$

we find

$$\sum_{k=1}^{\infty} \frac{h_k}{2^{2k} k} \binom{2k}{k} = \frac{\pi^2}{4}. \tag{17.17}$$

Next, for $0 < x < 1$, differentiating (17.8), then multiplying both sides by x, we obtain

$$\sum_{k=1}^{\infty} h_k x^{2k} = \frac{x}{2(1-x^2)} \ln \left(\frac{1+x}{1-x} \right). \tag{17.18}$$

Setting $x = 1/2$, we get

$$\sum_{k=1}^{\infty} \frac{h_k}{2^{2k}} = \frac{1}{3} \ln 3. \tag{17.19}$$

For $x = \sqrt{2}/2$,

$$\sum_{k=1}^{\infty} \frac{h_k}{2^k} = \frac{\sqrt{2}}{2} \ln(3 + 2\sqrt{2}). \tag{17.20}$$

17. Evaluations of Some Variant Euler Sums

We can now deduce a result similar to (17.15):

$$\sum_{k=1}^{\infty} \frac{h_k}{2^{2k}} F_{2k} = \frac{\sqrt{5}}{50} (10 \ln(5 + 2\sqrt{5}) + 3\sqrt{5} \ln 5 - 5 \ln 5). \qquad (17.21)$$

Finally, for $0 < x \le 1$, dividing both sides of (17.9) by x and integrating from 0 to x, we obtain

$$\sum_{k=1}^{\infty} \frac{h_k}{k^2} x^{2k} = \frac{1}{2} \int_0^x \frac{1}{t} \ln^2 \left(\frac{1+t}{1-t} \right) dt. \qquad (17.22)$$

Using the substitution $u = (1-x)/(1+x)$ and integration by parts, we get

$$\sum_{k=1}^{\infty} \frac{h_k}{k^2} x^{2k} = \int_{(1-x)/(1+x)}^{1} \frac{\ln^2 u}{1-u^2} du$$

$$= \frac{1}{2} \ln x \ln^2 \left(\frac{1-x}{1+x} \right) + \int_{(1-x)/(1+x)}^{1} \frac{\ln u}{u} \ln \left(\frac{1-u}{1+u} \right) du.$$

In view of (17.7), we have

$$\int_{(1-x)/(1+x)}^{1} \frac{\ln u}{u} \ln \left(\frac{1-u}{1+u} \right) du = -2 \sum_{k=0}^{\infty} \frac{1}{2k+1} \int_{(1-x)/(1+x)}^{1} u^{2k} \ln u \, du.$$

Since

$$\int u^{2k} \ln u \, du = \frac{1}{2k+1} u^{2k+1} \ln u - \frac{1}{(2k+1)^2} u^{2k+1} + C,$$

we find

$$\sum_{k=1}^{\infty} \frac{h_k}{k^2} x^{2k} = \frac{1}{2} \ln x \ln^2 \left(\frac{1-x}{1+x} \right) + 2 \ln \left(\frac{1-x}{1+x} \right) \sum_{k=0}^{\infty} \frac{1}{(2k+1)^2} \left(\frac{1-x}{1+x} \right)^{2k+1}$$

$$+ 2 \sum_{k=0}^{\infty} \frac{1}{(2k+1)^3} - 2 \sum_{k=0}^{\infty} \frac{1}{(2k+1)^3} \left(\frac{1-x}{1+x} \right)^{2k+1}. \qquad (17.23)$$

In order to proceed comfortably, we employ the classical *polylogarithm function*

$$\text{Li}_n(x) = \sum_{k=1}^{\infty} \frac{x^n}{k^n},$$

which is the *generating function* of the sequence $\{1/k^n\}$. The special case $n = 2$ is the *dilogarithm* introduced by Euler. Clearly, $\zeta(2) = \text{Li}_2(1)$ and

$$\sum_{k=0}^{\infty} \frac{x^n}{(2k+1)^n} = \frac{1}{2} (\text{Li}_n(x) - \text{Li}_n(-x)).$$

Appealing to (17.23) and

$$\sum_{k=0}^{\infty} \frac{1}{(2k+1)^3} = \sum_{k=0}^{\infty} \frac{1}{k^3} - \sum_{k=0}^{\infty} \frac{1}{(2k)^3} = \frac{7}{8} \zeta(3),$$

we finally obtain
$$\sum_{k=1}^{\infty} \frac{h_k}{k^2} x^{2k} = \frac{7}{4}\zeta(3) + \frac{1}{2}\ln x \ln^2\left(\frac{1-x}{1+x}\right)$$
$$+ \ln\left(\frac{1-x}{1+x}\right)\left(\text{Li}_2\left(\frac{1-x}{1+x}\right) - \text{Li}_2\left(\frac{x-1}{1+x}\right)\right)$$
$$- \left(\text{Li}_3\left(\frac{1-x}{1+x}\right) - \text{Li}_3\left(\frac{x-1}{1+x}\right)\right).$$

Taking the limit as x approaches 1, we get
$$\sum_{k=1}^{\infty} \frac{h_k}{k^2} = \frac{7}{4}\zeta(3). \tag{17.24}$$

For $x = 1/3$,
$$\sum_{k=1}^{\infty} \frac{h_k}{3^{2k} k^2} = \frac{7}{8}\zeta(3) - \frac{1}{2}\ln 3 \ln^3 2$$
$$+ \frac{1}{3}\ln^3 2 + \ln 2 \,\text{Li}_2(-1/2) + \text{Li}_3(-1/2),$$

where we have used
$$\text{Li}_2(1/2) = \frac{\pi^2}{12} - \frac{1}{2}\ln^2 2;$$
$$\text{Li}_3(1/2) = \frac{7}{8}\zeta(3) + \frac{1}{6}\ln^3 2 - \frac{\pi^2}{12}\ln 2.$$

Moreover, manipulating (17.22) yields
$$\sum_{k=1}^{\infty} \frac{h_k}{k^2}(1-x^{2k}) = \frac{1}{2}\int_x^1 \frac{1}{t}\ln^2\left(\frac{1+t}{1-t}\right) dt.$$

Appealing to
$$h_k = \sum_{i=1}^{k}\int_0^1 x^{2(i-1)}\,dt = \int_0^1 \left(\sum_{i=1}^{k} x^{2(i-1)}\right) dt = \int_0^1 \frac{1-x^{2k}}{1-x^2}\,dx$$

we have
$$\sum_{k=1}^{\infty} \frac{h_k^2}{k^2} = \sum_{k=1}^{\infty} \frac{h_k}{k^2}\int_0^1 \frac{1-x^{2k}}{1-x^2}\,dx$$
$$= \frac{1}{2}\int_0^1 \left(\frac{1}{1-x^2}\int_x^1 \frac{1}{t}\ln^2\left(\frac{1+t}{1-t}\right) dt\right) dx.$$

Interchanging the order of the integration, we get
$$\sum_{k=1}^{\infty} \frac{h_k^2}{k^2} = \frac{1}{2}\int_0^1 \left(\frac{1}{t}\ln^2\left(\frac{1+t}{1-t}\right)\int_0^t \frac{1}{1-x^2}\,dx\right) dt.$$
$$= \frac{1}{4}\int_0^1 \frac{1}{t}\ln^3\left(\frac{1+t}{1-t}\right) dt.$$

17. Evaluations of Some Variant Euler Sums

Using the substitution $x = (1-t)/(1+t)$ and the well-known fact that

$$\int_0^1 x^k \ln^3 x \, dx = -\frac{6}{(k+1)^3},$$

we find

$$\sum_{k=1}^\infty \frac{h_k^2}{k^2} = -\frac{1}{2} \int_0^1 \frac{\ln^3 x}{1-x^2} dx$$

$$= -\frac{1}{2} \sum_{k=0}^\infty \int_0^1 x^{2k} \ln^3 x \, dx$$

$$= 3 \sum_{k=0}^\infty \frac{1}{(2k+1)^4} = \frac{45}{16} \zeta(4). \tag{17.25}$$

Another path from (17.8) is via complex variables and the formula:

$$\frac{1}{i} \tan^{-1}(iz) = \tanh^{-1} z = \frac{1}{2} \ln\left(\frac{1+z}{1-z}\right).$$

Replacing x by ix in (17.8), we obtain

$$\sum_{k=1}^\infty \frac{(-1)^{k-1} h_k}{k} x^{2k} = (\tan^{-1} x)^2. \tag{17.26}$$

This series may be evaluated explicitly at the values $x = 2 - \sqrt{3}, \sqrt{3}/3, 1$:

$$\sum_{k=1}^\infty \frac{(-1)^{k-1} h_k (2-\sqrt{3})^{2k}}{k} = \frac{\pi^2}{144} = \frac{3}{72} \zeta(2), \tag{17.27}$$

$$\sum_{k=1}^\infty \frac{(-1)^{k-1} h_k}{3^k k} = \frac{\pi^2}{36} = \frac{1}{6} \zeta(2), \tag{17.28}$$

$$\sum_{k=1}^\infty \frac{(-1)^{k-1} h_k}{k} = \frac{\pi^2}{16} = \frac{3}{8} \zeta(2). \tag{17.29}$$

Similarly, differentiating and integrating (17.26), we derive the corresponding formulas

$$\sum_{k=1}^\infty (-1)^{k-1} h_k x^{2k} = \frac{x}{1+x^2} \tan^{-1} x, \tag{17.30}$$

$$\sum_{k=1}^\infty \frac{(-1)^{k-1} h_k}{k^2} x^{2k} = 2 \int_0^x \frac{(\tan^{-1} t)^2}{t} dt. \tag{17.31}$$

In particular, we find

$$\sum_{k=1}^\infty \frac{(-1)^{k-1} h_k}{3^k} = \frac{\sqrt{3}}{24} \pi, \tag{17.32}$$

$$\sum_{k=1}^\infty \frac{(-1)^{k-1} h_k}{k^2} = G\pi - \frac{7}{4} \zeta(3), \tag{17.33}$$

where G is Catalan's constant, given by
$$G = \sum_{k=0}^{\infty} \frac{(-1)^k}{(2k+1)^2}.$$

Finally, observing that
$$h_k = H_{2k} - \frac{1}{2} H_k, \tag{17.34}$$

we find from (17.24)
$$\sum_{k=1}^{\infty} \frac{1}{k^2} H_{2k} = \sum_{k=1}^{\infty} \frac{1}{k^2} h_k + \frac{1}{2} \sum_{k=1}^{\infty} \frac{1}{k^2} H_k = \frac{11}{4} \zeta(3). \tag{17.35}$$

Furthermore, using partial fraction yields
$$\sum_{i=1}^{\infty} \sum_{j=1}^{\infty} \frac{1}{ij(i+j)} = \sum_{i=1}^{\infty} \sum_{j=1}^{\infty} \frac{1}{i^2} \left(\frac{1}{j} - \frac{1}{i+j} \right)$$
$$= \sum_{i=1}^{\infty} \frac{1}{i^2} H_i = 2\zeta(3).$$

Similarly,
$$\sum_{i=1}^{\infty} \sum_{j=1}^{\infty} \frac{(-1)^{i+j}}{ij(i+j)} = \frac{1}{4} \zeta(3).$$

The difference gives
$$\sum_{i,j>1, i+j=\text{odd}} \frac{1}{ij(i+j)} = \frac{7}{8} \zeta(3).$$

Setting $i + j = 2k + 1$ and using partial fractions, we have
$$\sum_{i,j>1, i+j=\text{odd}} \frac{1}{ij(i+j)} = \sum_{k=1}^{\infty} \sum_{j=1}^{2k} \frac{1}{j(2k+1-j)(2k+1)}$$
$$= \sum_{k=1}^{\infty} \frac{1}{(2k+1)^2} \sum_{j=1}^{2k} \left(\frac{1}{j} + \frac{1}{2k+1-j} \right)$$
$$= \sum_{k=1}^{\infty} \frac{1}{(2k+1)^2} 2H_{2k}.$$

Thus,
$$\sum_{k=1}^{\infty} \frac{1}{(2k+1)^2} H_{2k} = \frac{7}{16} \zeta(3). \tag{17.36}$$

Subsequently, we have
$$\sum_{k=1}^{\infty} \frac{1}{(2k-1)^2} H_{2k} = \sum_{k=1}^{\infty} \frac{1}{(2k+1)^2} H_{2k} + \sum_{k=1}^{\infty} \frac{1}{(2k-1)^3} + \sum_{k=1}^{\infty} \frac{1}{2k(2k-1)^2}$$
$$= \frac{21}{16} \zeta(3) + \frac{1}{8} (\pi^2 - 8\ln 2). \tag{17.37}$$

17. Evaluations of Some Variant Euler Sums

From this and the known result

$$\sum_{k=1}^{\infty} \frac{1}{(2k-1)^2} H_k = \frac{1}{4}(\pi^2 - \pi^2 \ln 2 - 8\ln 2 + 7\zeta(3)), \tag{17.38}$$

we finally get

$$\sum_{k=1}^{\infty} \frac{1}{(2k-1)^2} h_k = \frac{7}{16}\zeta(3) + \frac{3}{4}\zeta(2)\ln 2. \tag{17.39}$$

There are many avenues for generalizing Euler sums. We close this chapter by posing some other new variant Euler sums that are generalizations of (17.38), (17.39) and (17.35) respectively. Define

$$\mathbb{S}(m,n) = \sum_{k=0}^{\infty} \frac{1}{(2k+1)^n} H_k^{(m)},$$

$$\mathbb{T}(m,n) = \sum_{k=0}^{\infty} \frac{1}{(2k+1)^n} h_k^{(m)},$$

$$\mathbb{U}(m,n,r) = \sum_{k=1}^{\infty} \frac{1}{k^n} H_{kr}^{(m)}.$$

It follows that

$$\mathbb{S}(m,n) = \sum_{k=1}^{\infty} \frac{1}{(2k)^n} H_{2k}^{(m)} + \sum_{k=0}^{\infty} \frac{1}{(2k+1)^n} H_{2k+1}^{(m)}$$

$$= 2^{-n}\mathbb{U}(m,n,2) + \mathbb{T}(m,n) + 2^{-n}\mathbb{S}(m,n).$$

One can compute a special case, a result similar to Euler's formula (17.2):

$$\mathbb{S}(1, 2p+1) = (2 - 2^{-(2p+1)})(S(1, 2p+1) - \zeta(2+2p))$$

$$- 2\lambda(2p+1)\ln 2 + \sum_{k=2}^{2p} 2^{1-k}\zeta(k)\lambda(2p+2-k), \tag{17.40}$$

where

$$\lambda(x) = \sum_{k=0}^{\infty} \frac{1}{(2k+1)^x} = (1 - 2^{-x})\zeta(x).$$

But, we have been unable to determine $\mathbb{S}(m,n)$ in closed form for $m \geq 2$. Future exploration is left to the interested reader.

Exercises

1. Show that

$$\sum_{i=1}^{\infty}\sum_{j=1}^{\infty} \frac{(i-1)!(j-1)!}{(i+j)!} = \zeta(2) \quad \text{and} \quad \sum_{i=0}^{\infty}\sum_{j=0}^{i} (-1)^j \binom{i}{j} \frac{H_{j+1}}{j+1} = \zeta(2).$$

2. Evaluate
$$\sum_{i,j=1, i \neq j}^{\infty} \frac{1}{i^2 - j^2}.$$

3. Show that, for $|x| < 1$,
$$\sum_{k=0}^{\infty} (-1)^k h_{k+1} x^{2k} = \frac{\arctan x}{x(1 + x^2)}.$$

4. For positive integers k, let $g_k = \sum_{0 \leq 2n+1 \leq k} 1/(2n + 1)$. Show that
$$e^x \sum_{k=1}^{\infty} \frac{(-2)^{k-1} x^k}{k! k} = \sum_{k=1}^{\infty} \frac{g_k x^k}{k!}.$$

Use this to prove:
$$g_n = \sum_{k=1}^{n} \binom{n}{k} \frac{(-2)^{k-1}}{k}.$$

5. Let $S(m, n)$ be defined by (17.3). Show that
$$S(m, n) + S(n, m) = \zeta(m)\zeta(n) + \zeta(m + n).$$

6. Show that $\text{Li}_2(x)$ satisfies the functional equation
$$\text{Li}_2(x) + \text{Li}_2(1 - x) = \zeta(2) - \ln x \ln(1 - x).$$

7. Let G be Catalan's constant. Prove that

(a) $\sum_{n=0}^{\infty} \frac{(-1)^n}{2n + 1} H_n = G - \frac{\pi}{2} \ln 2.$

(b) $\sum_{n=0}^{\infty} \frac{(-1)^n}{2n + 1} h_n = \frac{\pi}{8} \ln 2 - \frac{1}{2} G.$

(c) $\sum_{n=1}^{\infty} \frac{(-1)^{n+1}}{n^2} h_n = \pi G - \frac{7}{4} \zeta(3).$

(d) $\sum_{n=0}^{\infty} \frac{2^n}{(2n + 1) \binom{2n}{n}} h_{n+1} = 2G.$

8. Show that
$$S(2, 2) = \frac{1}{4\pi} \int_0^{2\pi} (\pi - t)^2 \ln^2(2 \sin(t/2)) dt.$$

9. **Ramanujan's Constant** $G(1)$ (see [1] Entry 11. p 255). Ramanujan claimed that
$$G(1) := \frac{1}{8} \sum_{k=1}^{\infty} \frac{1}{k^3} h_k = \frac{\pi}{4} \sum_{k=1}^{\infty} \frac{(-1)^k}{(4k + 1)^3} - \frac{\pi}{3\sqrt{3}} \sum_{k=1}^{\infty} \frac{1}{(2k + 1)^3}.$$

17. Evaluations of Some Variant Euler Sums

Disprove this identity. Show that

$$G(1) = \mathbb{S}(2, 2) + \frac{15}{16} \zeta(4).$$

10. Prove (17.40).

11. For $|x| < 1$, define

$$f(x) = \sum_{k=1}^{\infty} \frac{h_k}{2k - 1} x^{2k-1}.$$

   Show that

$$f\left(\frac{x}{2 - x}\right) = \frac{1}{8} \ln^2(1 - x) + \frac{1}{2} \operatorname{Li}_2(x).$$

12. Prove that, for $n > 1$,

$$\sum_{k=1}^{\infty} \frac{h_k}{k^{2n}} = \frac{1}{4} \left(\zeta(2n + 1) + \frac{1}{2} - 2 \sum_{i=1}^{n-1} \zeta(2i) \left[\zeta(2n + 1 - 2i) + \frac{1}{2} \right] \right).$$

References

[1] B. Berndt, *Ramanujan's Notebooks, Part I*, New York, Springer-Verlag, 1985.

[2] D. Borwein and J. Borwein, On some intriguing sums involving $\zeta(4)$, *Proc. Amer. Math. Soc.*, **123** (1995) 1191–1198.

[3] J. Borwein and D. Bailey, *Thirty-Two Goldbach Variations,* Available at users.cs.dal.ca/~jborwein/32goldbach.pdf

[4] J. Borwein, D. Bailey and R. Girgensohn, *Experimentation in Mathematics*, A. K. Peters, MA, 2004.

[5] H. Chen, Evaluations of some variant Euler sums, *J. Integer Seq.*, **9** (2006), Article 06.2.3. Available at www.cs.uwaterloo.ca/journals/JIS/VOL9/Chen/chen78.pdf

[6] V. S. Varadarajan, *Euler Through Time: A New Look at Old Themes*, AMS, 2006.

[7] E. Weisstein, Euler Sum, From MathWorld—A Wolfram Web Resource. Available at mathworld.wolfram.com/EulerSum.html

18

Interesting Series Involving Binomial Coefficients

In his wonderful paper [4], Lehmer defines a series to be *interesting* if the sum has a closed form in terms of well-known constants. Using the power series

$$\frac{2x \arcsin x}{\sqrt{1-x^2}} = \sum_{n=1}^{\infty} \frac{(2x)^{2n}}{n \binom{2n}{n}}, \tag{18.1}$$

he establishes a large class of interesting series, including

$$\sum_{n=1}^{\infty} \frac{1}{\binom{2n}{n}} = \frac{1}{2} + \frac{1}{6} + \frac{1}{20} + \frac{1}{70} + \cdots = \frac{1}{3} + \frac{2\sqrt{3}}{27} \pi, \tag{18.2}$$

$$\sum_{n=1}^{\infty} \frac{1}{n \binom{2n}{n}} = \frac{1}{2} + \frac{1}{12} + \frac{1}{60} + \frac{1}{280} + \cdots = \frac{\sqrt{3}}{9} \pi, \tag{18.3}$$

and

$$\sum_{n=1}^{\infty} \frac{1}{n^2 \binom{2n}{n}} = \frac{1}{2} + \frac{1}{24} + \frac{1}{180} + \frac{1}{1120} + \cdots = \frac{1}{18} \pi^2. \tag{18.4}$$

However, he points out that the series

$$\sum_{n=1}^{\infty} \frac{1}{n^3 \binom{2n}{n}} = \frac{1}{2} + \frac{1}{48} + \frac{1}{540} + \frac{1}{4480} + \cdots = 4 \int_0^{1/2} \frac{(\arcsin y)^2}{y} \, dy$$

is a "higher transcendent" and does not admit a simple closed form.

Motivated by such results, it is natural for us to ask when a given series involving the central binomial coefficients is interesting. In this chapter, we focus our attention on a class

201

of series that possess integral representations. First we replace the power series expansion (18.1) by an integral representation, namely

$$\sum_{n=1}^{\infty} \frac{x^n}{\binom{2n}{n}} = \int_0^1 \frac{x(1-t)}{(1-xt(1-t))^2}\, dt \quad \text{for } |x| < 4. \tag{18.5}$$

This, along with its integration, differentiation and specialization, will enable us to recover and extend many of Lehmer's related results by a different approach. Moreover, we will see that whether a series is interesting or not is closely linked to the partial fraction decomposition of the integrand. Next, we extend this technique to more general binomial series. Finally, in a spirit of experimental mathematics, we show how to search for rapidly converging series for π.

18.1 An integral representation and its applications

We begin with deriving the proposed integral representation (18.5). Instead of using the series (18.1), we turn to the binomial coefficients directly (see [2] and [3]). Recall that the β-function is defined by

$$\beta(m,n) = \int_0^1 t^{m-1}(1-t)^{n-1}\, dt$$

and for all positive integers m and n,

$$\beta(m,n) = \frac{(m-1)!(n-1)!}{(m+n-1)!}.$$

We have

$$\binom{2n}{n}^{-1} = \frac{n!\,n!}{(2n)!} = n\int_0^1 t^{n-1}(1-t)^n\, dt. \tag{18.6}$$

By the ratio test, we see that the series in (18.5) converges for $|x| < 4$. Next, interchanging the order of the summation and the integration yields

$$\sum_{n=1}^{\infty} \frac{x^n}{\binom{2n}{n}} = \int_0^1 \sum_{n=1}^{\infty} nx^n t^{n-1}(1-t)^n\, dt. \tag{18.7}$$

Appealing to the well-known series

$$\sum_{n=1}^{\infty} nz^{n-1} = \frac{1}{(1-z)^2} \quad \text{for } |z| < 1$$

and $t(1-t) \leq 1/4$ for $0 \leq t \leq 1$, we establish (18.5) from (18.7).

We now demonstrate some examples of interesting series via its specialization, integration, and differentiation of (18.5). Starting by setting $x = 1$ in (18.5), we recover (18.3)

18. Interesting Series Involving Binomial Coefficients

as

$$\sum_{n=1}^{\infty} \frac{1}{\binom{2n}{n}} = \int_0^1 \frac{(1-t)dt}{(t^2-t+1)^2} = \frac{1}{3} + \frac{2\sqrt{3}}{27}\pi.$$

For $x = -1$, letting $\phi = (1+\sqrt{5})/2$ be the Golden-Ratio, $\bar{\phi} = (1-\sqrt{5})/2$, we obtain

$$\sum_{n=1}^{\infty} \frac{(-1)^{n-1}}{\binom{2n}{n}} = \int_0^1 \frac{(1-t)dt}{(t^2-t-1)^2}$$

$$= \int_0^1 \frac{(1-t)dt}{[(t-\phi)(t-\bar{\phi})]^2}$$

$$= \frac{1}{5} + \frac{4\sqrt{5}}{25} \ln\phi.$$

In general, noting that

$$\int_0^1 \frac{x(1-t)}{(1-xt(1-t))^2} dt = -\frac{1}{2}\int_0^1 \frac{d(1-xt(1-t))}{(1-xt(1-t))^2} + \frac{x}{2}\int_0^1 \frac{dt}{(1-xt(1-t))^2}$$

$$= \frac{x}{2}\int_0^1 \frac{dt}{(1-xt(1-t))^2},$$

and taking into account that

$$\int \frac{dt}{(at^2+bt+c)^2} = \frac{2at+b}{(4ac-b^2)(at^2+bt+c)} + \frac{2a}{4ac-b^2}\int \frac{dt}{at^2+bt+c},$$

we find that, for $x > 0$,

$$\sum_{n=1}^{\infty} \frac{x^n}{\binom{2n}{n}} = \frac{x}{4-x} + \frac{4\sqrt{x}}{(4-x)^{3/2}} \arctan\left(\sqrt{\frac{x}{4-x}}\right). \tag{18.8}$$

Next, dividing (18.5) by x and then integrating from 0 to x, we have

$$\sum_{n=1}^{\infty} \frac{x^n}{n\binom{2n}{n}} = \int_0^1 \frac{x(1-t)}{1-xt(1-t)} dt = 2\sqrt{\frac{x}{4-x}} \arctan\left(\sqrt{\frac{x}{4-x}}\right). \tag{18.9}$$

In the same manner, from (18.9), we find that

$$\sum_{n=1}^{\infty} \frac{x^n}{n^2\binom{2n}{n}} = -\int_0^1 \frac{1}{t} \ln(1-xt(1-t)) dt = 2\left(\arctan\left(\sqrt{\frac{x}{4-x}}\right)\right)^2. \tag{18.10}$$

In return, setting $x = 1$ in (18.9) and (18.10), we arrive at (18.3) and (18.4) respectively.

If we divide both sides of (18.10) by x and integrate, we obtain

$$\sum_{n=1}^{\infty} \frac{x^n}{n^3 \binom{2n}{n}} = 2 \int_0^x \frac{\left(\arctan\left(\sqrt{\frac{t}{4-t}}\right)\right)^2}{t} \, dt.$$

This fails to give a closed form in terms of simple constants even for $x = 1$. However, integrating (18.10) and then setting $x = 1$, we have

$$\sum_{n=1}^{\infty} \frac{1}{n^2(n+1)\binom{2n}{n}} = \int_0^1 2\left(\arctan\left(\sqrt{\frac{x}{4-x}}\right)\right)^2 dx = -1 + \frac{\sqrt{3}}{3}\pi - \frac{1}{18}\pi^2.$$

We consider this to be one of the most interesting series deduced from (18.5).

Differentiating (18.5) gives

$$\sum_{n=1}^{\infty} \frac{nx^{n-1}}{\binom{2n}{n}} = \int_0^1 \frac{(1-t)}{(1-xt(1-t))^2} \, dt + \int_0^1 \frac{2xt(1-t)^2}{(1-xt(1-t))^3} \, dt$$

and setting $x = \pm 1$ respectively, we get

$$\sum_{n=1}^{\infty} \frac{n}{\binom{2n}{n}} = \frac{2}{3} + \frac{2}{27}\pi\sqrt{3},$$

$$\sum_{n=1}^{\infty} (-1)^{n-1} \frac{n}{\binom{2n}{n}} = \frac{6}{25} + \frac{4}{125}\sqrt{5} \ln \phi.$$

In general, for any positive integer k, applying Leibniz's rule for the k-th derivative of the product to (18.5), we have

$$\frac{d^k}{dx^k}(1 - xt(1-t))^{-2} = \frac{(k+1)!(t(1-t))^k}{(1-xt(1-t))^{k+2}}.$$

Appealing to

$$\int \frac{t^m \, dt}{(at^2 + bt + c)^n} = -\frac{t^{m-1}}{(2n-m-1)a(at^2+bt+c)^{n-1}}$$

$$- \frac{(n-m)b}{(2n-m-1)a} \int \frac{t^{m-1} \, dt}{(at^2+bt+c)^n}$$

$$+ \frac{(m-1)c}{(2n-m-1)a} \int \frac{t^{m-2} \, dt}{(at^2+bt+c)^n},$$

18. Interesting Series Involving Binomial Coefficients

we find that for $x = \pm 1$, respectively,

$$\sum_{n=1}^{\infty} \frac{n^k}{\binom{2n}{n}} = p_k + q_k \pi \sqrt{3};$$

$$\sum_{n=1}^{\infty} (-1)^{n-1} \frac{n^k}{\binom{2n}{n}} = r_k + s_k \sqrt{5} \ln \phi$$

for explicitly given rational numbers p_k, q_k, r_k and s_k.

Finally, we show how the method of integral representation can be extended to the series

$$\sum_{n=0}^{\infty} \frac{x^n}{(2n + 2k + 1) \binom{2n}{n}}$$

for all integers $k \geq 0$. These series are not covered in [4] and appear to be new.

As in the derivation of (18.5), using

$$\binom{2n}{n}^{-1} = (2n+1)\beta(n+1, n+1) = (2n+1) \int_0^1 t^n (1-t)^n \, dt,$$

we find

$$\sum_{n=0}^{\infty} \frac{x^n}{(2n+1) \binom{2n}{n}} = \int_0^1 \sum_{n=0}^{\infty} x^n t^n (1-t)^n \, dt = \int_0^1 \frac{dt}{1 - xt(1-t)}. \quad (18.11)$$

Setting $x = 1, -1, -1/2, 2$, respectively, we find that

$$\sum_{n=0}^{\infty} \frac{1}{(2n+1) \binom{2n}{n}} = \int_0^1 \frac{dt}{t^2 - t + 1} = \frac{2}{9} \pi \sqrt{3},$$

$$\sum_{n=0}^{\infty} \frac{(-1)^n}{(2n+1) \binom{2n}{n}} = \int_0^1 \frac{dt}{1 + t - t^2} = \frac{4}{5} \sqrt{5} \ln \phi,$$

$$\sum_{n=0}^{\infty} \frac{(-1)^n}{(2n+1) 2^n \binom{2n}{n}} = \int_0^1 \frac{2 \, dt}{2 + t - t^2} = \frac{4}{3} \ln 2,$$

$$\sum_{n=0}^{\infty} \frac{2^n}{(2n+1) \binom{2n}{n}} = \int_0^1 \frac{dt}{1 - 2t + 2t^2} = \frac{1}{2} \pi.$$

Next, we derive the closed form of

$$\sum_{n=0}^{\infty} \frac{x^n}{(2n+3)\binom{2n}{n}}$$

by transforming it into two known series. Indeed, since

$$\binom{2n}{n} = \frac{n+1}{2(2n+1)}\binom{2(n+1)}{n+1},$$

it follows that

$$\sum_{n=0}^{\infty} \frac{x^n}{(2n+3)\binom{2n}{n}} = \sum_{n=0}^{\infty} \frac{2(2n+1)}{(2n+3)(n+1)\binom{2(n+1)}{n+1}}.$$

Replacing $n+1$ by n gives

$$\sum_{n=0}^{\infty} \frac{x^n}{(2n+3)\binom{2n}{n}} = \sum_{n=1}^{\infty} \frac{2(2n-1)x^{n-1}}{(2n+1)n\binom{2n}{n}}.$$

By the partial fraction decomposition

$$\frac{2n-1}{(2n+1)n} = \frac{4}{2n+1} - \frac{1}{n},$$

we obtain from (18.9) and (18.11)

$$\sum_{n=0}^{\infty} \frac{x^n}{(2n+3)\binom{2n}{n}} = -\frac{8}{x} + \frac{4(8-x)}{x^2}\sqrt{\frac{x}{4-x}} \arctan\left(\sqrt{\frac{x}{4-x}}\right). \quad (18.12)$$

Setting $x = 1, -1, -1/2, 2$ respectively yields

$$\sum_{n=0}^{\infty} \frac{1}{(2n+3)\binom{2n}{n}} = -8 + \frac{14}{9}\pi\sqrt{3},$$

$$\sum_{n=0}^{\infty} \frac{(-1)^n}{(2n+3)\binom{2n}{n}} = 8 - \frac{36}{5}\sqrt{5}\ln\phi,$$

$$\sum_{n=0}^{\infty} \frac{(-1)^n}{(2n+3)2^n\binom{2n}{n}} = 16 - \frac{68}{3}\ln 2,$$

$$\sum_{n=0}^{\infty} \frac{2^n}{(2n+3)\binom{2n}{n}} = -4 + \frac{3}{2}\pi.$$

18. Interesting Series Involving Binomial Coefficients

Similarly, we have

$$\sum_{n=0}^{\infty} \frac{x^n}{(2n+5)\binom{2n}{n}} = -\frac{16}{9x^2}(x+24) \\ + \frac{4}{3x^3}(128 - 16x - x^2)\sqrt{\frac{x}{4-x}} \arctan\left(\sqrt{\frac{x}{4-x}}\right). \quad (18.13)$$

In particular, we find that

$$\sum_{n=0}^{\infty} \frac{1}{(2n+5)\binom{2n}{n}} = -\frac{400}{9} + \frac{74}{9}\pi\sqrt{3},$$

$$\sum_{n=0}^{\infty} \frac{(-1)^n}{(2n+5)\binom{2n}{n}} = -\frac{368}{9} + \frac{572}{15}\sqrt{5}\ln\phi,$$

$$\sum_{n=0}^{\infty} \frac{(-1)^n}{(2n+5)\,2^n\binom{2n}{n}} = -\frac{1504}{9} + \frac{724}{3}\ln 2,$$

$$\sum_{n=0}^{\infty} \frac{2^n}{(2n+5)\binom{2n}{n}} = -\frac{104}{9} + \frac{23}{6}\pi.$$

In general, for any integer $k \geq 0$, we successively find that

$$\sum_{n=0}^{\infty} \frac{1}{(2n+2k+1)\binom{2n}{n}} = r_{1k} + s_{1k}\pi\sqrt{3},$$

$$\sum_{n=0}^{\infty} \frac{(-1)^n}{(2n+2k+1)\binom{2n}{n}} = r_{2k} + s_{2k}\sqrt{5}\ln\phi,$$

$$\sum_{n=0}^{\infty} \frac{(-1)^n}{(2n+2k+1)\,2^n\binom{2n}{n}} = r_{3k} + s_{3k}\ln 2,$$

$$\sum_{n=0}^{\infty} \frac{2^n}{(2n+2k+1)\binom{2n}{n}} = r_{4k} + s_{4k}\pi,$$

for appropriate rational numbers $r_{ik}, s_{ik} (1 \leq i \leq 4)$.

In summary, these derivations indicate that
- A given series has a closed form if it yields an integral representation;
- The closed form is in terms of well-known constants if the integrand displays a sufficiently simple factorization.

18.2 Some Extensions

In this section, we extend the integral representation of (18.5) to the more general series involving binomial coefficients

$$\sum_{n=1}^{\infty} \frac{x^n}{\binom{kn}{ln}} \qquad (18.14)$$

where k and l are positive integers with $k > l$.

In view of (18.6), we have

$$\binom{kn}{ln}^{-1} = ln \int_0^1 t^{ln-1}(1-t)^{(k-l)n}\, dt.$$

Therefore,

$$\sum_{n=1}^{\infty} \frac{x^n}{\binom{kn}{ln}} = l \int_0^1 \frac{xt^{l-1}(1-t)^{k-l}}{(1-xt^l(1-t)^{k-l})^2}\, dt. \qquad (18.15)$$

Observing that

$$\int_0^1 \frac{dt}{2-t} = \ln 2, \quad \int_0^1 \frac{dt}{1+t^2} = \frac{\pi}{4},$$

$$\int_0^1 \frac{dt}{1-t+t^2} = \frac{2}{9}\pi\sqrt{3} \quad \text{and} \quad \int_0^1 \frac{dt}{1+t-t^2} = \frac{4}{5}\sqrt{5}\ln\phi,$$

we see that there exists a closed form evaluation for (18.15) in terms of $\{\pi, \ln 2, \pi\sqrt{3}, \sqrt{5}\ln\phi\}$, as long as the denominator of the integrand $1 - xt^l(1-t)^{k-l}$ has factors of the form

$$\{t-2, 1+t^2, 1-t+t^2, 1+t-t^2\}.$$

With this as a guideline, as with the help of Mathematica, we can search for interesting series. In the following we display two examples of partial fraction decompositions and the corresponding closed forms. We use Mathematica to carry out the partial fraction decompositions and integrations.

Example 1. For $k = 3, l = 1$, (18.15) becomes

$$\sum_{n=1}^{\infty} \frac{x^n}{\binom{3n}{n}} = \int_0^1 \frac{x(1-t)^2}{(1-xt(1-t)^2)^2}\, dt. \qquad (18.16)$$

When $x = 1/2$, $1 - xt(1-t)^2 = (t^2+1)(t-2)/2$. Since

$$\frac{1-t^2}{(1-t(1-t)^2/2)^2} = \frac{4}{25(t-2)^2} + \frac{8}{125(t-2)} - \frac{8(3t-4)}{25(1+t^2)^2} - \frac{4(2t+9)}{125(1+t^2)},$$

we obtain

$$\sum_{n=1}^{\infty} \frac{1}{2^n \binom{3n}{n}} = \frac{2}{25} - \frac{6}{125}\ln 2 + \frac{11}{250}\pi. \qquad (18.17)$$

18. Interesting Series Involving Binomial Coefficients

Applying $x \frac{d}{dx}$ to (18.16) shows that

$$\sum_{n=1}^{\infty} \frac{nx^n}{\binom{3n}{n}} = \int_0^1 \left[\frac{2x^2 t(1-t)^4}{(1-xt(1-t)^2)^3} + \frac{x(1-t)^2}{(1-xt(1-t)^2)^2} \right] dt$$

from which we get, for $x = 1/2$,

$$\sum_{n=1}^{\infty} \frac{n}{2^n \binom{3n}{n}} = \frac{81}{625} - \frac{18}{3125} \ln 2 + \frac{79}{3125} \pi. \tag{18.18}$$

Similarly, we have

$$\sum_{n=1}^{\infty} \frac{n^2}{2^n \binom{3n}{n}} = \frac{561}{3125} + \frac{42}{15625} \ln 2 + \frac{673}{31250} \pi.$$

In particular, by solving (18.17) and (18.18) for π, we recover Gosper's identity:

$$\sum_{n=0}^{\infty} \frac{50n - 6}{2^n \binom{3n}{n}} = \pi. \tag{18.19}$$

Example 2. When $k = 4, l = 2$, (18.15) becomes

$$\sum_{n=1}^{\infty} \frac{x^n}{\binom{4n}{2n}} = 2 \int_0^1 \frac{xt(1-t)^2}{(1-xt^2(1-t)^2)^2} dt. \tag{18.20}$$

Since

$$t^2(1-t)^2 - 1 = (t^2 - t - 1)(t^2 - t + 1)$$

and

$$\frac{t(1-t^2)}{(1-t^2(1-t)^2)^2} = \frac{t-1}{4(t^2-t-1)^2} + \frac{1-t}{4(t^2-t+1)^2},$$

substituting $x = 1$ into (18.20) yields

$$\sum_{n=1}^{\infty} \frac{1}{\binom{4n}{2n}} = \frac{1}{15} + \frac{1}{27} \pi \sqrt{3} - \frac{2}{25} \sqrt{5} \ln \phi.$$

Similarly,

$$\sum_{n=1}^{\infty} \frac{n}{\binom{4n}{2n}} = 2 \int_0^1 \left[\frac{2t^3(1-t)^4}{(1-t^2(1-t)^2)^3} + \frac{t(1-t)^2}{(1-t^2(1-t)^2)^2} \right] dt$$

$$= \frac{8}{75} + \frac{1}{54} \pi \sqrt{3} - \frac{1}{125} \sqrt{5} \ln \phi;$$

$$\sum_{n=1}^{\infty} \frac{n^2}{\binom{4n}{2n}} = 2 \int_0^1 \left[\frac{6t^5(1-t)^6}{(1-t^2(1-t)^2)^4} + \frac{6t^3(1-t)^4}{(1-t^2(1-t)^2)^3} + \frac{t(1-t)^2}{(1-t^2(1-t)^2)^2} \right] dt$$

$$= \frac{11}{75} + \frac{5}{324} \pi \sqrt{3} - \frac{1}{250} \sqrt{5} \ln \phi.$$

18.3 Searching for new formulas for π

Based on Gosper's identity (18.19), an algorithm was developed for computing the nth decimal of π without computing the earlier ones (see `fabrice.bellard.free.fr/pi`). This is a radical idea, since all previous algorithms for the nth digit of π required the computation of all previous digits.

Now, we present a proof from the perspective of experimental mathematics: suppose we wish to see if π can be expressed as

$$\pi = \sum_{n=0}^{\infty} \frac{an+b}{2^n \binom{3n}{n}}.$$

Rewriting

$$\binom{3n}{n}^{-1} = (3n+1) \int_0^1 t^{2n}(1-t)^n \, dt,$$

we feed the general sum to Mathematica

`Sum[(a n + b)(3 n + 1)x^n, {n, 0, Infinity}]`

`(-b - 4ax - bx - 2ax^2 + 2bx^2)/(-1 + x)^3`

`% /. x -> t^2(1 - t)/2`

`(-b - 2a(1 - t)t^2 - 1/2b(1 - t)t^2 -`
  `1/2a(1 - t)^2t^4 + 1/2b(1 - t)^2t^4)/(-1 + 1/2(1 - t)t^2)^3`

`Simplify[%]`
`(4(ax^2(4 - 4x + x^2 - 2x^3 + x^4) -`
  `b(-2 - x^2 + x^3 + x^4 - 2x^5 + x^6)))/(2 - x^2 + x^3)^3`

`Integrate[%, {t, 0, 1}]`

`(2a(405 + 79\[Pi] - 18Log[2]) + 25 b (270 + 11\[Pi] - 12Log[2]))/6250`

We regroup the coefficients of π and $\ln 2$

`Collect[%, {Pi, Log[2]}]`

`(810 a + 6750b)/6250 + ((158 a + 275b)\[Pi])/6250 + ((-36a - 300b)Log[2])/6250`

18. Interesting Series Involving Binomial Coefficients

Now we simply search for values of a and b that cause all but the second summand to vanish and the second to equal π. This can be done by

Solve[{810a + 6750b == 0, 158a + 275b == 6250, 36a + 300b == 0}, {a, b}]

{{a -> 50, b -> -6}}

This recaptures (18.19).

We then ask if there are other such formulas in the form of

$$\pi = \sum_{n=0}^{\infty} \frac{p_m(n)}{b^n \binom{kn}{ln}}, \qquad (18.21)$$

where $p_m(n)$ is an mth degree polynomial in n and b is an integer. Using the same idea as before, and appealing to

$$\binom{kn}{ln}^{-1} = (kl+1) \int_0^1 t^{ln}(1-t)^{(k-l)n}\, dt,$$

we have

$$\text{RHS of (18.21)} = \int_0^1 \sum_{n=0}^{\infty} (kl+1) p_m(n) \left(\frac{x^l(1-x)^{k-l}}{b} \right)^n dx.$$

Since

$$\sum_{n=0}^{\infty} (kl+1) p_m(n) x^n = \frac{q(x)}{(1-x)^{k+2}},$$

where $q(x)$ has degree $m+1$. Substituting $x = t^l(1-t)^{k-l}/b$, we obtain

$$\text{RHS of (18.21)} = \int_0^1 \frac{P(t)\, dt}{(t^l(1-t)^{k-l} - b)^{k+2}}, \qquad (18.22)$$

where $P(t)$ is a polynomial of degree $k(m+1)$.

Now, we want the integral in (18.22) to equal π. A decent way to generate π is to have either $\arctan x$ or $\arctan(1-x)$ after integration. This would imply that

$$t^l(1-t)^{k-l} - b$$

must have the factor $t^2 + 1$ or $(t-1)^2 + 1$; that is, it must have a zero at i or $1+i$, thereby restricting k, l and b. Experimenting with Mathematica, we can find formulas for π in the following cases (there could be many more, see [1])

k	l	b	m
3	1	2	1 or 2
4	2	1	2
8	4	-4	4

For example, we have

$$\pi = \sum_{n=1}^{\infty} \frac{-150n^2 + 230n - 36}{2^n \binom{3n}{n}},$$

$$\pi = \frac{27}{4}\sqrt{3} \sum_{n=1}^{\infty} \frac{-10n^2 + 15n - 2}{\binom{4n}{2n}}.$$

It is interesting to see that these results discovered experimentally indeed are based on the rigorous mathematics of the *Hermite-Ostrogradski Formula* in [5]: Given a rational function $P(x)/Q(x)$ with $Q(x) = \prod_{i=1}^n q_i(x)^{\alpha_i}$, where $q_i(x)$ are either linear or irreducible quadratic factors, then

$$\int \frac{P(x)}{Q(x)} dx = \frac{P_1(x)}{Q_1(x)} + \int \frac{P_2(x)}{Q_2(x)} dx,$$

where

$$Q_1(x) = \prod_{i=1}^n q_i(x)^{\alpha_i - 1}; \quad Q_2(x) = \prod_{i=1}^n q_i(x).$$

This formula says what is intuitively clear — in a partial fraction decomposition, the repeated factors of the decomposition give us the rational part and the factors without repetition give us the transcendental part. This enables us to perform the experiment on (18.22): if

$$\int_0^1 \frac{P(x)}{Q(x)^{k+2}} dx = \frac{R(x)}{Q^{k+1}(x)} + r(x),$$

we would expect that $R(0) = R(1) = 0$ and $r(x)\big|_0^1 = \pi$.

In the last two decades, computer algebra systems have become an active tool in mathematical research. They are often used to experiment with various examples, to formulate credible conjectures and to decide if a potential result leans in the desired direction. With the continued advance of computing power and accessibility, it is expected that computer-aided research will play a more significant role.

Exercises

1. Prove that

$$\sum_{k=0}^n \frac{1}{\binom{n}{k}} = \frac{n+1}{2^n} \sum_{k=0}^n \frac{2^k}{k+1}, \quad \sum_{k=0}^n \frac{1}{\binom{2n}{2k}} = \frac{2n+1}{2^{2n+1}} \sum_{k=0}^{2n+1} \frac{2^k}{k+1}.$$

2. Let m be a positive integer. Show that

$$\sum_{n=1}^{\infty} \frac{1}{n^2 \binom{2mn}{mn}} = -\frac{m}{2} \int_0^1 \frac{\ln[1 - t^m(1-t)^m]dt}{t(1-t)},$$

$$\sum_{n=1}^{\infty} \frac{(-1)^n}{n^2 \binom{2mn}{mn}} = -\frac{m}{2} \int_0^1 \frac{\ln[1 + t^m(1-t)^m]dt}{t(1-t)}.$$

18. Interesting Series Involving Binomial Coefficients

3. Let G be Catalan's constant which is defined by $G = \sum_{n=0}^{\infty} \frac{(-1)^n}{(2n+1)^2}$. Let

$$a(n) = \sum_{k=0}^{\infty} \frac{1}{(2k+n)2^{4k}} \binom{2k}{k}^2.$$

Prove that $a(n)$ satisfies the recurrence

$$a(0) = 2\ln 2 - \frac{4G}{\pi}, \quad a(1) = \frac{4G}{\pi},$$

$$(n-1)^2 a(n) = (n-2)^2 a(n-2) + \frac{2}{\pi}, \quad \text{for } n \geq 2.$$

4. Let G be Catalan's constant. Prove that

(a) $\displaystyle\sum_{n=0}^{\infty} \frac{1}{(2n+1)^2 \binom{2n}{n}} = \frac{8G}{3} - \frac{\pi \ln(2+\sqrt{3})}{3}.$

(b) $\displaystyle\sum_{n=0}^{\infty} \frac{(-1)^n}{(2n+1)^2 \binom{2n}{n}} = \frac{\pi^2}{6} - 3\ln^2(\phi).$

(c) $\displaystyle\sum_{n=0}^{\infty} \frac{2^{2n}}{(2n+1)^2 \binom{2n}{n}} = 2G.$

(d) $\displaystyle\sum_{n=0}^{\infty} \frac{(-1)^n 2^{2n}}{(2n+1)^2 \binom{2n}{n}} = \frac{\pi^2}{8} - \frac{1}{2}\ln^2(1+\sqrt{2}).$

5. Prove that

(a) $\displaystyle\sum_{n=1}^{\infty} \frac{2^{4n}}{n^3 \binom{2n}{n}^2} = 8\pi G - 14\zeta(3).$

(b) $\displaystyle\sum_{n=0}^{\infty} \frac{2^{4n}}{(2n+1)^3 \binom{2n}{n}^2} = \frac{7}{2}\zeta(3) - \pi G.$

6. **Ramanujan Problem.** Given

$$\sum_{j=0}^{\infty} \binom{2j}{j} \frac{x^j}{4^j} = \frac{1}{\sqrt{1-x}} \quad -1 \leq x < 1.$$

Determine

$$\sum_{j=0}^{\infty} \binom{2j}{j} \frac{1}{(2j+1)^2 4^j}.$$

7. Prove that

$$\sum_{n=1}^{\infty} \frac{1}{n\, 2^n \binom{3n}{n}} = \frac{1}{10}\pi - \frac{1}{5}\ln 2,$$

$$\sum_{n=1}^{\infty} \frac{1}{n^2\, 2^n \binom{3n}{n}} = \frac{1}{24}\pi^2 - \ln 2.$$

8. Show that

$$\ln 2 = \frac{1}{6}\sum_{n=1}^{\infty} \frac{-575n^2 + 965n - 273}{2^n \binom{3n}{n}},$$

$$\ln 2 = -\frac{37}{24} + \frac{1}{24}\sum_{n=1}^{\infty} \frac{(-1)^n(-728n + 17)}{4^n \binom{3n}{n}},$$

$$\ln 2 = \frac{3}{16}\sum_{n=1}^{\infty} \frac{-882n^2 + 1295n - 296}{4^n \binom{4n}{2n}}.$$

9. Show that

$$\pi = \frac{1}{3^2 5^2 7^2}\sum_{n=0}^{\infty} \frac{p_4(n)}{(-4)^n \binom{8n}{4n}},$$

where

$$p_4(n) = -89286 + 3875948n - 34970134n^2 + 110202472n^3 - 115193600n^4.$$

10. **Monthly Problem 11356** [2008, 365]. Prove that, for any positive integer n,

$$\sum_{k=0}^{n} \frac{\binom{n}{k}^2}{(2k+1)\binom{2n}{2k}} = \frac{2^{4n}(n!)^4}{(2n)!(2n+1)!}.$$

11. It is known that

$$\sum_{n=1}^{\infty} \frac{H_n}{n^2} = 2\zeta(3) = 2\int_0^1 \frac{\ln x \ln(1-x)}{1-x}\, dx.$$

Prove that

$$\sum_{n=1}^{\infty} \frac{H_n}{n^p} = (-1)^p \zeta(p+1)\Gamma(p) + \frac{(-1)^p}{\Gamma(p)}\int_0^1 \frac{\ln^{p-1} x \ln(1-x)}{1-x}\, dx.$$

18. Interesting Series Involving Binomial Coefficients

12. **Open Problems**: Determine the exact sum of

$$\sum_{n=0}^{\infty} \frac{1}{(2n+1)^3 \binom{2n}{n}}, \quad \sum_{n=0}^{\infty} \frac{(-1)^n}{(2n+1)^3 \binom{2n}{n}}, \quad \sum_{n=0}^{\infty} \frac{(-1)^n 2^n}{(2n+1)^3 \binom{2n}{n}}.$$

References

[1] G. Almkvist, C. Krattenthaler and J. Petersson, Some new formulas for π, *Experimental Math.* **12** (2000) 441–456.

[2] A. Chen, Accelerated series for π and $\ln 2$, *Pi Mu Epsilon J.*, **12** (2008) 529–534.

[3] J. Borwein, D. Bailey and R. Girgensohn, *Experimentation in Mathematics*, A. K. Peters, MA, 2004.

[4] D. H. Lehmer, Interesting series involving the central binomial coefficient, *Amer. Math. Monthly*, **92** (1985) 449–457.

[5] T. N. Subramaniam and Donald E. G. Malm, How to integrate rational functions, *Amer. Math. Monthly*, **99** (1992) 762–772.

19

Parametric Differentiation and Integration

> *A mathematician is a person to whom $\int_0^\infty e^{-x^2}\,dx = \sqrt{\pi}/2$ is as obvious as $1 + 1 = 2$.*
> — W. Thomson

In this chapter, we present an integration method that evaluates integrals via differentiation and integration with respect to a parameter. This approach has been a favorite tool of applied mathematicians and theoretical physicists. In his autobiography [3], eminent physicist Richard Feynman mentioned how he frequently used this approach when confronted with difficult integrations associated with mathematics and physics problems. He referred to this approach as "a different box of tools". However, most modern texts either ignore this subject or provide only few examples. In the following, we illustrate this method with the help of some selected examples, most of them being improper integrals. These examples will show that the parametric differentiation and integration technique requires only the mathematical maturity of calculus, and often provides a straightforward method to evaluate difficult integrals which conventionally require the more sophisticated method of contour integration.

To illustrate the basic idea, we begin with two examples.

Example 1
Show that
$$\int_0^\infty \frac{\sin x}{x}\,dx = \frac{\pi}{2}. \tag{19.1}$$

This integral is generally evaluated by using contour integration and thus requires the theory of complex functions. Here we consider the parametric integral

$$I(p) = \int_0^\infty e^{-px} \frac{\sin x}{x}\,dx, \quad (p > 0).$$

Differentiating under the integral sign yields

$$I'(p) = -\int_0^\infty e^{-px} \sin x\,dx = -\frac{1}{1+p^2},$$

which has solutions
$$I(p) = -\arctan p + C,$$
where C is an arbitrary constant. Since $\lim_{p\to\infty} I(p) = 0$, it follows that $C = \pi/2$ and
$$I(p) = \frac{\pi}{2} - \arctan p,$$
from which (19.1) follows by setting $p = 0$.

Example 2

Show that the probability integral
$$G = \int_0^\infty e^{-x^2}\,dx = \frac{\sqrt{\pi}}{2}. \tag{19.2}$$

One clever way of evaluating this integral resides in the introduction of polar coordinates
$$G^2 = \int_0^\infty \int_0^\infty e^{-(x^2+y^2)}\,dx\,dy.$$

Here, we offer a modification that uses the double integrals but applies parametric integration. Notice that if $p > 0$, the change of variable $x = pt$ yields
$$G = p\int_0^\infty e^{-p^2 t^2}\,dt.$$

Multiplying both sides by e^{-p^2} and then integrating p from 0 to ∞ leads to
$$G \cdot \int_0^\infty e^{-p^2}\,dp = G^2 = \int_0^\infty pe^{-p^2}\,dp \int_0^\infty e^{-p^2 t^2}\,dt.$$

Exchanging the order of the right-hand side integrations gives
$$G^2 = \int_0^\infty \int_0^\infty pe^{-(1+t^2)p^2}\,dp\,dt = \frac{1}{2}\int_0^\infty \frac{dt}{1+t^2} = \frac{\pi}{4},$$
from which (19.2) follows easily.

These two examples suggest a natural question: how should one introduce a parameter within the integrand? In many integrals, especially in the formulas of integrals, parameters are already present. For example, differentiating the formula
$$\int_0^1 x^p\,dx = \frac{1}{p+1}, \qquad (p \neq -1),$$
k times with respect to p gives
$$\int_0^1 x^p (\ln x)^k\,dx = \frac{(-1)^k k!}{(p+1)^{k+1}}$$
as a new integral formula. However, there also exist integrals containing no parameter, like (19.1) and (19.2). In such cases, a parameter is generally introduced either by substitution, as we did in Example 2, or by specialization, where the given integral is viewed as a special case of a parametric integral. See Example 3 below. The foregoing examples also require some justification of the following three operations:

19. Parametric Differentiation and Integration

1. Differentiation with respect to a parameter under the integral sign;
2. Integration with respect to a parameter under the integral sign;
3. Exchanging the limit and integral operations.

Many theorems have been devoted to determining when the order of these processes may be interchanged (see [1] and [5]). For proper integrals, our discussion and illustrations will be based on the following *Leibniz's rule* for differentiation and *Fubini's theorem* for interchanging orders of integration.

Let I be an integral whose integrand $f(x, p)$ contains a parameter p, namely

$$I(p) = \int_a^b f(x, p)\, dx.$$

Leibniz's Rule. *If f and its partial derivative f_p are continuous on the rectangle $R = \{(x, p) : x \in [a, b], p \in [p_1, p_2]\}$, then $I(p)$ is differentiable in (p_1, p_2) and*

$$I'(p) = \int_a^b f_p(x, p)\, dx.$$

Fubini's Theorem. *If f is continuous on the rectangle $R = \{(x, p) : x \in [a, b], p \in [p_1, p_2]\}$, then for any $t \in [p_1, p_2]$,*

$$\lim_{p \to t} \int_a^b f(x, p)\, dx = \int_a^b \lim_{p \to t} f(x, p)\, dx$$

and

$$\int_{p_1}^t I(p)\, dp = \int_{p_1}^t dp \int_a^b f(x, p)\, dx = \int_a^b dx \int_{p_1}^t f(x, p)\, dp.$$

Now, we single out two integrals and demonstrate how to apply these two theorems.

Example 3

To evaluate

$$\int_0^1 \frac{\arctan x}{x\sqrt{1 - x^2}}\, dx,$$

we introduce a parametric integral

$$T(p) = \int_0^1 \frac{\arctan(px)}{x\sqrt{1 - x^2}}\, dx.$$

Differentiating with respect to p by Leibniz's rule gives

$$T'(p) = \int_0^1 \frac{dx}{(1 + p^2 x^2)\sqrt{1 - x^2}}.$$

The substitution $x = \sin\theta$ yields

$$T'(p) = \int_0^{\pi/2} \frac{d\theta}{1 + p^2 \sin^2\theta} = \frac{1}{\sqrt{1 + p^2}} \arctan(\sqrt{1 + p^2}\tan\theta)\Big|_0^{\pi/2} = \frac{\pi}{2} \frac{1}{\sqrt{1 + p^2}}.$$

Appealing to $T(0) = 0$, we have
$$T(p) = \frac{\pi}{2} \ln(p + \sqrt{1+p^2}).$$

In particular,
$$\int_0^1 \frac{\arctan x}{x\sqrt{1-x^2}} dx = T(1) = \frac{\pi}{2} \ln(1+\sqrt{2}).$$

Example 4

To compute
$$I = \int_0^{\pi/2} \ln\left(\frac{a+b\sin x}{a-b\sin x}\right) \cdot \frac{dx}{\sin x}, \qquad (a > b > 0),$$

observing that
$$\frac{1}{\sin x} \ln\left(\frac{a+b\sin x}{a-b\sin x}\right) = 2ab \int_0^1 \frac{dy}{a^2 - b^2 y^2 \sin^2 x},$$

we get
$$I = 2ab \int_0^{\pi/2} dx \int_0^1 \frac{dy}{a^2 - b^2 y^2 \sin^2 x}.$$

For $a > b > 0$, $1/(a^2 - b^2 y^2 \sin^2 x)$ is continuous on $[0, \pi/2] \times [0, 1]$. By Fubini's theorem, interchanging the order of integrations, we obtain
$$I = 2ab \int_0^1 dy \int_0^{\pi/2} \frac{dx}{a^2 - b^2 y^2 \sin^2 x}$$
$$= 2ab \int_0^1 \frac{\pi \, dy}{2a\sqrt{a^2 - b^2 y^2}}$$
$$= \pi \arcsin\left(\frac{b}{a}\right).$$

The following example warns us that switching the order of integrations without justification may change the answer.

Example 5

Consider
$$f(x, y) = \frac{y^2 - x^2}{(x^2 + y^2)^2} \quad \text{on } [0, 1] \times [0, 1].$$

We have
$$\int_0^1 f(x, y) dx = \frac{x}{x^2 + y^2}\Big|_0^1 = \frac{1}{1+y^2}$$

and so
$$\int_0^1 dy \int_0^1 f(x, y) \, dx = \frac{\pi}{4}.$$

However,
$$\int_0^1 dx \int_0^1 f(x, y) \, dy = -\frac{\pi}{4}.$$

19. Parametric Differentiation and Integration

Here the discontinuity of $f(x, y)$ at $(0, 0)$ causes the discrepancy.

For an improper integral

$$I(p) = \int_a^\infty f(x, p)\, dx,$$

to avoid the absurdity that appeared in Example 5, uniform convergence and absolute integrability are usual requirements. We say that $I(p)$ converges uniformly on $[p_1, p_2]$ if for every $\epsilon > 0$ and $p \in [p_1, p_2]$ there is a constant $A \geq a$, independent of p, such that $x \geq A$ implies

$$\left| \int_a^\infty f(x, p)\, dx - \int_a^A f(x, p)\, dx \right| = \left| \int_A^\infty f(x, p)\, dx \right| < \epsilon.$$

There is a very convenient test for uniform convergence, due to Weierstrass.

M-Test. *If $f(x, p)$ is integrable on every finite interval and there is an integrable function $\phi(x)$ on $[0, \infty)$ such that*

$$|f(x, p)| \leq \phi(x), \quad \text{for all } x \geq a,\, p \in [p_1, p_2],$$

then $I(p)$ converges uniformly on $[p_1, p_2]$.

When the integrand is in product form, the following criterion is sometimes useful.

Dini Test. *If*

$$\left| \int_a^A f(x, p)\, dx \right|$$

is uniformly bounded in A and p, $g(x, p)$ is monotonic in x and $g(x, p) \to 0$ uniformly in p as $x \to \infty$, then

$$\int_a^\infty f(x, p) g(x, p)\, dx$$

converges uniformly on $[p_1, p_2]$.

As an immediate consequence of the Dini test, if $\int_a^\infty f(x)\, dx$ converges, then

$$\int_a^\infty e^{-xp} f(x)\, dx \quad \text{and} \quad \int_a^\infty e^{-x^2 p} f(x)\, dx$$

are uniformly convergent for $p > 0$.

When the range of parameter p also becomes infinite, in order to switch the order of integration, besides the uniform convergence of the integrals

$$\int_a^\infty f(x, p)\, dx \quad \text{and} \quad \int_b^\infty f(x, p)\, dp,$$

Fubini's theorem further requires that one of the double integrals

$$\int_b^\infty \int_a^\infty |f(x, p)|\, dx\, dp \quad \text{and} \quad \int_a^\infty \int_b^\infty |f(x, p)|\, dp\, dx$$

exists. Combining with the M-test, we have the following theorem to justify the desired steps in our subsequent examples.

Extended Fubini's Theorem. *Let $f(x, y)$ be continuous on $(a, \infty) \times (b, \infty)$. Suppose the improper integrals*

$$\int_b^\infty f(x, y)\,dy \quad \text{and} \quad \int_a^\infty f(x, y)\,dx$$

exist and converge uniformly for x and y restricted to every finite interval, respectively. In addition, suppose that, for $s, t > a$,

$$\left|\int_s^t f(x, y)\,dx\right| \le M(y)$$

and $\int_b^\infty M(y)\,dy$ exists. Then

$$\int_a^\infty dx \int_b^\infty f(x, y)\,dy = \int_b^\infty dy \int_a^\infty f(x, y)\,dx.$$

As a nice application of this theorem, the integral

$$\frac{1}{x} = \int_0^\infty e^{-xy}\,dy$$

enables us to evaluate (19.1) in one line:

$$\int_0^\infty \sin x\,dx \int_0^\infty e^{-xy}\,dy = \int_0^\infty dy \int_0^\infty e^{-xy} \sin x\,dx = \int_0^\infty \frac{dy}{1+y^2} = \frac{\pi}{2}.$$

Now, we turn to more selected examples, in which most of the justifications of applications of Leibniz's rule or Fubini's theorem are straightforward — the verifications are left to the reader.

Example 6

Let $\alpha, \beta > 0$. Evaluate

$$\int_0^\infty e^{-\alpha^2 x^2 - \frac{\beta^2}{x^2}}\,dx. \tag{19.3}$$

Let $y = \alpha x$, $p = \alpha\beta$. Then

$$\int_0^\infty e^{-\alpha^2 x^2 - \frac{\beta^2}{x^2}}\,dx = \frac{1}{\alpha} \int_0^\infty e^{-y^2 - \frac{p^2}{y^2}}\,dy.$$

If

$$J(p) = \int_0^\infty e^{-y^2 - \frac{p^2}{y^2}}\,dy,$$

then

$$J'(p) = -2 \int_0^\infty \frac{p}{y^2} e^{-y^2 - \frac{p^2}{y^2}}\,dy,$$

and the substitution $z = p/y$ yields

$$J'(p) = -2 \int_0^\infty e^{-z^2 - \frac{p^2}{z^2}}\,dz = -2 J(p).$$

19. Parametric Differentiation and Integration

Appealing to (19.2), we find

$$J(p) = J(0)e^{-2p} = \left(\int_0^\infty e^{-y^2}\,dy\right)e^{-2p} = \frac{\sqrt{\pi}}{2}e^{-2p}$$

and so

$$\int_0^\infty e^{-\alpha^2 x^2 - \frac{\beta^2}{x^2}}\,dx = \frac{\sqrt{\pi}}{2\alpha}e^{-2\alpha\beta}.$$

Remark. In general, by substitution, one can establish that

$$\int_0^\infty f\left[\left(Ax - \frac{B}{x}\right)^2\right]dx = \frac{1}{A}\int_0^\infty f(y^2)\,dy, \quad (A > 0, B > 0).$$

A related Putnam problem is 1968-B4, which asks us to prove

$$\int_{-\infty}^\infty f\left(x - \frac{1}{x}\right)dx = \int_{-\infty}^\infty f(x)\,dx.$$

Example 7

Let $\alpha, \beta > 0$. Show that

$$H(\beta) = \int_0^\infty e^{-\alpha x^2}\cos\beta x\,dx = \frac{1}{2}\sqrt{\frac{\pi}{\alpha}}e^{-\beta^2/4\alpha}. \tag{19.4}$$

Differentiating with respect to β and then integrating by parts gives

$$H'(\beta) = -\int_0^\infty xe^{-\alpha x^2}\sin\beta x\,dx$$

$$= \frac{1}{2\alpha}\int_0^\infty \sin\beta x\,d(e^{-\alpha x^2})$$

$$= \frac{1}{2\alpha}e^{-\alpha x^2}\sin\beta x\Big|_0^\infty - \frac{\beta}{2\alpha}\int_0^\infty e^{-\alpha x^2}\cos\beta x\,dx$$

$$= -\frac{\beta}{2\alpha}H(\beta).$$

Moreover, by (19.3),

$$H(0) = \int_0^\infty e^{-\alpha x^2}\,dx = \frac{1}{\sqrt{\alpha}}\int_0^\infty e^{-x^2}\,dx = \frac{1}{2}\sqrt{\frac{\pi}{\alpha}}.$$

Consequently, the desired integral follows by solving the initial value problem. It is interesting to note that the similar integral

$$\int_0^\infty e^{-\alpha x^2}\sin\beta x\,dx$$

has no elementary closed form.

Example 8

Let $\alpha, \beta > 0$. Evaluate the *Laplace integrals*

$$L_1(\beta) = \int_0^\infty \frac{\cos \beta x}{\alpha^2 + x^2}\, dx \quad \text{and} \quad L_2(\beta) = \int_0^\infty \frac{x \sin \beta x}{\alpha^2 + x^2}\, dx.$$

Applying Leibniz's rule to $L_1(\beta)$ gives

$$L_1'(\beta) = -\int_0^\infty \frac{x \sin \beta x}{\alpha^2 + x^2}\, dx.$$

We caution that here $L_1'(\beta)$ can not be directly differentiated under the integral sign anymore, because the resulting integral is divergent. Indeed, differentiating $L'(\beta)$ under the integral sign yields

$$-\int_0^\infty \frac{x^2 \cos \beta x}{\alpha^2 + x^2}\, dx,$$

which can be rewritten as

$$-\int_0^{\pi/2\beta} \frac{x^2 \cos \beta x}{\alpha^2 + x^2}\, dx + \sum_{n=1}^\infty (-1)^{n+1} \int_{(2n-1)\pi/2\beta}^{(2n+1)\pi/2\beta} \frac{x^2 |\cos \beta x|}{\alpha^2 + x^2}\, dx.$$

Since

$$\int_{(2n-1)\pi/2\beta}^{(2n+1)\pi/2\beta} \frac{x^2 |\cos \beta x|}{\alpha^2 + x^2}\, dx$$

$$\geq \left(\min_{(2n-1)\pi/2\beta \leq x \leq (2n+1)\pi/2\beta} \frac{x^2}{\alpha^2 + x^2}\right) \int_{(2n-1)\pi/2\beta}^{(2n+1)\pi/2\beta} |\cos \beta x|\, dx$$

$$= \frac{2((2n-1)\pi/2\beta)^2}{(\alpha^2 + ((2n-1)\pi/2\beta)^2)\beta} \to \frac{2}{\beta}, \quad \text{as } n \to \infty,$$

the series, hence the corresponding integral, is divergent. On the other hand, by observing that

$$\frac{\alpha^2 \sin \beta x}{x(\alpha^2 + x^2)} = \frac{\sin \beta x}{x} - \frac{x \sin \beta x}{\alpha^2 + x^2}$$

and

$$\int_0^\infty \frac{\sin \beta x}{x}\, dx = \int_0^\infty \frac{\sin x}{x}\, dx = \frac{\pi}{2},$$

we have

$$L_1'(\beta) = -\frac{\pi}{2} + \int_0^\infty \frac{\alpha^2 \sin \beta x}{x(\alpha^2 + x^2)}\, dx.$$

Hence

$$L_1''(\beta) = \int_0^\infty \frac{\alpha^2 \cos \beta x}{\alpha^2 + x^2}\, dx = \alpha^2 L_1(\beta),$$

which leads to

$$L_1(\beta) = c_1 e^{\alpha \beta} + c_2 e^{-\alpha \beta},$$

19. Parametric Differentiation and Integration

where c_1 and c_2 are arbitrary constants. Since

$$L_1(\beta) \leq \int_0^\infty \frac{dx}{\alpha^2 + x^2} = \frac{\pi}{2\alpha},$$

$L_1(\beta)$ is uniformly bounded. Therefore,

$$c_1 = 0, \quad c_2 = L_1(0) = \frac{\pi}{2\alpha}$$

and consequently

$$L_1(\beta) = \frac{\pi}{2\alpha} e^{-\alpha\beta} \quad \text{and} \quad L_2(\beta) = -L_1'(\beta) = \frac{\pi}{2} e^{-\alpha\beta}.$$

The value of L_1 also can be captured by using the parametric integration. Indeed, noticing

$$\frac{1}{\alpha^2 + x^2} = \int_0^\infty e^{-t(\alpha^2 + x^2)} \, dt,$$

we have

$$L_1 = \int_0^\infty \cos \beta x \, dx \int_0^\infty e^{-t(\alpha^2 + x^2)} \, dt.$$

Interchanging the integral order yields

$$L_1 = \int_0^\infty e^{-\alpha^2 t} \, dt \int_0^\infty e^{-tx^2} \cos \beta x \, dx.$$

(19.4) implies the inner integral

$$\int_0^\infty e^{-tx^2} \cos \beta x \, dx = \frac{1}{2} \sqrt{\frac{\pi}{t}} e^{-\beta^2/4t}.$$

Finally, appealing to (19.3), we arrive at

$$L_1 = \frac{\sqrt{\pi}}{2} \int_0^\infty e^{-\alpha^2 t - \beta^2/4t} \frac{dt}{\sqrt{t}}$$

$$= \sqrt{\pi} \int_0^\infty e^{-\alpha^2 s^2 - \beta^2/4s^2} \, ds \quad (\text{let } t = s^2)$$

$$= \frac{\pi}{2\alpha} e^{-\alpha\beta}.$$

Example 9

Evaluate the Fresnel integrals

$$F_c = \int_0^\infty \cos(x^2) \, dx \quad \text{and} \quad F_s = \int_0^\infty \sin(x^2) \, dx.$$

It is standard to substitute x for x^2, so

$$F_c = \frac{1}{2} \int_0^\infty \frac{\cos x}{\sqrt{x}} \, dx \quad \text{and} \quad F_s = \frac{1}{2} \int_0^\infty \frac{\sin x}{\sqrt{x}} \, dx.$$

Recall that
$$\frac{1}{\sqrt{x}} = \frac{2}{\sqrt{\pi}} \int_0^\infty e^{-xy^2}\, dy.$$

Then
$$\begin{aligned}
F_c &= \frac{1}{\sqrt{\pi}} \int_0^\infty \cos x\, dx \int_0^\infty e^{-xy^2}\, dy \\
&= \frac{1}{\sqrt{\pi}} \int_0^\infty dy \int_0^\infty e^{-xy^2} \cos x\, dx \\
&= \frac{1}{\sqrt{\pi}} \int_0^\infty \frac{dy}{1+y^4} = \frac{1}{\sqrt{\pi}} \frac{\pi}{2\sqrt{2}} = \frac{1}{2}\sqrt{\frac{\pi}{2}}.
\end{aligned}$$

Similarly,
$$F_s = \frac{1}{2}\sqrt{\frac{\pi}{2}}.$$

Example 10

Evaluate
$$F = \int_0^\infty \frac{e^{-ax} - e^{-bx}}{x}\, dx, \quad (a,b > 0).$$

Recall that
$$\int_a^b e^{-yx}\, dy = \frac{e^{-ax} - e^{-bx}}{x}.$$

This yields
$$F = \int_0^\infty dx \int_a^b e^{-yx}\, dy = \int_a^b dy \int_0^\infty e^{-yx}\, dx = \int_a^b \frac{1}{y}\, dy = \ln\frac{b}{a}.$$

Remarks. Here F belongs to the family of *Frullani integrals* [2, pp. 406–407], which are defined by
$$\int_0^\infty \frac{f(ax) - f(bx)}{x}\, dx \quad (a,b > 0).$$
In general, if $f(x)$ is continuous on $[0, \infty)$ and
$$f(+\infty) = \lim_{x \to \infty} f(x)$$
exists, then
$$\int_0^\infty \frac{f(ax) - f(bx)}{x}\, dx = (f(0) - f(+\infty)) \ln\frac{b}{a}.$$
If $\lim_{x \to \infty} f(x)$ has no finite limit, but
$$\int_A^\infty \frac{f(x)}{x}\, dx$$
exists, then
$$\int_0^\infty \frac{f(ax) - f(bx)}{x}\, dx = f(0) \ln\frac{b}{a}.$$

19. Parametric Differentiation and Integration

Similarly, if f is not continuous at $x = 0$, but

$$\int_0^A \frac{f(x)}{x} dx$$

exists, then

$$\int_0^\infty \frac{f(ax) - f(bx)}{x} dx = f(\infty) \ln \frac{b}{a}.$$

Finally, we conclude this chapter with two comments. First, sometimes one may need to deal with integrals with more than one parameter. For example, to evaluate

$$\int_0^\infty \frac{e^{-px} \cos qx - e^{-ax} \cos bx}{x} dx, \quad (p, a > 0),$$

it is instructive to check that differentiating under the integral sign with one parameter or using parametric integration via

$$\frac{1}{x} = \int_0^\infty e^{-xt} dt$$

fails to yield an explicit value. However, let $I(p, q)$ denote the integral to be computed. Differentiating $I(p, q)$ with respect to p and q respectively gives

$$\frac{\partial I}{\partial p} = -\int_0^\infty e^{-px} \cos qx \, dx = -\frac{p}{p^2 + q^2},$$

$$\frac{\partial I}{\partial q} = -\int_0^\infty e^{-px} \sin qx \, dx = -\frac{q}{p^2 + q^2},$$

which have solutions

$$I(p, q) = -\frac{1}{2} \ln(p^2 + q^2) + C,$$

where constant C is independent of p and q. Since $I(a, b) = 0$, we obtain that

$$C = \frac{1}{2} \ln(a^2 + b^2)$$

and so

$$I(p, q) = \frac{1}{2} \ln \frac{a^2 + b^2}{p^2 + q^2}.$$

Second, we present two examples of what can go wrong when differentiating under the integral sign or interchanging the order of integrations is not valid. In 1851 Cauchy obtained the result

$$\int_0^\infty \sin(x^2) \cos(px) \, dx = \frac{1}{2} \sqrt{\frac{\pi}{2}} \left[\cos\left(\frac{p^2}{4}\right) - \sin\left(\frac{p^2}{4}\right) \right].$$

He then differentiated under the integral sign with respect to p yielding

$$\int_0^\infty x \sin(x^2) \sin(px) \, dx = \frac{p}{4} \sqrt{\frac{\pi}{2}} \left[\sin\left(\frac{p^2}{4}\right) + \cos\left(\frac{p^2}{4}\right) \right]. \quad (19.5)$$

This formula has subsequently been reproduced and still appears in standard tables today. However, Talvila in [4] recently proved that the integral in (19.5) is divergent! Next, direct calculation shows

$$\int_1^\infty dy \int_1^\infty \frac{y^2 - x^2}{(x^2 + y^2)^2} dx = -\frac{\pi}{4} \quad \text{and} \quad \int_1^\infty dx \int_1^\infty \frac{y^2 - x^2}{(x^2 + y^2)^2} dy = \frac{\pi}{4}.$$

Since both

$$\int_1^\infty dy \int_1^\infty \frac{|y^2 - x^2|}{(x^2 + y^2)^2} dx \quad \text{and} \quad \int_1^\infty dx \int_1^\infty \frac{|y^2 - x^2|}{(x^2 + y^2)^2} dy$$

diverge, Fubini's theorem does not apply to the given function.

Exercises

1. Show that
$$\lim_{y \to 0} \int_0^1 \frac{2xy^2}{(x^2 + y^2)^2} \neq \int_0^1 \left(\lim_{y \to 0} \frac{2xy^2}{(x^2 + y^2)^2} \right) dx.$$

2. Show that
$$\int_0^\infty \frac{\sin px}{x} dx$$
is uniformly convergent for $p \geq p_0 > 0$, but fails to converge uniformly for $p \geq 0$.

3. Evaluate
$$\int_0^1 \frac{x-1}{\ln x} dx.$$
Hint: Consider $\int_0^1 \frac{x^p - 1}{\ln x} dx$.

4. **Putnam Problem 2005-A5.** Evaluate
$$\int_0^1 \frac{\ln(x+1)}{x^2 + 1} dx.$$

There are many ways to solve this problem. Try to introduce a parametric integral and then differentiate it.

5. Let $\alpha, \beta, k > 0$. Evaluate
   (a) $\int_0^\infty \frac{1 - \cos \alpha x}{x} e^{-kx} dx.$
   (b) $\int_0^\infty \frac{\sin \alpha x}{x} \cdot \frac{\sin \beta x}{x} e^{-kx} dx.$

6. Let $a, b > 0$. Prove that
   (a) $\int_0^\infty \frac{\ln(1 + a^2 x^2)}{b^2 + x^2} dx = \frac{\pi}{b} \ln(1 + ab).$
   (b) $\int_0^\infty \frac{\arctan ax}{x(1 + x^2)} dx = \frac{\pi}{2} \ln(1 + a).$

19. Parametric Differentiation and Integration

(c) $\displaystyle\int_0^\infty \frac{\arctan ax \cdot \arctan bx}{x^2}\, dx = \frac{\pi}{2} \ln \frac{(a+b)^{a+b}}{a^a \cdot b^b}.$

7. For $a, b > 0$, show that

$$\int_0^\infty \frac{\cos ax - \cos bx}{x^2}\, dx = \frac{\pi}{2}(b-a).$$

8. Evaluate

$$\int_0^\infty x^{-1/2} e^{-ax - px^{-1}}\, dx.$$

Remark. For $a = p = 1985$, this is the **Putnam Problem 1985-B5**. In contrast to Bernau's answer for the Putnam problem, here differentiating under the integral sign with respect to p yields an alternative solution.

9. For $p > q > 0$, evaluate the integral

$$\int_0^\infty \frac{e^{px} - e^{qx}}{x(e^{px}+1)(e^{qx}+1)}\, dx.$$

10. Evaluate

$$\int_0^\infty e^{-x^2} \cos(a^2/x^2)\, dx \quad \text{and} \quad \int_0^\infty e^{-x^2} \sin(a^2/x^2)\, dx.$$

11. Prove that

$$\int_0^\infty \frac{xe^{-x^2}\, dx}{\sqrt{x^2 + a^2}} = \frac{a}{\sqrt{\pi}} \int_0^\infty \frac{e^{-x^2}\, dx}{x^2 + a^2} \quad (a > 0).$$

12. Evaluate

$$\int_0^\infty \frac{e^{-x^2}\, dx}{(x^2 + a^2)^2}.$$

When $a^2 = 1/2$, this is **Monthly Problem 4212** [1946, 397; 1947, 601–603].

13. **Monthly Problem 3766** [1936, 50; 1938, 56–58]. Evaluate

$$\int_0^\infty e^{-x} \ln^2 x\, dx.$$

14. Let $J_0(x)$ be the zero-order Bessel function. Evaluate

$$\int_0^\infty e^{-sx} J_0(x)\, dx \quad (s > 0).$$

15. For $p > 0$ and $0 < s < 1$, evaluate

$$\int_0^\infty \frac{\cos px}{x^s}\, dx.$$

Hint: use

$$\frac{1}{x^s} = \frac{1}{\Gamma(s)} \int_0^\infty y^{s-1} e^{-xy}\, dy.$$

16. **Monthly Problem 11101** [2004, 626; 2006, 270–271]. Evaluate
$$\int_0^\infty a \arctan\left(\frac{b}{\sqrt{a^2+x^2}}\right) \frac{dx}{\sqrt{a^2+x^2}}$$
in closed form for $a, b, > 0$.

17. **Monthly Problem 11113** [2004, 822; 2006, 573–574]. Evaluate
$$I_k(a,b) = \int_0^\infty \int_0^\infty \frac{e^{-k\sqrt{x^2+y^2}} \sin(ax)\sin(bx)}{xy\sqrt{x^2+y^2}} \, dx\, dy$$
in closed form for $a, b, k > 0$.

18. **Monthly Problem 11225** [2006, 459; 2007, 750]. Find
$$\lim_{n\to\infty} \frac{1}{n} \int_0^n \frac{x\ln(1+x/n)}{1+x} \, dx.$$

References

[1] T. M. Apostol, *Mathematical Analysis*, 2nd edition, Addison-Wesley, 1974.

[2] H. Jeffreys and B. S. Jeffreys, *Methods of Mathematical Physics*, 3rd ed, Cambridge University Press, Cambridge, England, 1988.

[3] R. P. Feynman, *"Surely You're Joking, Mr. Feynman!": Adventures of a curious character*, Bantam Books, New York, 1985.

[4] E. Talvila, Some divergent trigonometric integrals, *Amer. Math. Monthly*, **108** (2001) 432–436.

[5] ——, Necessary and sufficient conditions for differentiating under the integral sign, *Amer. Math. Monthly*, **108** (2001) 544–548.

20
Four Ways to Evaluate the Poisson Integral

In general, it is difficult to decide whether or not a given function can be integrated via elementary methods. In light of this, it is quite surprising that the value of the Poisson integral

$$I(x) = \int_0^\pi \ln(1 - 2x \cos \theta + x^2) \, d\theta$$

can be determined precisely. Even more surprising is that we can do so for every value of the parameter x. In this chapter, using four different methods, we show that

$$I(x) = \begin{cases} 0, & \text{if } |x| < 1; \\ 2\pi \ln |x|, & \text{if } |x| > 1. \end{cases}$$

Our integral is one of several known as the Poisson integral. All are related in some way to Poisson's integral formula, which recovers an analytic function on the disk from its boundary values, a relationship we mention below. However, none of our methods involves complex analysis at all. The first one uses Riemann sums and relies on a trigonometric identity. The second method is based on a functional equation and involves a sequence of integral substitutions. The third method uses parametric differentiation and the half angle substitution. Finally, we finish with an approach based on infinite series. It is interesting to see how wide a range of mathematical topics this chapter exploits. These evaluations are suitable for an advanced calculus class and provide a very nice application of Riemann sums, functional equations, parametric differentiation and infinite series.

We begin with three elementary observations:

1. $I(0) = 0$.
2. $I(-x) = I(x)$.
3. $I(x) = 2\pi \ln |x| + I(1/x)$, $(x \neq 0)$.

To verify these results, first note that for $|x| < 1$,

$$(1 - |x|)^2 \leq 1 - 2x \cos \theta + x^2 \leq (1 + |x|)^2.$$

Thus, taking the logarithm and integrating with respect to θ from 0 to π, we find

$$2\pi \ln(1 - |x|) \leq I(x) \leq 2\pi \ln(1 + |x|).$$

Letting $x \to 0$, we have that $I(0) = 0$. Next, applying the substitution $\theta = \pi - \alpha$ in $I(x)$, we have

$$I(x) = \int_0^\pi \ln(1 + 2x \cos \alpha + x^2) \, d\alpha = I(-x).$$

Finally, if $x \neq 0$, we have

$$I(x) = \int_0^\pi \ln\left[x^2\left(1 - \frac{2}{x}\cos\theta + \frac{1}{x^2}\right)\right] d\theta$$

$$= \int_0^\pi \ln x^2 \, d\theta + I(1/x) = 2\pi \ln|x| + I(1/x).$$

In light of the third fact, the main formula follows easily once we show that $I(x) = 0$ for $|x| < 1$. This will be the goal of the next four sections.

20.1 Using Riemann Sums

Since

$$1 - 2x \cos \theta + x^2 \geq (1 - |x|)^2, \qquad \text{for } |x| < 1,$$

the integrand is continuous and integrable. Partition the interval $[0, \pi]$ into n equal subintervals by the partition points

$$\left\{x_k = \frac{k\pi}{n} : 1 \leq k \leq n\right\}.$$

We have the corresponding Riemann sum for $I(x)$, namely

$$\mathcal{R}_n = \frac{\pi}{n} \sum_{k=1}^n \ln\left(1 - 2x \cos\left(\frac{k\pi}{n}\right) + x^2\right)$$

$$= \frac{\pi}{n} \ln\left[(1+x)^2 \prod_{k=1}^{n-1}\left(1 - 2x \cos\left(\frac{k\pi}{n}\right) + x^2\right)\right]. \tag{20.1}$$

Recall that (see (5.2))

$$\prod_{k=1}^{n-1}\left(1 - 2x \cos\left(\frac{k\pi}{n}\right) + x^2\right) = \frac{x^{2n} - 1}{x^2 - 1}. \tag{20.2}$$

Substituting the identity (20.2) into (20.1), we have

$$\mathcal{R}_n = \frac{\pi}{n} \ln\left(\frac{x+1}{x-1}(x^{2n} - 1)\right).$$

Since $|x| < 1$, $x^{2n} \to 0$ as $n \to \infty$. Hence, we obtain

$$I(x) = \lim_{n \to \infty} \mathcal{R}_n = \lim_{n \to \infty} \frac{\pi}{n} \ln\left(\frac{x+1}{x-1}(x^{2n} - 1)\right) = 0.$$

Remark. The above method relies on the trigonometric identity (20.2), which is of interest in its own right.

20.2 Using A Functional Equation

The functional equation we have in mind is

$$I(x) = I(-x) = \frac{1}{2} I(x^2). \tag{20.3}$$

Applying the identity

$$(1 - 2x \cos \theta + x^2)(1 + 2x \cos \theta + x^2) = 1 - 2x^2 \cos(2\theta) + x^4,$$

we obtain

$$I(x) + I(-x) = \int_0^\pi \ln\left[(1 - 2x \cos \theta + x^2)(1 + 2x \cos \theta + x^2)\right] d\theta$$

$$= \int_0^\pi \ln(1 - 2x^2 \cos 2\theta + x^4) \, d\theta.$$

Setting $\alpha = 2\theta$, we have

$$I(x) + I(-x) = \frac{1}{2} \int_0^{2\pi} \ln(1 - 2x^2 \cos \alpha + x^4) \, d\alpha$$

$$= \frac{1}{2} I(x^2) + \frac{1}{2} \int_\pi^{2\pi} \ln(1 - 2x^2 \cos \alpha + x^4) \, d\alpha.$$

Using the substitution $\alpha = 2\pi - t$ in the last integral shows that it is exactly the same as the first integral. Since the two terms on the left are the same (recalling that $I(x) = I(-x)$), we obtain (20.3) as desired. Applying equation (20.3) repeatedly, we find that

$$I(x) = \frac{1}{2} I(x^2) = \frac{1}{2^2} I(x^4) = \cdots = \frac{1}{2^n} I(x^{2^n}).$$

Again we assume that $|x| < 1$, so that $x^{2^n} \to 0$ as $n \to \infty$ and consequently

$$I(x) = \lim_{n \to \infty} \frac{1}{2^n} I(x^{2^n}) = 0.$$

Remark. Equation (20.3) holds for any x. In particular, we have that $I(0) = 0$ and $I(\pm 1) = 0$. The latter equation leads to an added bonus:

$$\int_0^{\pi/2} \ln(\sin \theta) \, d\theta = \int_0^{\pi/2} \ln(\cos \theta) \, d\theta = -\frac{\pi}{2} \ln 2,$$

since

$$I(1) = \int_0^\pi \ln(2 - 2 \cos \theta) \, d\theta = 2\pi \ln 2 + 4 \int_0^{\pi/2} \ln(\sin \theta) \, d\theta;$$

$$I(-1) = \int_0^\pi \ln(2 + 2 \cos \theta) \, d\theta = 2\pi \ln 2 + 4 \int_0^{\pi/2} \ln(\cos \theta) \, d\theta.$$

These two integrals are improper. To show convergence, for example, using integration by parts, we have

$$\int_0^{\pi/2} \ln(\sin\theta)\, d\theta = \lim_{\epsilon\to 0} \int_\epsilon^{\pi/2} \ln(\sin\theta)\, d\theta$$

$$= \lim_{\epsilon\to 0} \epsilon \ln(\sin\epsilon) - \lim_{\epsilon\to 0} \int_\epsilon^{\pi/2} \frac{\theta \cos\theta}{\sin\theta}\, d\theta$$

$$= -\int_0^{\pi/2} \theta \cot\theta\, d\theta.$$

Since $\theta \cot\theta$ is Riemann integrable on $[0, \pi/2]$, $\int_0^{\pi/2} \ln(\sin\theta)\, d\theta$ converges.

20.3   Using Parametric Differentiation

Since $I(x)$ is differentiable for $|x| < 1$, we apply Leibniz's rule to $I(x)$ to find

$$I'(x) = \int_0^\pi \frac{-2\cos\theta + 2x}{1 - 2x\cos\theta + x^2}\, d\theta.$$

Clearly, $I'(0) = 0$. We now show that $I'(x) = 0$ for $x \neq 0$. First, we prove that

$$\int_0^\pi \frac{1 - x^2}{1 - 2x\cos\theta + x^2}\, d\theta = \pi. \tag{20.4}$$

The integrand in (20.4) is called Poisson's kernel. It is used to derive solutions to the two-dimensional Laplace's equation on the unit circle, and it also plays an important role in summation of Fourier series. Computing the value of this integral is often used to show the usefulness of the residue theorem – a relatively advanced tool. We give a more straightforward method using the half angle substitution. Setting $t = \tan(\theta/2)$, we have

$$\int \frac{1 - x^2}{1 - 2x\cos\theta + x^2}\, d\theta = 2(1 - x^2) \int \frac{dt}{(1-x)^2 + (1+x)^2 t^2}$$

$$= 2\arctan\left(\frac{1+x}{1-x} t\right) + C = 2\arctan\left(\frac{1+x}{1-x} \tan(\theta/2)\right) + C.$$

The fundamental theorem of calculus gives

$$\int_0^\pi \frac{1 - x^2}{1 - 2x\cos\theta + x^2}\, d\theta = \lim_{\theta\to\pi} 2\arctan\left(\frac{1+x}{1-x} \tan(\theta/2)\right) = \pi.$$

Using (20.4) and $x \neq 0$, we get

$$I'(x) = \frac{1}{x} \int_0^\pi \left(1 - \frac{1 - x^2}{1 - 2x\cos\theta + x^2}\right) d\theta = 0.$$

Thus, we have $I'(x) = 0$ for $|x| < 1$ and so $I(x)$ is a constant. Since $I(0) = 0$, we have shown that $I(x) \equiv 0$ for all $|x| < 1$.

20.4 Using Infinite Series

We first show that

$$\ln(1 - 2x\cos\theta + x^2) = -2\sum_{n=1}^{\infty} \frac{x^n}{n}\cos n\theta, \tag{20.5}$$

where the series converges uniformly for $|x| < 1$. Once we establish this, integrating (20.5) with respect to θ from 0 to π, will show that $I(x) = 0$ once again.

To prove (20.5) and to keep the evaluation at an elementary level, instead of using the Fourier series, we start with

$$\frac{1-x^2}{1-2x\cos\theta+x^2} = \frac{1-x^2}{1-x(e^{i\theta}+e^{-i\theta})+x^2} = \frac{1-x^2}{(1-xe^{i\theta})(1-xe^{-i\theta})},$$

where we have used the relation $2\cos\theta = e^{i\theta} + e^{-i\theta}$. Now, decomposing into partial fractions yields

$$\frac{1-x^2}{1-2x\cos\theta+x^2} = -1 + \frac{1}{1-xe^{i\theta}} + \frac{1}{1-xe^{-i\theta}}.$$

Thus, the geometric series expansion leads to

$$\frac{1-x^2}{1-2x\cos\theta+x^2} = 1 + 2\sum_{n=1}^{\infty} x^n \cos n\theta. \tag{20.6}$$

The series (20.6) converges uniformly since $\sum_{n=1}^{\infty} |x|^n$ converges for $|x| < 1$. Removing 1 from the right-hand side of (20.6) and then dividing by x, we have

$$\frac{2\cos\theta - 2x}{1-2x\cos\theta+x^2} = 2\sum_{n=1}^{\infty} x^{n-1} \cos n\theta. \tag{20.7}$$

Since the series (20.7) converges uniformly, integrating from 0 to x term by term, we have established (20.5) as desired.

Remark. As a bonus, we have another proof of (20.4) that follows from integrating the series (20.6) from 0 to π term by term.

We have seen a variety of evaluations of the Poisson integral. The interested reader is encouraged to investigate additional approaches.

Exercises

1. Use

$$\ln(1-z) = -\sum_{n=1}^{\infty} \frac{z^n}{n}, \qquad (|z| < 1)$$

to prove (20.5) and

$$\arctan\left(\frac{x\sin\theta}{1-x\cos\theta}\right) = \sum_{n=1}^{\infty} \frac{\sin n\theta}{n} x^n, \qquad (|x| < 1).$$

2. Show that
$$\int_0^1 x^{-x}\,dx = \sum_{n=1}^{\infty} n^{-n}.$$

3. Let f be a continuous function. Prove that
$$\int_0^{2\pi} f(a\cos\theta + b\sin\theta)\,d\theta = 2\int_0^{\pi} f(\sqrt{a^2+b^2}\cos\phi)\,d\phi.$$

4. Show that
$$3^{3z} \cdot \frac{\Gamma(z)\Gamma(z+1/3)\Gamma(z+2/3)}{\Gamma(3z)}$$
is periodic of period 1. Hence it is a constant. Determine the constant.

5. Show that
$$I(\theta) = \int_0^1 \frac{\ln(1-2x\cos\theta+x^2)}{x}\,dx$$
satisfies the functional equation
$$I\left(\frac{\theta}{2}\right) + I\left(\pi - \frac{\theta}{2}\right) = \frac{1}{2}I(\theta),$$
and then use this equation to evaluate the integral.

6. If f is a real-valued continuous function on R and satisfies the functional equation
$$f(x)\cdot f(y) = f(\sqrt{x^2+y^2}), \qquad \text{for all } x \text{ and } y,$$
show that
$$f(x) = 0 \quad \text{or} \quad f(x) = e^{\alpha x^2},$$
where α is an arbitrary constant. Use this to prove that
$$\int_0^{\infty} e^{-x^2}\cos(px)\,dx = \frac{\sqrt{\pi}}{2} e^{-p^2/4}.$$

7. Let
$$F(p) = \int_0^{\infty} e^{-pt^2}\cos x^2\,dx \quad \text{and} \quad G(p) = \int_0^{\infty} e^{-pt^2}\sin x^2\,dx.$$
Show that F and G satisfy
$$F^2(p) - G^2(p) = \frac{p\pi}{4(1+p^2)} \quad \text{and} \quad 2F(p)G(p) = \frac{\pi}{4(1+p^2)},$$
and solve the simultaneous quadratic equations for $F(p)$ and $G(p)$.

8. **Putnam Problem 1990-B1.** Find all continuously differentiable functions f on the real line such that, for all x,
$$f^2(x) = \int_0^x [f^2(t) + (f'(t))^2]\,dt + 1990.$$

20. Four Ways to Evaluate the Poisson Integral

9. For all p, show that
$$\int_0^{\pi/2} \frac{dx}{1+\tan^p x} = \frac{\pi}{4}.$$
   Remark. For $p = \sqrt{2}$, this is **Putnam Problem 1980-A3**.

10. Define the parametric integral
$$I_n(z) = \int_0^\pi \cos(zt) \cos^n t\, dt.$$
   (a) Establish the reduction formula $(n^2 - z^2) I_n(z) = n(n-1) I_{n-2}(z)$.
   (b) Show that
$$\lim_{n\to\infty} \frac{I_n(z)}{I_n(0)} = 1.$$
   (c) Deduce the Euler sine product formula.

11. The *theta function* is defined by
$$\theta(s) = \sum_{-\infty}^\infty e^{-n^2 \pi s} = 1 + 2e^{-\pi s} + 2e^{-4\pi s} + 2e^{-9\pi s} + \cdots, \qquad s > 0.$$
   Prove that $\theta(1/s) = \sqrt{s}\, \theta(s)$; that is,
$$\sum_{-\infty}^\infty e^{-n^2 \pi / s} = \sqrt{s} \sum_{-\infty}^\infty e^{-n^2 \pi s}.$$

12. (P. Bracken) Let α and β be positive real numbers with $\alpha\beta = \pi$, and let y be a real number. Prove that
$$\frac{1}{2} + \sum_{k=1}^\infty e^{-\alpha k} \cos(\alpha y k) = \frac{1}{\alpha} \sum_{j=-\infty}^\infty \frac{1}{1 + (y + 2\beta j)^2}.$$

13. **Monthly Problem 11036** [2003, 743; 2005, 569-572]. For $0 \leq a \leq \sqrt{3}$, evaluate
$$I(a) = \int_{-1}^1 \frac{\ln(1 + x^2 - x\sqrt{a^2 + x^2})}{\sqrt{1-x^2}}\, dx$$
   in closed form.

14. **Monthly Problem 11072** [2004, 259; 2005, 845-846]. Evaluate the integral
$$I(a, k) = \int_0^\infty \int_0^\infty \exp(-k\sqrt{x^2 + y^2}) \frac{\sin ax}{x} \frac{\sin ay}{y}\, dy\, dx.$$

15. Using
$$\frac{1}{\sin t} = \frac{1}{t} + \sum_{k=0}^\infty (-1)^k \left[\frac{1}{t - k\pi} + \frac{1}{t + k\pi}\right],$$
   prove that
$$\int_0^\infty \frac{\sin x}{x}\, dx = \frac{\pi}{2}.$$

16. Evaluate
$$\int_0^\infty \frac{\ln|\cos x|}{x^2} dx.$$

17. Let $B_n(t)$ be the Bernoulli polynomials. Show that
$$\zeta(2n+1) = (-1)^{n+1} \frac{(2\pi)^{2n+1}}{(2n+1)!} \int_0^{1/2} B_{2n+1}(t) \cot(\pi t) dt.$$

18. Let
$$I_n = \int_0^{\pi/2} (\ln \sin t)^n dt.$$

(a) Define
$$D(s) = \sum_{k=0}^\infty \frac{(2k-1)!!}{(2n)!!} \cdot \frac{1}{(2k+1)^s}.$$

Show that
$$D(n+1) = \frac{(-1)^n}{n!} \int_0^{\pi/2} (\ln \sin t)^n dt.$$

(b) Find $I(n)$ explicitly for $n = 2, 3, 4$.

(c) Show that
$$I_n = \frac{(-1)^n}{2} \pi \begin{vmatrix} \sigma_1 & -1 & 0 & 0 & \cdots & 0 \\ \sigma_2 & \sigma_1 & -2 & 0 & \cdots & 0 \\ \sigma_3 & \sigma_2 & \sigma_1 & -3 & \cdots & 0 \\ \cdots & \cdots & \cdots & \cdots & \cdots & \cdots \\ \sigma_n & \sigma_{n-1} & \sigma_{n-2} & \sigma_{n-4} & \cdots & \sigma_1 \end{vmatrix},$$

where
$$\sigma_k = \frac{1}{1^k} - \frac{1}{2^k} + \frac{1}{3^k} - \frac{1}{4^k} + \cdots = (1 - 2^{1-k})\zeta(k).$$

(d) Generalize these results to
$$I_n(p, q) = \int_0^{\pi/2} (\ln \sin^p t \ln \cos^q t)^n dt.$$

19. **Monthly Problem 11041** [2003, 843; 2005, 655–657]. Let
$$\alpha(k) = \sum_{n=0}^\infty \frac{1}{(2n+1)^k}, \quad \beta(k) = \sum_{n=0}^\infty \frac{(-1)^n}{(2n+1)^k}.$$

For $|x| < 1$, define
$$f(x) = \sum_{k=1}^\infty \alpha(2k+1)x^{2k}, \quad g(x) = \sum_{k=1}^\infty \beta(2k)x^{2k-1}.$$

Show that
$$f(x) + \frac{1}{2} \ln 2 = \frac{1}{2}\left(f\left(\frac{x+1}{2}\right) + f\left(\frac{x-1}{2}\right)\right)$$

and
$$g(x) = \frac{1}{2}\left(f\left(\frac{x+1}{2}\right) - f\left(\frac{x-1}{2}\right)\right).$$

20. Four Ways to Evaluate the Poisson Integral

20. For positive integer $p \geq 2$, let

$$A(p) = 2\pi^{p-1} \sum_{k=1}^{\infty} \frac{\alpha(2k)}{(2k)(2k+1)\cdots(2k+p-1)},$$

$$B(p) = \frac{\pi^{p-1}}{2^{p-1}} \sum_{k=0}^{\infty} \frac{\beta(2k+1)}{(2k+1)(2k+2)\cdots(2k+p)},$$

where $\alpha(k)$ and $\beta(k)$ are defined as in Exercise 19 above. Prove that

$$A(p) = \frac{2^{p-1}}{(p-1)!} \int_0^{\pi/2} x^{p-1} \cot x \, dx, \quad B(p) = \frac{1}{2 \cdot (p-1)!} \int_0^{\pi/2} \frac{x^{p-1}}{\sin x} \, dx.$$

In particular,

$$B(2) = \frac{\pi}{2} \sum_{k=0}^{\infty} \frac{\beta(2k+1)}{(2k+1)(2k+2)} = \frac{1}{2} \int_0^{\pi/2} \frac{x}{\sin x} \, dx = G.$$

References

[1] T. M. Apostol, *Mathematical Analysis*, 2nd edition, Addison-Wesley, 1974.

[2] E. W. Weisstein, "Definite integral." From *Mathworld* — A Wolfram Web Resource. mathworld.wolfram.com/about/author.html

21

Some Irresistible Integrals

I could never resist an integral. — G. H. Hardy

Over the years, we have seen a great many interesting integral problems and clever solutions published in the Monthly. Most of them contain mathematical ingenuities. In this chapter, we record eight of these intriguing integrals. They are so striking and so elegant that we can not resist including an account of them in this book. I still remember how exciting I found them on first encounter. In presenting the combination of approaches required to evaluate these integrals, I have tried to follow the most interesting route to the results and endeavored to highlight connections to other problems and to more advanced topics. At the time this chapter was written, the Monthly had not yet published solutions of about half of the problems. I have referenced the solutions published up to that time and look forward to those forthcoming.

21.1 Monthly Problem 10611 [1997, 665; 1999, 75]

Find the largest value of a and the smallest value b for which the inequalities

$$\frac{1+\sqrt{1-e^{-ax^2}}}{2} < \frac{1}{\sqrt{2\pi}} \int_{-\infty}^{x} e^{-y^2/2} dy < \frac{1+\sqrt{1-e^{-bx^2}}}{2}$$

hold for all $x > 0$.

The inequalities give tight bounds for the *normal distribution*. In the following, we show that $a = 1/2$ and $b = 2/\pi$ are the best possible constants for which the stated inequalities hold. First, by (19.2) we have

$$\int_{-\infty}^{0} e^{-y^2/2} dy = \int_{0}^{\infty} e^{-y^2/2} dy = \sqrt{\frac{\pi}{2}}$$

and so the stated inequalities are equivalent to

$$\frac{\sqrt{1-e^{-ax^2}}}{2} < f(x) < \frac{\sqrt{1-e^{-bx^2}}}{2}, \qquad (21.1)$$

where
$$f(x) = \frac{1}{\sqrt{2\pi}} \int_0^x e^{-y^2/2} dy.$$

Next, if the second inequality of (21.1) holds for all $x > 0$, then the series expansions give
$$0 < \frac{\sqrt{1 - e^{-bx^2}}}{2} - f(x) = \left(\frac{\sqrt{b}}{2} - \frac{1}{\sqrt{2\pi}}\right) x + O(x^3)$$

as $x \to 0$, which implies that $b \geq 2/\pi$. Similarly, if the first inequality of (21.1) holds for all $x > 0$, appealing to

$$\frac{\sqrt{1 - e^{-ax^2}}}{2} = \frac{1}{2} - \frac{1}{4} e^{-ax^2} - \frac{1}{16} e^{-2ax^2} + O(e^{-3ax^2})$$

$$f(x) = \frac{1}{2} - \frac{1}{\sqrt{2\pi}} \int_x^\infty e^{-y^2/2} dy$$

$$= \frac{1}{2} + \frac{1}{\sqrt{2\pi}} \int_x^\infty \frac{1}{y} d(e^{-y^2/2})$$

$$= \frac{1}{2} - \frac{1}{\sqrt{2\pi} x} e^{-x^2/2} + O\left(\frac{e^{-x^2/2}}{x^2}\right),$$

we have
$$0 < f(x) - \frac{\sqrt{1 - e^{-ax^2}}}{2} = \frac{1}{4} e^{-ax^2} - \frac{1}{\sqrt{2\pi} x} e^{-x^2/2} + O(e^{-3ax^2}) + O\left(\frac{e^{-x^2/2}}{x^2}\right)$$

as $x \to \infty$. Dividing each side by $e^{-x^2/2}$ yields $a \leq 1/2$.

Finally, we show that (21.1) holds for all $x > 0$ when $a = 1/2$ and $b = 2/\pi$. To this end, we consider
$$f^2(x) = \frac{1}{2\pi} \int_0^x \int_0^x e^{-(y^2+z^2)/2} dy dz.$$

Let
$$D = [0, x] \times [0, x], \quad D_1 = \{(y, z) : 0 \leq y, 0 \leq z, y^2 + z^2 \leq x^2\}$$
$$D_2 = \{(y, z) : 0 \leq y, 0 \leq z, y^2 + z^2 \leq (4/\pi)x^2\}.$$

See Figure 21.1. We have
$$\frac{1}{2\pi} \iint_{D_1} e^{-(y^2+z^2)/2} dy dz \leq \frac{1}{2\pi} \iint_D e^{-(y^2+z^2)/2} dy dz$$
$$\leq \frac{1}{2\pi} \iint_{D_2} e^{-(y^2+z^2)/2} dy dz, \quad (21.2)$$

where the first inequality holds because $D_1 \subset D$; the second because D and D_2 have the same area and $e^{-(y^2+z^2)/2} \leq e^{-(2/\pi)x^2}$ for $(y, z) \in D - D_2$ while $e^{-(y^2+z^2)/2} \geq e^{-(2/\pi)x^2}$ for $(y, z) \in D_2 - D$. Evaluating the outer integrals in (21.2) via polar coordinates, we obtain
$$\frac{1 - e^{-x^2/2}}{4} < f^2(x) < \frac{1 - e^{-2x^2/\pi}}{4},$$

which is equivalent to (21.1).

21. Some Irresistible Integrals

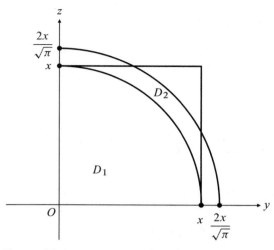

Figure 21.1. Square D and quarter circles D_1 and D_2

21.2 Monthly Problem 11206 [2006, 180]

Evaluate

$$\lim_{n\to\infty} \frac{1}{n} \sum_{k=1}^{n} \left\{\frac{n}{k}\right\}^2$$

where $\{x\}$ denotes $x - \lfloor x \rfloor$, the fractional part of x.

The problem posed originally in the February, 2006, Monthly was to evaluate

$$\lim_{n\to\infty} \sum_{k=1}^{n} \left\{\frac{n}{k}\right\}^2.$$

The crucial factor $1/n$ before the summation was missing. Since $\{1/x\}^2$ is bounded and continuous on $(0, 1)$ except at each reciprocal of a positive integer, $\{1/x\}^2$ is Riemann integrable. Therefore,

$$\lim_{n\to\infty} \frac{1}{n} \sum_{k=1}^{n} \left\{\frac{n}{k}\right\}^2 = \int_0^1 \left\{\frac{1}{x}\right\}^2 dx.$$

Now, we show that

$$\int_0^1 \left\{\frac{1}{x}\right\}^2 dx = \ln(2\pi) - \gamma - 1. \qquad (21.3)$$

where γ is Euler's constant. To see this, for $x \in (1/(n+1), 1/n)$, we have

$$n < 1/x < n+1, \quad \left\{\frac{1}{x}\right\} = \frac{1}{x} - n,$$

and so

$$\int_0^1 \left\{\frac{1}{x}\right\}^2 dx = \sum_{n=1}^{\infty} \int_{1/(n+1)}^{1/n} \left(\frac{1}{x} - n\right)^2 dx$$

$$= \sum_{n=1}^{\infty} \left(1 - 2n \ln \frac{n+1}{n} + \frac{n}{n+1}\right).$$

Let S_N be the partial sum, namely

$$S_N = \sum_{n=1}^{N} \left(1 - 2n \ln \frac{n+1}{n} + \frac{n}{n+1}\right).$$

Since

$$\sum_{n=1}^{N} \left(1 + \frac{n}{n+1}\right) = \sum_{n=1}^{N} \left(2 - \frac{1}{n+1}\right) = 2N - \sum_{n=1}^{N} \frac{1}{n+1},$$

$$\sum_{n=1}^{N} n \ln \frac{n+1}{n} = (\ln 2 - \ln 1) + 2(\ln 3 - \ln 2) + \cdots + N(\ln(N+1) - \ln N)$$

$$= N \ln(N+1) - \ln(N!),$$

we find

$$S_N = 2N - \sum_{n=1}^{N} \frac{1}{n+1} - 2N \ln(N+1) + 2 \ln(N!).$$

Recall the asymptotic formula of the harmonic numbers

$$H_k = \ln k + \gamma + O(1/k)$$

and Stirling's formula

$$k! \sim \sqrt{2\pi k} \left(\frac{k}{e}\right)^k.$$

We obtain

$$S_N = \ln(2\pi) - \gamma + 1 - (2N+1) \ln \frac{1+N}{N} + O(1/N),$$

from which (21.3) follows by letting $N \to \infty$.

Using similar paths, one can evaluate more challenge integrals like

$$\int_0^1 \left\{\frac{1}{x}\right\}^2 \left\{\frac{1}{1-x}\right\}^2 dx = 4\ln(2\pi) - 4\gamma - 5,$$

which appeared in *Pi Mu Epsilon Journal* as Problem 1151, 2006.

21.3 Monthly Problem 11275 [2007, 165]

Find

$$I_1 := \int_{y=0}^{\infty} \int_{x=y}^{\infty} \frac{(x-y)^2 \ln((x+y)/(x-y))}{xy \sinh(x+y)} dxdy.$$

Substitution of $x = uy$ gives

$$I_1 = \int_{y=0}^{\infty} \int_{1}^{\infty} \frac{y(u-1)^2 \ln((u+1)/(u-1))}{u \sinh(y(u+1))} dudy.$$

21. Some Irresistible Integrals

Interchanging the order of the integration yields

$$I_1 = \int_1^\infty \frac{(u-1)^2 \ln((u+1)/(u-1))}{u} \left(\int_0^\infty \frac{y}{\sinh(y(u+1))} dy \right) du.$$

Substitution of $t = e^{-(1+u)y}$ leads to

$$\int_0^\infty \frac{y}{\sinh(y(u+1))} dy = -\frac{2}{(1+u)^2} \int_0^1 \frac{\ln t}{1-t^2} dt$$

$$= -\frac{2}{(1+u)^2} \int_0^1 \sum_{k=0}^\infty t^{2k} \ln t \, dt$$

$$= \frac{2}{(1+u)^2} \sum_{k=0}^\infty \frac{1}{(2k+1)^2}$$

$$= \frac{\pi^2}{4(1+u)^2},$$

where we have used the facts that

$$\int_0^1 t^{2k} \ln t \, dt = \frac{1}{(2k+1)^2} \quad \text{and} \quad \sum_{k=0}^\infty \frac{1}{(2k+1)^2} = \frac{3}{4} \sum_{k=1}^\infty \frac{1}{k^2} = \frac{\pi^2}{8}.$$

Therefore

$$I_1 = \frac{\pi^2}{4} \int_1^\infty \frac{(u-1)^2 \ln((u+1)/(u-1))}{u(1+u)^2} du.$$

Finally, substituting $s = (u-1)/(u+1)$, we get

$$I_1 = -\frac{\pi^2}{2} \int_0^1 \frac{s^2 \ln s}{1-s^2} ds = -\frac{\pi^2}{2} \int_0^1 \left(\frac{\ln s}{1-s^2} - \ln s \right) ds = \frac{\pi^2(\pi^2 - 8)}{16}.$$

21.4 Monthly Problem 11277
[2007, 259; 2008, 758–759]

Evaluate the double integral

$$I_2 := \int_0^{\pi/2} \int_0^{\pi/2} \frac{\log(2 - \sin\theta \cos\phi) \sin\theta}{2 - 2\sin\theta \cos\phi + \sin^2\theta \cos^2\phi} d\theta d\phi.$$

Bailey and Borwein [2] recently evaluated its numerical value to exceedingly high precision and found the exact value via the *Inverse Symbolic Calculator*, an online numerical constant recognition tool available at

oldweb.cecm.sfu.ca/projects/ISC/ISCmain.html

They then established an analytical evaluation via double series expansions and Wallis's formula. We now calculate this integral based on a lemma stated below. The lemma enables us to convert a class of double integrals into single integrals, which leads to a straightforward evaluation of I_2. We begin with the lemma.

Lemma 21.1. *If f is integrable on $[0, 1]$, then*

$$\int_0^{\pi/2}\int_0^{\pi/2} f(\sin\theta\cos\phi)\sin\theta\,d\theta d\phi = \frac{\pi}{2}\int_0^1 f(x)\,dx. \qquad (21.4)$$

Proof. Let S be the unit sphere in the first octant. The double integral can then be viewed as a surface integral over S with the angle θ measured from the z-axis. Thus, $x = \sin\theta\cos\phi, y = \sin\theta\sin\phi, z = \cos\theta$ and

$$\|\mathbf{T}_\theta \times \mathbf{T}_\phi\| = |\sin\theta|.$$

This yields

$$\int_0^{\pi/2}\int_0^{\pi/2} f(\sin\theta\cos\phi)\sin\theta\,d\theta d\phi = \iint_S f(x)\,dS.$$

On the other hand, viewing the x-axis as the polar axis and ϕ as the angle, we have

$$\iint_S f(x)\,dS = \int_0^{\pi/2}\int_0^1 f(x)dx d\phi = \frac{\pi}{2}\int_0^1 f(x)\,dx,$$

which proves (21.4).

Clearly, replacing θ by $\pi/2 - \theta$ in (21.4), we also find

$$\int_0^{\pi/2}\int_0^{\pi/2} f(\cos\theta\cos\phi)\cos\theta\,d\theta d\phi = \frac{\pi}{2}\int_0^1 f(x)\,dx.$$

Now we are ready to use the lemma to evaluate I_2. Setting $f(x) = \ln(2-x)/(2-2x+x^2)$, by (21.4), I_2 becomes

$$I_2 = \frac{\pi}{2}\int_0^1 \frac{\ln(2-x)}{2-2x+x^2}\,dx.$$

Substituting $x = 1 - \tan\alpha$, we obtain

$$I_2 = \frac{\pi}{2}\int_0^{\pi/4} \ln(1+\tan\alpha)\,d\alpha$$

$$= \frac{\pi}{2}\int_0^{\pi/4} \ln\left(\frac{\sqrt{2}\cos(\pi/4-\alpha)}{\cos\alpha}\right)\,d\alpha$$

$$= \frac{\pi}{2}\left(\frac{\pi}{8}\ln 2 + \int_0^{\pi/4} \ln\cos(\pi/4-\alpha)\,d\alpha - \int_0^{\pi/4} \ln\cos\alpha\,d\alpha\right)$$

$$= \frac{\pi^2}{16}\ln 2,$$

where the last two integrals are equal by symmetry about $\alpha = \pi/8$.

21.5 Monthly Problem 11322 [2007, 835]

Let N be a positive integer. Prove that

$$\int_0^1\int_0^1 \frac{(x(1-x)y(1-y))^N}{(1-xy)(-\ln(xy))}\,dydx = \sum_{n=N+1}^\infty \int_n^\infty \left(\frac{N!}{t(t+1)\cdots(t+N)}\right)^2\,dt. \qquad (21.5)$$

21. Some Irresistible Integrals

Let I_3 denote the left-hand integral in (21.5). Expanding $1/(1-xy)$ in a geometric series and using the monotone convergence theorem yields

$$I_3 = \int_0^1 \int_0^1 \sum_{k=0}^\infty \frac{(xy)^k (x(1-x)y(1-y))^N}{(-\ln(xy))} \, dy \, dx$$

$$= \sum_{k=0}^\infty \int_0^1 \int_0^1 \frac{(xy)^k (x(1-x)y(1-y))^N}{(-\ln(xy))} \, dy \, dx.$$

In view of the fact that

$$\int_{k+N}^\infty (xy)^t \, dt = \frac{(xy)^{k+N}}{(-\ln(xy))}, \quad \text{for } 0 < xy < 1,$$

we have

$$I_3 = \sum_{k=0}^\infty \int_0^1 \int_0^1 \left(\int_{k+N}^\infty (xy)^t (1-x)^N (1-y)^N \, dt \right) dy \, dx.$$

Since the integrand is positive, exchanging the order of integration gives

$$I_3 = \sum_{k=0}^\infty \int_{k+N}^\infty \left(\int_0^1 \int_0^1 (xy)^t (1-x)^N (1-y)^N \, dy \, dx \right) dt.$$

Using Euler's *beta function*, we have

$$\int_0^1 x^t (1-x)^N \, dx = B(t+1, N+1) = \frac{\Gamma(t+1)\Gamma(N+1)}{\Gamma(t+N+2)}.$$

Repeatedly using $\Gamma(x+1) = x\Gamma(x)$ yields

$$\frac{\Gamma(t+1)}{\Gamma(t+N+2)} = \frac{1}{(t+N+1)(t+N)\cdots(t+1)}.$$

Appealing to $\Gamma(N+1) = N!$, we have

$$I_3 = \sum_{k=0}^\infty \int_{k+N}^\infty \left(\frac{N!}{(t+N+1)(t+N)\cdots(t+1)} \right)^2 dt.$$

Replacing $t+1$ by t and then setting $n = k+N+1$, we arrive at

$$I_3 = \sum_{n=N+1}^\infty \int_n^\infty \left(\frac{N!}{t(t+1)\cdots(t+N)} \right)^2 dt$$

as desired.

21.6 Monthly Problem 11329 [2007, 925]

Let $f(t) = 2^{-t} \ln \Gamma(t)$, and let γ be Euler's constant. Derive the following integral identities:

$$\int_0^\infty f(t) \, dt = 2 \int_0^1 f(t) \, dt - \frac{\gamma + \ln\ln 2}{\ln 2},$$

$$\int_0^\infty t \, f(t) \, dt = 2 \int_0^1 (t+1) f(t) \, dt - \frac{(\gamma + \ln\ln 2)(1 + 2\ln 2) - 1}{\ln^2 2}.$$

Using that $\Gamma(t+1) = t\Gamma(t)$, we see that $f(t)$ satisfies

$$2f(t+1) = f(t) + 2^{-t} \ln t. \qquad (21.6)$$

Hence

$$\int_0^\infty f(t)\,dt = \int_0^1 f(t)\,dt + \int_0^\infty f(t+1)\,dt$$
$$= \int_0^1 f(t)\,dt + \frac{1}{2}\left(\int_0^\infty f(t)\,dt + \int_0^\infty 2^{-t} \ln t\,dt\right),$$

and so

$$\int_0^\infty f(t)\,dt = 2\int_0^1 f(t)\,dt + \int_0^\infty 2^{-t} \ln t\,dt.$$

Recall the classical formula (for example, see [3, Formula 4.331.1])

$$\int_0^\infty e^{-\alpha t} \ln t\,dt = -\frac{\gamma + \ln \alpha}{\alpha}. \qquad (21.7)$$

Setting $\alpha = \ln 2$, we obtain the first desired integral identity.

Differentiating (21.7) with respect to the parameter α yields

$$\int_0^\infty t e^{-\alpha t} \ln t\,dt = \frac{1 - (\gamma + \ln \alpha)}{\alpha^2}.$$

Thus, using (21.6) and the first integral identity, we obtain

$$\int_0^\infty t f(t)\,dt = 2\int_0^1 (t+1) f(t)\,dt + \int_0^\infty (t+2) 2^{-t} \ln t\,dt$$
$$= 2\int_0^1 (t+1) f(t)\,dt - \frac{(\gamma + \ln \ln 2)(1 + 2\ln 2) - 1}{\ln^2 2}.$$

Remark: This problem is related to the Laplace transform

$$L(\alpha) = \int_0^\infty e^{-\alpha t} \psi(t+1)\,dt,$$

where $\psi(t) = \frac{d}{dt} \ln \Gamma(t)$ is the *digamma function*. Integrating by parts yields

$$L(\alpha) = -\gamma - \ln \alpha + \alpha \int_0^\infty e^{-\alpha t} \ln \Gamma(t)\,dt.$$

In particular,

$$L(\ln 2) = -\gamma - \ln(\ln 2) + \ln 2 \int_0^\infty f(t)\,dt.$$

Refer to [4] for details.

21.7 Monthly Problem 11331 [2007, 926]

Show that if k is a positive integer, then

$$\int_0^\infty \left(\frac{\ln(1+t)}{t}\right)^{k+1} dt = (k+1) \sum_{j=1}^k a_j \zeta(j+1), \qquad (21.8)$$

where ζ denotes the Riemann zeta function and a_j is the coefficient of x^j in $x \prod_{n=1}^{k-1}(1-nx)$.

Here we have made a slight modification: originally, the factor $(k+1)$ before the summation in (21.8) was missing. Let I_4 denote the desired integral in (21.8). The substitution $t = (1-x)/x$ yields

$$I_4 = \int_0^1 \frac{(\ln(\frac{1}{x}))^{k+1} x^{k-1}}{(1-x)^{k+1}} dx.$$

Using

$$\frac{1}{(1-x)^{k+1}} = \sum_{n=0}^\infty \binom{n+k}{k} x^n,$$

we deduce

$$I_4 = \sum_{n=0}^\infty \binom{n+k}{k} \int_0^1 \left(\ln\left(\frac{1}{x}\right)\right)^{k+1} x^{n+k-1} dx.$$

Setting $s = \ln\left(\frac{1}{x}\right)$, we have

$$\int_0^1 \left(\ln\left(\frac{1}{x}\right)\right)^{k+1} x^{n+k-1} dx = \int_0^\infty s^{k+1} e^{-(n+k)s} ds = \frac{(k+1)!}{(n+k)^{k+2}},$$

and so

$$I_4 = \sum_{n=0}^\infty \binom{n+k}{k} \frac{(k+1)!}{(n+k)^{k+2}} = (k+1) \sum_{n=0}^\infty \frac{(n+1)(n+2)\cdots(n+k-1)(n+k)}{(n+k)^{k+2}}.$$

Let

$$t \prod_{j=1}^{k-1}(1-jt) = \sum_{j=1}^k a_j t^j. \qquad (21.9)$$

Replacing t by $1/x$ and then multiplying by x^{k+1} on both sides gives

$$x \prod_{j=1}^{k-1}(x-j) = \sum_{j=1}^k a_j x^{k+1-j}.$$

Thus, we have

$$(n+1)(n+2)\cdots(n+k-1)(n+k) = (n+k) \prod_{j=0}^{k-1}(n+k-j) = \sum_{j=1}^k a_j (n+k)^{k+1-j}$$

and
$$I_4 = (k+1) \sum_{n=0}^{\infty} \sum_{j=1}^{k} \frac{a_j}{(n+k)^{j+1}}$$
$$= (k+1) \sum_{j=1}^{k} a_j \sum_{n=0}^{\infty} \frac{1}{(n+k)^{j+1}}$$
$$= (k+1) \sum_{j=1}^{k} a_j \left(\sum_{n=1}^{\infty} \frac{1}{n^{j+1}} - \sum_{i=1}^{k-1} \frac{1}{i^{j+1}} \right)$$
$$= (k+1) \sum_{j=1}^{k} a_j \zeta(j+1) - (k+1) \sum_{i=1}^{k-1} \frac{1}{i} \left(\sum_{j=1}^{k} \frac{a_j}{i^j} \right).$$

Noticing that $1/i$ ($i = 1, 2, \ldots, k-1$) are solutions of (21.9) implies
$$\sum_{j=1}^{k} \frac{a_j}{i^j} = 0, \quad \text{for } i = 1, 2, \ldots, k-1.$$

Finally, we find that
$$I_4 = (k+1) \sum_{j=1}^{k} a_j \zeta(j+1)$$
as desired.

21.8 Monthly Problem 11418 [2009, 276]

Find
$$I_5 := \int_{-\infty}^{\infty} \frac{t^2 \text{sech}^2 t}{a - \tanh t} \, dt$$
for complex a with $|a| > 1$.

Substitution of $x = \tanh t$ gives
$$I_5 = \int_{-1}^{1} \frac{\text{arctanh}^2 x}{a - x} \, dx.$$

Recall that
$$\text{arctanh } x = \frac{1}{2} \ln \left(\frac{1+x}{1-x} \right).$$

Substitution of $s = (1+x)/(1-x)$ gives
$$I_5 = \frac{1}{2} \int_0^{\infty} \frac{\ln^2 s \, ds}{(s+1)[(a-1)s + a + 1]}.$$

Rewrite I_5 in the form of
$$I_5 = \frac{1}{2} \left(\int_0^1 \frac{\ln^2 s \, ds}{(s+1)[(a-1)s + a + 1]} + \int_1^{\infty} \frac{\ln^2 s \, ds}{(s+1)[(a-1)s + a + 1]} \right).$$

21. Some Irresistible Integrals

Substituting s by $1/s$ in the second integral yields

$$I_5 = \frac{1}{2}\left(\int_0^1 \frac{\ln^2 s}{(s+1)[(a-1)s+a+1]} + \int_0^1 \frac{\ln^2 s}{(s+1)[(a+1)s+a-1]}\right).$$

The partial fraction decompositions

$$\frac{1}{(s+1)[(a-1)s+a+1]} = \frac{1}{2}\left(\frac{1}{s+1} - \frac{a-1}{(a-1)s+a+1}\right) \quad \text{and}$$

$$\frac{1}{(s+1)[(a+1)s+a-1]} = \frac{1}{2}\left(-\frac{1}{s+1} + \frac{a+1}{(a+1)s+a-1}\right)$$

lead to

$$I_5 = \frac{1}{4}\int_0^1 \left(\frac{a+1}{(a+1)s+a-1} - \frac{a-1}{(a-1)s+a+1}\right)\ln^2 s \, ds.$$

To proceed further, we recall the *trilogarithm function* defined by

$$\text{Li}_3(x) := \sum_{n=1}^{\infty} \frac{x^k}{n^3}.$$

Let $\alpha = (a+1)/(a-1)$. Appealing to

$$\int_0^1 s^n \ln^k s \, ds = (-1)^k \frac{k!}{(n+1)^{k+1}},$$

we find that

$$\int_0^1 \frac{(a+1)\ln^2 s}{(a+1)s+a-1} ds = \alpha \int_0^1 \frac{\ln^2 s \, ds}{\alpha s + 1}$$

$$= \alpha \sum_{n=0}^{\infty} (-\alpha)^n \int_0^1 s^n \ln^2 s \, ds$$

$$= 2\alpha \sum_{n=0}^{\infty} \frac{(-\alpha)^n}{(n+1)^3} = -2\text{Li}_3(-\alpha).$$

Similarly,

$$\int_0^1 \frac{(a-1)\ln^2 s}{(a-1)s+a+1} ds = \int_0^1 \frac{\ln^2 s}{s+\alpha} = -2\text{Li}_3\left(-\frac{1}{\alpha}\right).$$

Therefore

$$I_5 = -\frac{1}{2}\left(\text{Li}_3(-\alpha) - \text{Li}_3\left(-\frac{1}{\alpha}\right)\right).$$

Finally, by the trilogarithmic identity
(see mathworld.wolfram.com/Trilogarithm.html)

$$\text{Li}_3(-\alpha) - \text{Li}_3\left(-\frac{1}{\alpha}\right) = -\frac{1}{6}(\ln^3 \alpha + \pi^2 \ln \alpha),$$

we obtain

$$I_5 = \frac{1}{12}\left(\ln^3\left(\frac{a+1}{a-1}\right) + \pi^2 \ln\left(\frac{a+1}{a-1}\right)\right). \tag{21.10}$$

Symbolically, it is interesting to see that Mathematica 6.0 displays

Assuming[a > 1, {Integrate[t^2*Sech[t]^2/(a - Tanh[t]), {t, -Infinity, Infinity}]}]

{-(1/12) Log[(-1 + a)/(1 + a)] (\[Pi]^2 + Log[(-1 + a)/(1 + a)]^2)}

which is equivalent to (21.10).

Exercises

1. Prove that
$$\int_0^\infty \frac{dx}{(1+x^2)(1+x^\alpha)}$$
   is independent of α.

2. Let a be a positive number. Prove that
$$\int_1^a \frac{\ln x \, dx}{x^2 + ax + a} = \frac{\ln a}{2} \int_1^a \frac{dx}{x^2 + ax + a}.$$

3. **Putnam Problem 1983-B5.** Let $\{a\}$ be the fractional part of a. Determine
$$\lim_{n \to \infty} \frac{1}{n} \int_1^n \left\{\frac{n}{x}\right\} dx.$$

4. Let $\lfloor a \rfloor$ denote the greatest integer $\leq a$. Determine
$$\lim_{n \to \infty} \frac{1}{n} \sum_{k=1}^n \left(\left\lfloor \frac{2\sqrt{n}}{\sqrt{k}} \right\rfloor - 2 \left\lfloor \frac{\sqrt{n}}{\sqrt{k}} \right\rfloor\right).$$

5. Let $n = 1, 3, 4$. Evaluate
$$\int_0^1 \left\{\frac{1}{x}\right\}^n dx.$$

6. Prove that
$$\int_0^1 \frac{\ln x \ln(1-x)}{(1+x)^2} dx = \frac{\pi^2}{24} - \frac{1}{2} \ln^2 2.$$

7. Prove that
$$\int_0^1 \int_0^1 \frac{x-1}{(1-xy)\ln(xy)} dx\,dy = \gamma,$$
$$\int_0^1 \int_0^1 \frac{x-1}{(1+xy)\ln(xy)} dx\,dy = \ln(4/\pi). \qquad \text{(J. Sondow)}$$

8. **Monthly Problem 11148** [2005, 366; 2007, 80]. Show that
$$\int_0^\infty \frac{(x^8 - 4x^6 + 9x^4 - 5x^2 + 1)\,dx}{x^{12} - 10x^{10} + 37x^8 - 42x^6 + 26x^4 - 8x^2 + 1} = \frac{\pi}{2}.$$

9. **Monthly Problem 11152** [2005, 567; 2006, 945]. Evaluate
$$\int_0^1 \frac{\ln(\cos(\pi x/2))}{x(1+x)} dx.$$

21. Some Irresistible Integrals

10. **Monthly Problem 11159** [2005; 567; 2007, 167]. For $|a| < \pi/2$, evaluate
$$\int_0^{\pi/2}\int_0^{\pi/2} \frac{\cos\theta\, d\theta d\phi}{\cos(a\cos\theta\cos\phi)}.$$

11. **Monthly Problem 11234** [2006, 568; 2008, 169–170]. Let $a_1,\ldots,a_n$ and $b_1,\ldots,b_{n-1}$ be real numbers with $a_1 < b_1 < a_2 < \cdots < a_{n-1} < b_{n-1} < a_n$, and let h be an integrable function from $\mathbb{R}$ to $\mathbb{R}$. Show that
$$\int_{-\infty}^{\infty} h\left(\frac{(x-a_1)\cdots(x-a_n)}{(x-b_1)\cdots(x-b_{n-1})}\right) dx = \int_{-\infty}^{\infty} h(x)\, dx.$$

Comment. A more detailed analysis of this type of integrals appeared in Glasser's "A remarkable property of definite integrals" (*Math. Comp.*, **60** (1983) 561–563).

12. **Hadamard Product.** Let $f(z) = \sum_{n=0}^{\infty} a_n z^n$ and $g(z) = \sum_{n=0}^{\infty} b_n z^n$ be analytical functions. Let $h(\xi) = \sum_{n=0}^{\infty} a_n b_n \xi^n$. Then
$$h(\xi) = \frac{1}{2\pi}\int_0^{2\pi} f(ze^{i\theta})g(\zeta e^{-i\theta})\, d\theta,$$
where $\xi = z\zeta$.

13. **Sequence of Definite Integrals.** It is known that $n! = (-1)^n 2^{n+1}\int_0^1 x\ln^n x\, dx$. Let
$$I_n = \int_0^1 \frac{1}{(x^2+x+1)^n}\, dx.$$
Show that $I_n = a_n + b_n\sqrt{3}\pi$ and determine the closed forms of a_n and b_n.

14. **Putnam Problem 1985-A5.** Let
$$I_m = \int_0^{2\pi} \cos(x)\cos(2x)\cdots\cos(mx)\, dx.$$
For which integers m, $1 \le m \le 10$, is $I_m \ne 0$?

15. **Open Problem.** Determine the maximum of
$$I_k = \int_0^{2\pi} \cos(n_1 x)\cos(n_2 x)\cdots\cos(n_k x)\, dx$$
where $n_1, n_2, \ldots, n_k$ are positive integers.

16. **Open Problem.** Let $n \in \mathbb{N}, a > 1$. Find
$$I(a) = \int_0^{\infty} \frac{\ln^n x\, dx}{(x+1)(x+a)}$$
in closed form. Along the way you may use the functional equation
$$\mathrm{Li}_n(-z) + (-1)^n \mathrm{Li}_n(-1/z) = -\frac{(2\pi i)^n}{n!} B_n\left(\frac{1}{2} + \frac{\ln z}{2\pi i}\right),$$
where $B_n(x)$ is the Bernoulli polynomial.

References

[1] T. M. Apostol, *Mathematical Analysis*, 2nd edition, Addison-Wesley, 1974.

[2] D. H. Bailey and J. M. Borwein, "Solution to Monthly Problem 11410". Available at `crd.lbl.gov/~dhbailey/dhbpapers/amm-11275.pdf`

[3] I. S. Gradshteyn and I. M. Ryzhik, *Table of Integrals, Series, and Products*, 6th edition, edited by A. Jeffrey and D. Zwillinger. Academic Press, New York, 2000.

[4] M. Glasser and D. Manna, On the Laplace transform of the psi-function, *Contemporary Mathematics*. Available at `locutus.cs.dal.ca:8088/archive/00000375/01/ContempMathFinal.pdf`

[5] E. W. Weisstein, "Definite integral." From *Mathworld* — A Wolfram Web Resource. Available at `mathworld.wolfram.com/about/author.html`

[6] Wikipedia, Polylogarithm. Available at `en.wikipedia.org/wiki/Polylogarithm`

Solutions to Selected Problems

Here the solutions do not include any Monthly and Putnam problems. For Monthly problems, please refer to the journal. The second pair of numbers indicates when and where the solution has been published. For Putnam problems and solutions before 2000, see

1. *The William Lowell Putnam Mathematical Competition. Problems and Solutions: 1938–1964*, by A. M. Gleason, R. E. Greenwood and L. M. Kelly.

2. *The William Lowell Putnam Mathematical Competition. Problems and Solutions: 1965–1984*, by G. L. Alexanderson, L. F. Klosinski and L. C. Larson.

3. *The William Lowell Putnam Mathematical Competition 1985–2000: Problems, Solutions and Commentary*, by K. S. Kedlaya, B. Poonen and R. Vakil.

All three are currently in print, and should be available for purchase through the MAA online bookstore (www.maa.org/EBUSPPRO/). For the problems and solutions after 2000, see

www.unl.edu/amc/a-activities/a7-problems/putnamindex.shtml.

Chapter 1

4. (a) We present three distinct proofs as follows:

First Proof: For any positive numbers a and b, the AM-GM inequality yields

$$\frac{4a^2}{a+b} + (a+b) \geq 4a,$$

which implies

$$\frac{a^2}{a+b} = \frac{1}{4}\frac{4a^2}{a+b} \geq \frac{1}{4}(3a-b).$$

Repeatedly using this inequality gives

$$\sum_{k=1}^{n} \frac{a_k^2}{a_k + a_{k+1}} \geq \frac{1}{4}(3a_1 - a_2) + \frac{1}{4}(3a_2 - a_3) + \cdots + \frac{1}{4}(3a_n - a_1)$$

$$= \frac{1}{2}(a_1 + a_2 + \cdots + a_n) = \frac{1}{2}.$$

Second Proof: Let $a_{n+1} = a_1$. The Cauchy-Schwarz inequality yields

$$\left(\sum_{k=1}^{n} a_k\right)^2 = \left(\sum_{k=1}^{n} \frac{a_k}{\sqrt{a_k + a_{k+1}}} \cdot \sqrt{a_k + a_{k+1}}\right)^2$$

$$\leq \sum_{k=1}^{n} \frac{a_k^2}{a_k + a_{k+1}} \cdot \sum_{k=1}^{n} (a_k + a_{k+1})$$

$$= 2 \sum_{k=1}^{n} \frac{a_k^2}{a_k + a_{k+1}} \cdot \sum_{k=1}^{n} a_k,$$

from which and $\sum_{k=1}^{n} a_k = 1$ the desired inequality follows.

Third Proof: Let $a_{n+1} = a_1$. The key observation is

$$\sum_{k=1}^{n} \frac{a_k^2}{a_k + a_{k+1}} = \frac{1}{2} \sum_{k=1}^{n} \frac{a_k^2 + a_{k+1}^2}{a_k + a_{k+1}},$$

which is equivalent to

$$\sum_{k=1}^{n} \frac{a_k^2}{a_k + a_{k+1}} = \sum_{k=1}^{n} \frac{a_{k+1}^2}{a_k + a_{k+1}}.$$

This immediately follows from

$$\sum_{k=1}^{n} \frac{a_k^2 - a_{k+1}^2}{a_k + a_{k+1}} = \sum_{k=1}^{n} \frac{(a_k + a_{k+1})(a_k - a_{k+1})}{a_k + a_{k+1}} = \sum_{k=1}^{n} (a_k - a_{k+1}) = 0.$$

Now, the original inequality is equivalent to

$$\sum_{k=1}^{n} \frac{a_k^2 + a_{k+1}^2}{a_k + a_{k+1}} \geq 1.$$

To prove this, notice that

$$\frac{a_k^2 + a_{k+1}^2}{a_k + a_{k+1}} \geq \frac{a_k + a_{k+1}}{2},$$

which follows directly after clearing out denominators and rearranging in the form $(a_k - a_{k+1})^2 \geq 0$. Therefore,

$$\sum_{k=1}^{n} \frac{a_k^2 + a_{k+1}^2}{a_k + a_{k+1}} \geq \sum_{k=1}^{n} \frac{a_k + a_{k+1}}{2} = \sum_{k=1}^{n} a_k = 1.$$

Remark. One may use a similar argument to prove that, if $\sum_{k=1}^{n} a_k = 1$ and $a_{n+1} = a_1$, then for any positive integer m,

$$\sum_{k=1}^{n} \frac{a_k^m}{a_k^{m-1} + a_k^{m-2} a_{k+1} + a_k^{m-3} a_{k+1}^2 + \cdots + a_{k+1}^{m-1}} \geq \frac{1}{m}.$$

4.(e) By the AM-GM inequality,

$$\sqrt{1 + a_0 + \cdots + a_{k-1}} \sqrt{a_k + \cdots + a_n} \leq \frac{1}{2}(1 + a_0 + \cdots + a_{k-1} + a_k + \cdots + a_n) = 1.$$

Hence, for all $1 \leq k \leq n$,

$$b_k := \frac{a_k}{\sqrt{1 + a_0 + \cdots + a_{k-1}} \sqrt{a_k + \cdots + a_n}} \geq a_k,$$

Solutions to Selected Exercises

and therefore
$$\sum_{k=1}^{n} b_k \geq \sum_{k=1}^{n} a_k = 1.$$

This proves the first inequality. To prove the second inequality, for $0 \leq k \leq n$, define
$$\theta_k = \arcsin(a_0 + a_1 + \cdots + a_k).$$

Clearly,
$$0 = \theta_0 < \theta_1 < \cdots < \theta_n = \arcsin 1 = \frac{\pi}{2}.$$

Notice that
$$a_k = \sin \theta_k - \sin \theta_{k-1}$$
$$= 2 \cos \frac{\theta_k + \theta_{k-1}}{2} \sin \frac{\theta_k - \theta_{k-1}}{2},$$

and
$$\cos \theta_{k-1} = \sqrt{1 - \sin^2 \theta_{k-1}}$$
$$= \sqrt{1 - (a_0 + a_1 + \cdots + a_{k-1})^2}$$
$$= \sqrt{1 + a_0 + \cdots + a_{k-1}} \sqrt{a_k + \cdots + a_n}.$$

By the well-known inequality
$$\sin \theta < \theta, \qquad \theta \in (0, \pi/2),$$

we have
$$a_k < \cos \frac{\theta_k + \theta_{k-1}}{2} (\theta_k - \theta_{k-1}) \leq \cos \theta_{k-1} (\theta_k - \theta_{k-1}),$$

and therefore
$$\sum_{k=1}^{n} b_k = \sum_{k=1}^{n} \frac{a_k}{\cos \theta_{k-1}} \leq \sum_{k=1}^{n} (\theta_k - \theta_{k-1}) = \frac{\pi}{2}.$$

Remark. Observe that
$$b_k = \frac{a_k}{\sqrt{1 - (a_0 + a_1 + \cdots + a_{k-1})^2}} = \frac{(a_0 + \cdots + a_k) - (a_0 + \cdots + a_{k-1})}{\sqrt{1 - (a_0 + a_1 + \cdots + a_{k-1})^2}}.$$

The proposed sum can be viewed as a Riemann sum for the function $1/\sqrt{1-x^2}$ on the interval $[0, 1]$ when the partition is
$$\{0 = a_0, a_0 + a_1, a_0 + a_1 + a_2, \ldots, a_0 + a_1 + \cdots + a_n = 1\}$$

It follows at once that
$$1 \cdot (1-0) \leq \sum_{k=1}^{n} b_k \leq \int_0^1 \frac{dx}{\sqrt{1-x^2}} = \frac{\pi}{2}.$$

7. There are two distinct cases. First, if $x^2 \geq a_1(a_1 - 1)$, since $\sum_{k=1}^{n} \frac{1}{a_k} \leq 1$, we have
$$\left(\sum_{k=1}^{n} \frac{1}{a_k^2 + x^2} \right)^2 \leq \left(\sum_{k=1}^{n} \frac{1}{2a_k |x|} \right)^2$$
$$= \frac{1}{4x^2} \left(\sum_{k=1}^{n} \frac{1}{a_k} \right)^2 \leq \frac{1}{4x^2}$$
$$\leq \frac{1}{2} \cdot \frac{1}{a_1(a_1 - 1) + x^2}.$$

Next, if $x^2 < a_1(a_1 - 1)$, applying the Cauchy-Schwarz inequality,

$$\left(\sum_{k=1}^{n} \frac{1}{a_k^2 + x^2}\right)^2 \leq \left(\sum_{k=1}^{n} \frac{1}{a_k}\right)\left(\sum_{k=1}^{n} \frac{a_k}{(a_k^2 + x^2)^2}\right)$$

$$\leq \sum_{k=1}^{n} \frac{a_k}{(a_k^2 + x^2)^2}.$$

Notice that $a_{k+1} \geq a_k + 1$ for $k = 1, 2, \ldots, n-1$. We have

$$\frac{2a_k}{(a_k^2 + x^2)^2} \leq \frac{2a_k}{(a_k^2 + x^2 + 1/4)^2 - a_k^2}$$

$$= \frac{1}{(a_k - 1/2)^2 + x^2} - \frac{1}{(a_k + 1/2)^2 + x^2}$$

$$\leq \frac{1}{(a_k - 1/2)^2 + x^2} - \frac{1}{(a_{k+1} - 1/2)^2 + x^2}.$$

Similarly, we have

$$\frac{2a_n}{(a_n^2 + x^2)^2} \leq \frac{1}{(a_n - 1/2)^2 + x^2} - \frac{1}{(a_n + 1/2)^2 + x^2} \leq \frac{1}{(a_n - 1/2)^2 + x^2}.$$

Thus, telescoping yields

$$\sum_{k=1}^{n} \frac{a_k}{(a_k^2 + x^2)^2} \leq \frac{1}{2}\sum_{k=1}^{n-1}\left[\frac{1}{(a_k - 1/2)^2 + x^2} - \frac{1}{(a_{k+1} - 1/2)^2 + x^2}\right] + \frac{1}{2}\frac{1}{(a_n - 1/2)^2 + x^2}$$

$$= \frac{1}{2}\frac{1}{(a_1 - 1/2)^2 + x^2} \leq \frac{1}{2}\frac{1}{a_1(a_1 - 1) + x^2}$$

as expected.

8. Let

$$s_0 = 0, \quad s_k = a_1 + a_2 + \cdots + a_k, \quad (k = 1, 2, \ldots, n).$$

Then

$$s_k = \frac{1}{2}[(a_1 + a_k) + (a_2 + a_{k-1}) + \cdots + (a_k + a_1)] \geq \frac{1}{2}ka_{k+1}.$$

Appealing to Abel's summation formula

$$\sum_{k=1}^{n} a_k b_k = a_n \sum_{k=1}^{n} b_k - \sum_{k=1}^{n-1}(a_{k+1} - a_k)\sum_{i=1}^{k} b_i,$$

we have

$$\sum_{k=1}^{n} \frac{a_k}{k} = \sum_{k=1}^{n} \frac{s_k - s_{k-1}}{k}$$

$$= \frac{1}{n}s_n + \sum_{k=1}^{n-1}\left(\frac{1}{k} - \frac{1}{k+1}\right)s_k$$

$$\geq \frac{1}{n}s_n + \frac{1}{2}s_1 + \frac{1}{2}\sum_{k=1}^{n-1}\left(\frac{1}{k} - \frac{1}{k+1}\right)ka_{k+1}$$

$$= \frac{1}{n}s_n + \frac{1}{2}s_1 + \frac{1}{2}\sum_{k=1}^{n-1}\frac{a_{k+1}}{k+1}$$

$$= \frac{1}{n}s_n + \frac{1}{2}\sum_{k=1}^{n}\frac{a_k}{k},$$

Solutions to Selected Exercises

and so

$$\sum_{k=1}^{n} \frac{a_k}{k} \geq \frac{2}{n} s_n = \frac{2}{n}(s_{n-1} + a_n)$$

$$\geq \frac{2}{n}\left(\frac{n-1}{2}a_n + a_n\right) = \frac{n+1}{n}a_n > a_n.$$

Chapter 2

4 Define $f_k(x) = a_k x - b_k \ln x$, $E = (0, \infty)$. Since $f'_k(x) = a_k - b_k/x$, $f''_k(x) = b_k/x^2$, $f_k(x)$ has a unique critical point at $x_{kc} = b_k/a_k$, where it has minimum value $f_k(x_{kc}) = b_k - b_k \ln(b_k/a_k)$. Similarly, the function $f(x) = \sum_{k=1}^{n} f_k(x)$ has a unique critical point $x_c = (\sum_{k=1}^{n} a_k)/(\sum_{k=1}^{n} b_k)$ and its minimum value is

$$f(x_c) = \sum_{k=1}^{n} b_k \left(1 - \ln \frac{\sum_{k=1}^{n} b_k}{\sum_{k=1}^{n} a_k}\right).$$

By using (2.1) we obtain

$$\sum_{k=1}^{n} a_k \ln \frac{b_k}{a_k} \geq \left(\sum_{k=1}^{n} b_k\right) \ln \frac{\sum_{k=1}^{n} b_k}{\sum_{k=1}^{n} a_k},$$

which is equivalent to the required inequality.

6. Define

$$f_k(x) = p_k \left(\frac{a_k}{x} - \frac{1-a_k}{1-x} - \ln \frac{1-x}{x}\right), \quad E = (0, 1/2].$$

Notice that $f'_k(x) = p_k(a_k - x)(2x - 1)/(x^2(1-x)^2)$. We find that f_k has a minimum at $x_{kc} = a_k$ and its value is $f_k(x_{kc}) = \ln[a_k/(1-a_k)]^{p_k}$. On the other hand, we have

$$f(x) = \sum_{k=1}^{n} f_k(x) = \left(\frac{A}{x} - \frac{1-A}{1-x} - \ln \frac{1-x}{x}\right),$$

which has a minimum at $x_c = A$ and its value is $f(A) = \ln(A/A')$. Now, applying (2.1) yields

$$\ln \prod_{k=1}^{n} \left(\frac{a_k}{1-a_k}\right)^{p_k} \leq \ln \frac{A}{A'},$$

which is equivalent to the desired inequality.

7. Let $f(x) = 1/(1 + e^x)$. Since

$$f''(x) = \frac{e^x(e^x - 1)}{(1 + e^x)^3} = \begin{cases} > 0, & x \in (0, \infty), \\ < 0, & x \in (-\infty, 0), \end{cases}$$

we see that f is convex in $(0, \infty)$ and concave in $(-\infty, 0)$. Therefore, for $x_k = \pm \ln(1 - a_k)/a_k$ $(1 \leq k \leq n)$, by *Jensen's inequality*,

$$f\left(\sum_{k=1}^{n} p_k x_k\right) = \begin{cases} \leq \sum_{k=1}^{n} p_k f(x_k), & x_k \in (0, \infty), \\ \geq \sum_{k=1}^{n} p_k f(x_k), & x_k \in (-\infty, 0). \end{cases}$$

This proves the proposed inequalities.

Chapter 3

7. We show the first inequality based on the following two lemmas. The first lemma yields a rational bound for $\ln t$.

Lemma 1. For $t \geq 1$ we have
$$\ln t \leq \frac{(t^2-1)(t^2+4t+1)}{4t(t^2+t+1)}. \tag{1}$$

Proof. Let
$$f(t) = \frac{(t^2-1)(t^2+4t+1)}{4t(t^2+t+1)} - \ln t.$$
Note that $f(1) = 0$, so it suffices to show that $f'(t) > 0$ for $t > 1$. This follows from the fact that
$$f'(t) = \frac{(t-1)^4(t+1)^2}{4t^2(t^2+t+1)^2} > 0, \qquad \text{for } t > 1.$$

Lemma 2. For $t \geq 1$ we have
$$\frac{\ln t}{t-1} - \ln\left(\frac{t+\sqrt{t}+1}{3t}\right) \geq 1. \tag{2}$$

Proof. Note that left-hand side becomes continuous if we define it to have the value 1 at $t = 1$. Thus, it suffices to show that it has a nonnegative derivative for $t > 1$, in other words,
$$-\frac{\ln t}{(t-1)^2} + \frac{1}{(t-1)} - \frac{1+2\sqrt{t}}{2\sqrt{t}(t+1)+2t} \geq 0.$$
In fact, it suffices to show this with t replaced by t^2, and this reduces to (1).

To prove the first inequality, we may assume $b > a$ and set $t = b/a$ in (2). We find that
$$\frac{a(\ln a - \ln b)}{a-b} + \ln b - 1 > \ln\left(\frac{a+\sqrt{ab}+b}{3}\right),$$
and the first inequality follows upon exponentiation. Also note that the strict inequality holds unless $a = b$.

Remark. The constants $2/3$ and $2/e$ in the double inequality indeed are the best possible. In fact, since $G(a,b) < I(a,b) < A(a,b)$, one may assume that
$$I(a,b) > sA(a,b) + (1-s)G(a,b), \qquad \text{for } 0 < s < 1.$$
This is equivalent to
$$\frac{a(\ln a - \ln b)}{a-b} + \ln b - 1 > \ln[2s(a+b) + (1-s)\sqrt{ab}].$$
The analogue of $f'(t)$ in Lemma 1 becomes
$$-\frac{(t-1)^2(1+t)^2(-s+s^2-4t+8st-2s^2t-st^2+s^2t^2)}{t^2(s+2t-2st+st^2)^2} \geq 0.$$
Letting $t \to 1$ yields $s \leq 2/3$. Similarly, one can show that if
$$I(a,b) < \beta A(a,b) + (1-\beta)G(a,b)$$
then $\beta \geq 2/e$.

10. First, since $S_\alpha(a,b)$ is strictly increasing in α, for $0 < a < b, \alpha \geq 2$, we have

$$\frac{b^\alpha - a^\alpha}{b - a} \geq \alpha \left(\frac{a+b}{2}\right)^{\alpha-1}$$

with equality if and only if $\alpha = 2$. Next, let $A = a^x, B = b^x$. Then $a^{x+y} = A^{(x+y)/x}$, $b^{x+y} = B^{(x+y)/x}$. Applying the inequality above implies

$$\frac{B^{(x+y)/x} - A^{(x+y)/x}}{B - A} \geq \frac{x+y}{x}\left(\frac{A+B}{2}\right)^{y/x}.$$

Recall that $M_p(a,b)$ is strictly increasing in p. Hence

$$\left(\frac{a^x + b^x}{2}\right)^{1/x} \geq \frac{a+b}{2}$$

for $x \geq 1$ with equality if and only if $x = 1$. Finally, we obtain

$$\frac{b^{x+y} - a^{x+y}}{b^x - a^x} \geq \frac{x+y}{x} A(a,b)^y$$

for $x, y \geq 1$ with equality if and only if $x = y = 1$.

Remark. When $0 < x, y < 1$, the inequality is reversed, namely

$$\frac{b^{x+y} - a^{x+y}}{b^x - a^x} \leq \frac{x+y}{x} A(a,b)^y.$$

But for $x, y \geq 0$, one can prove that

$$\frac{b^{x+y} - a^{x+y}}{b^x - a^x} \geq \frac{x+y}{x} G(a,b)^y.$$

11. We show that $L(a,b) \leq M_{1/3}(a,b)$ by using *Simpson's 3/8 rule*:

$$\int_c^d f(x)\,dx = \frac{d-c}{8}\left(f(c) + 3f\left(\frac{2c+d}{3}\right) + 3f\left(\frac{c+2d}{3}\right) + f(d)\right) - \frac{(d-c)^5}{6480} f^{(4)}(\xi),$$

where $\xi \in (c,d)$. Let $f(x) = e^x$. Replacing c and d with $\ln a$ and $\ln b$, respectively yields the proposed inequality.

13. (b) Recall the Gauss quadrature formula with two knots, namely,

$$\int_0^1 f(t)\,dt = \frac{1}{2} f\left(\frac{1}{2} + \frac{1}{2\sqrt{3}}\right) + \frac{1}{2} f\left(\frac{1}{2} - \frac{1}{2\sqrt{3}}\right) + \frac{1}{4320} f^{(4)}(\xi), \quad 0 < \xi < 1.$$

Let $f(t) = \ln[ta + (1-t)b]$. Then

$$\ln I(a,b) = \frac{1}{2} \ln\left(\frac{2}{3} A^2(a,b) + \frac{1}{3} G^2(a,b)\right) - \frac{(b-a)^4}{720[\xi a + (1-\xi)b]^4},$$

which implies

$$\exp\left(\frac{1}{360}\left(\frac{b-a}{\max(a,b)}\right)^4\right) < \frac{2A^2/3 + G^2/3}{I^2} < \exp\left(\frac{1}{360}\left(\frac{b-a}{\min(a,b)}\right)^4\right).$$

This yields the required inequality.

Remark. Stimulated by this result, for $p \geq 2$, one may prove that the double inequality

$$\alpha A^p(a,b) + (1-\alpha) G^p(a,b) < I^p(a,b) < \beta A^p(a,b) + (1-\beta) G^p(a,b),$$

holds for all positive numbers $a \neq b$ if and only if $\alpha \leq (2/e)^p$ and $\beta \geq 2/3$.

Chapter 4

1. Let $f(x) = x\ln(1+x), g(x) = x - \ln(1+x)$ and let

$$\frac{f_1(x)}{g_1(x)} := \frac{f'(x)}{g'(x)} = \frac{(1+x)\ln(1+x) + x}{x}.$$

Since

$$\frac{f_1'(x)}{g_1'(x)} = 2 + \ln(1+x)$$

is increasing, by the LMR, $f_1(x)/g_1(x)$ is increasing. An appeal to the LMR again establishes the assertion.

6. Apply the LMR to

$$\left(\frac{1-\cos x}{x} - \frac{2}{\pi}\right) \Big/ \left(\frac{\pi}{2} - x\right).$$

7. Rewrite the inequality as

$$\frac{2}{\pi} < \cot(\pi x/2) \Big/ \left(\frac{1-x^2}{x}\right) < \frac{\pi}{4}.$$

Then apply the LMR.

10. (a) We prove a stronger result: Let $\sum_{k=1}^{n} a_k \sin(2k-1)x \geq 0$, $0 < x < \pi$. Then

$$\sum_{k=1}^{n} \frac{a_k \sin kx}{k} > 0, \qquad 0 < x < \pi.$$

Notice that

$$\frac{d}{dt}\left(\frac{\sin(mt)}{m\sin^m t}\right) = -\frac{\sin(m-1)t}{\sin^{m+1} t}.$$

For $m = 2k$, integrating this equation from x to $\pi/2$ yields

$$\frac{\sin(2kx)}{2k \sin^{2k} x} = \int_x^{\pi/2} \frac{\sin(2k-1)t}{\sin^{2k+1} t}\, dt.$$

Replacing x by $x/2$ yields

$$\frac{\sin kx}{k} = 2\int_{x/2}^{\pi/2} \left(\frac{\sin(x/2)}{\sin t}\right)^{2k} \frac{\sin(2k-1)t}{\sin t}\, dt.$$

Thus

$$\sum_{k=1}^{n} \frac{a_k \sin kx}{k} = 2\int_{x/2}^{\pi/2} \sum_{k=1}^{n} a_n \left(\frac{\sin(x/2)}{\sin t}\right)^{2n} \frac{\sin(2n-1)t}{\sin t}\, dt.$$

The proposed inequality follows from $\sum_{k=1}^{n} a_k r^{2k-1} \sin(2k-1)t > 0$ for $0 < r < 1$ since $\sum_{k=1}^{n} a_k \sin(2k-1)t \geq 0$. When all $a_k = 1$, we have

$$\sum_{k=1}^{n} \sin(2k-1)x = \frac{\sin^2(nx)}{\sin x} > 0.$$

Solutions to Selected Exercises

Chapter 5

2. Let $P = e^{i\theta_0}$. Recall that the vertices of a regular n-gon are given by ω^k ($1 \le k \le n$). Thus,

$$\sum_{k=1}^{n} |P - \omega^k|^2 = \sum_{k=1}^{n} (P - \omega^k)\overline{(P - \omega)}$$

$$= \sum_{k=1}^{n} (|P|^2 - (P\overline{\omega} + \overline{P}\omega) + |\omega|^2)$$

$$= 2\sum_{k=1}^{n} \left(1 - \cos\left(\theta_0 + \frac{2k\pi}{n}\right)\right)$$

$$= 2n.$$

7. Repeatedly using the double angle formula $\sin 2\theta = 2\sin\theta\cos\theta$, we obtain

$$\sin x = 2\sin\frac{x}{2}\cos\frac{x}{2}$$

$$= 2^2 \sin\frac{x}{2^2}\cos\frac{x}{2}\cos\frac{x}{2^2}$$

$$= 2^3 \sin\frac{x}{2^3}\cos\frac{x}{2}\cos\frac{x}{2^2}\cos\frac{x}{2^3}$$

$$\vdots$$

$$= 2^n \sin\frac{x}{2^n}\cos\frac{x}{2}\cos\frac{x}{2^2}\cdots\cos\frac{x}{2^n}.$$

But from elementary calculus we have that $\lim_{t \to 0} \sin t/t = 1$, and therefore

$$\lim_{n \to \infty} 2^n \sin\frac{x}{2^n} = x.$$

This proves the identity (a). Next, let

$$P_n = \prod_{k=1}^{n} \cos\frac{x}{2^k} \quad \text{and} \quad Q_n = x\cos^2\frac{x}{2} + x\sum_{k=1}^{n}\sin^2\frac{x}{2^{k+1}}P_k.$$

Since

$$\sin x = \cos\frac{x}{2}\left(x\cos\frac{x}{2} + x\sin^2\frac{x}{4} - x\cos^2\frac{x}{4} + 2\sin\frac{x}{2}\right),$$

we have

$$\sin x = Q_1 - xP_2\cos\frac{x}{4} + 2P_1\sin\frac{x}{2}.$$

By induction,

$$\sin x = Q_n - xP_{n+1}\cos\frac{x}{2^{n+1}} + 2^n P_n \sin\frac{x}{2^n}.$$

Appealing to that $|P_n| \le 1$ and

$$\lim_{n \to \infty}\left(-xP_{n+1}\cos\frac{x}{2^{n+1}} + 2^n P_n \sin\frac{x}{2^n}\right) = \lim_{n \to \infty} P_n\left(-x\cos^2\frac{x}{2^{n+1}} + 2^n \sin\frac{x}{2^n}\right) = 0,$$

we establish that

$$\sin x = \lim_{n \to \infty} Q_n$$

which is equivalent to the desired identity (b).

12. Replacing x by $(x+iy)\pi/2$ in the infinite product of cosine (5.29) yields

$$\cos\frac{(x+iy)\pi}{2} = \prod_{k=1}^{\infty}\left(1 - \frac{x^2 - y^2 + 2xyi}{(2k-1)^2}\right).$$

On the other hand,

$$\cos\frac{(x+iy)\pi}{2} = \cos\frac{x\pi}{2}\cosh\frac{y\pi}{2} - i\sin\frac{x\pi}{2}\sinh\frac{y\pi}{2}.$$

It follows from (5.32) that

$$\sum_{k=1}^{\infty}\arctan\left(\frac{2xy}{(2k-1)^2 - x^2 + y^2}\right) = \arctan\left(\tan\frac{\pi x}{2}\tanh\frac{\pi y}{2}\right) \pmod{2\pi}.$$

Chapter 6

4. For $n \geq 2$, we have

$$\sum_{k=1}^{n-1} H_k = \sum_{k=1}^{n-1}\sum_{i=1}^{k}\frac{1}{i} \qquad \text{(interchange the order of sums)}$$

$$= (n-1)\cdot 1 + (n-2)\cdot\frac{1}{2} + \cdots + [n-(n-2)]\cdot\frac{1}{n-2} + [n-(n-1)]\cdot\frac{1}{n-1}$$

$$= nH_{n-1} - (n-1) = nH_n - n.$$

This identity can also be established by using Abel's summation formula. Next, we have

$$H_k^2 - H_{k-1}^2 = \left(1 + \frac{1}{2} + \cdots + \frac{1}{k}\right)^2 - \left(1 + \frac{1}{2} + \cdots + \frac{1}{k-1}\right)^2$$

$$= \frac{1}{k^2} + 2\cdot\frac{1}{k}\left(1 + \frac{1}{2} + \cdots + \frac{1}{k-1}\right)$$

$$= \frac{1}{k^2} + \frac{2}{k}\left(H_k - \frac{1}{k}\right)$$

$$= 2\cdot\frac{H_k}{k} - \frac{1}{k^2}.$$

Telescoping yields

$$H_n^2 - H_1^2 = 2\left(\frac{H_2}{2} + \frac{H_3}{3} + \cdots + \frac{H_n}{n}\right) - \left(\frac{1}{2^2} + \frac{1}{3^2} + \cdots + \frac{1}{n^2}\right).$$

Thus, we get

$$H_n^2 = 2\left(\frac{H_2}{2} + \frac{H_3}{3} + \cdots + \frac{H_n}{n}\right) + \left(1 - \frac{1}{2^2} - \frac{1}{3^2} + \cdots - \frac{1}{n^2}\right)$$

$$> 2\left(\frac{H_2}{2} + \frac{H_3}{3} + \cdots + \frac{H_n}{n}\right) + \left(1 - \frac{1}{1\cdot 2} - \frac{1}{2\cdot 3} + \cdots - \frac{1}{(n-1)\cdot n}\right)$$

$$= 2\left(\frac{H_2}{2} + \frac{H_3}{3} + \cdots + \frac{H_n}{n}\right) + \frac{1}{n}$$

as desired. Note that telescoping is used in the last summation again.

Solutions to Selected Exercises

8. Appealing to
$$H_n = \int_0^1 (1 + t + \cdots + t^{n-1})\, dt = \int_0^1 \frac{1-t^n}{1-t}\, dt,$$
we have
$$\sum_{n=1}^\infty a_n H_n x^n = \sum_{n=1}^\infty a_n x^n \int_0^1 \frac{1-t^n}{1-t}\, dt$$
$$= \sum_{n=1}^\infty \int_0^1 \frac{a_n x^n - a_n(tx)^n}{1-t}\, dt$$
$$= \int_0^1 \frac{f(x) - f(tx)}{1-t}\, dt.$$

9. Rewrite R_n as
$$R_n = \sum_{k=1}^n \left(\frac{1}{k} - \int_k^{k+1} \frac{1}{x}\, dx\right) = \sum_{k=1}^n \int_0^1 \frac{t}{k(k+t)}\, dt.$$
Thus
$$\gamma - R_n = \sum_{k=n+1}^\infty \int_0^1 \frac{t}{k(k+t)}\, dt$$
$$= \sum_{k=n+1}^\infty \int_0^1 t\left(\frac{1}{k(k+t)} - \frac{1}{k(k+1)}\right) dt + \int_0^1 t\, dt \sum_{k=n+1}^\infty \frac{1}{k(k+1)}$$
$$= \sum_{k=n+1}^\infty \int_0^1 \frac{t(1-t)}{k(k+1)(k+t)}\, dt + \frac{1}{n+1} \int_0^1 t\, dt.$$
Define
$$R_1(n) = \sum_{k=n+1}^\infty \int_0^1 \frac{t(1-t)}{k(k+1)(k+t)}\, dt, \quad a_1 = \int_0^1 t\, dt.$$
Then
$$\gamma - R_n = R_1(n) + \frac{a_1}{n+1}.$$
Repeating this process, we have
$$R_1(n) = \sum_{k=n+1}^\infty \int_0^1 \frac{t(1-t)(2-t)}{k(k+1)(k+2)(k+t)}\, dt + \frac{1}{(n+1)(n+2)} \int_0^1 t(1-t)\, dt.$$
Let the sum be $R_2(n)$ and $a_2 = 1/2 \int_0^1 t(1-t)\, dt$. Then
$$\gamma - R_n = R_2(n) + \frac{a_1}{n+1} + \frac{a_2}{(n+1)(n+2)}.$$
By induction, for $N \geq 2$, we have
$$\gamma - R_n = \sum_{k=1}^N \frac{a_k}{(n+1)(n+2)\cdots(n+k)} + R_N(n),$$
where
$$R_N(n) = \sum_{k+1}^\infty \int_0^1 \frac{t(1-t)\cdots(N-t)}{k(k+1)\cdots(k+N)(k+t)}\, dt;$$

$$a_k = \frac{1}{k} \int_0^1 t(1-t)\cdots(k-1-t)\,dt.$$

The required answer follows from two estimates:

$$\frac{1}{6k}(k-2)! \le a_k \le \frac{1}{6k}(k-1)!;$$

$$\frac{1}{6(n+2)k(k+1)} \le \binom{n+k}{k} R_k(n) \le \frac{1}{6n(k+1)}.$$

13. Note that $n! = \int_0^\infty x^n e^{-x}\,dx$. The substitution of $x = \sqrt{n}t + n$ yields

$$n! = n^n \sqrt{n}\, e^{-n} \int_{-\sqrt{n}}^\infty \left(1 + \frac{t}{\sqrt{n}}\right)^n e^{-\sqrt{n}t}\,dt.$$

Let the integrand be $f_n(t)$. Note that

$$f_n(t) = \left(1 + \frac{t}{\sqrt{n}}\right)^n e^{-\sqrt{n}t} = e^{-\sqrt{n}t + n\ln(1+t/\sqrt{n})} = e^{-t^2/2 + O(1/\sqrt{n})}.$$

Hence, the expected limit follows from the *dominated convergence theorem*

$$\lim_{n\to\infty} \int_{-\sqrt{n}}^\infty f_n(t)\,dt = \int_{-\infty}^\infty \lim_{n\to\infty} f_n(t)\,dt = \int_{-\infty}^\infty e^{-t^2/2}\,dt = \sqrt{2\pi}.$$

To justify this, we show that $\int_\mathbb{R} \sup_n |f_n|\,dx$ is bounded. Indeed, for $t \ge 0$, $f_n(t)$ is decreasing in n and so $\sup_n |f_n| = (1+t)e^{-t}$, and that is integrable. For $t \le 0$, $f_n(t)$ is increasing in n, which shows that in this case the sup occurs as $n \to \infty$, and so has the value $e^{-t^2/2}$. This is also integrable.

14. Let $f(x) = 1/x$. Then $f^{(k)}(x) = (-1)^k k! x^{-(k+1)}$. The Euler-Maclaurim summation formula gives

$$H_n = \ln n + \frac{1}{2}(1 + 1/n) + \sum_{k=1}^m \frac{B_{2k}}{(2k)!}\left((-1)^{2k-1}\frac{(2k-1)!}{n^{2k}} - (-1)^{2k-1}(2k-1)!\right) + R_n(m)$$

$$= \ln n + \frac{1}{2}(1 + 1/n) + \sum_{k=1}^m \frac{B_{2k}}{2k}\left(1 - \frac{1}{n^{2k}}\right) + R_n(m).$$

Thus

$$\gamma = \lim_{n\to\infty}(H_n - \ln n) = \frac{1}{2} + \sum_{k=1}^m \frac{B_{2k}}{2k} + R_\infty(m),$$

and so

$$H_n = \ln n + \gamma + \frac{1}{2n} - \sum_{k=1}^m \frac{B_{2k}}{2k}\frac{1}{n^{2k}} + R_n(m) - R_\infty(m).$$

This implies the desired formula.

Chapter 7

2. The limit is 2. For $n \ge 3$ we have

$$\frac{n}{2^n}\sum_{k=1}^n \frac{2^k}{k} = \frac{n}{2^n}\sum_{k=0}^{n-1}\frac{2^{n-k}}{n-k}$$

$$= \sum_{k=0}^{n-1}\frac{1}{2^k}\frac{n}{n-k} \quad \left(\text{using } \frac{n}{n-k} = 1 + \frac{k}{n-k}\right)$$

$$= \sum_{k=0}^{n-1}\frac{1}{2^k} + \sum_{k=1}^{n-1}\frac{k}{2^k(n-k)}.$$

Since

$$\sum_{k=1}^{n-1} \frac{k}{2^k(n-k)} \le \frac{1}{2(n-1)} + \sum_{k=2}^{n-1} \frac{k}{(k^2-k)(n-k)}$$

$$= \frac{1}{2(n-1)} + \frac{2}{n-1} \sum_{k=1}^{n-2} \frac{1}{k}$$

$$\le \frac{1}{2(n-1)} + \frac{2}{n-1} (\ln n + \gamma + O(1/n)) \to 0, \text{ as } n \to \infty$$

we obtain

$$L = \lim_{n \to \infty} \sum_{k=0}^{n-1} \frac{1}{2^k} = 2.$$

4. Consider the sequence $\{a_n\}$ defined by

$$\frac{1}{2n + a_n} = |A_n| = \frac{1}{n+1} - \frac{1}{n+2} + \frac{1}{n+3} - \cdots,$$

or $a_n = |A_n|^{-1} - 2n$. For every $n \ge 1$, we have

1. $1 + \frac{1}{\sqrt{(n+1)^2+1}+n+1} < a_n < 1 + \frac{1}{\sqrt{n^2+1}+n}$;
2. a_n is strictly decreasing and converges to 1;
3. The best constants are $a = a_1 = 1/(1 - \ln 2) - 2 = 1.258891\ldots$ and $b = 1$.

5. The best possible constants are $\alpha = 11/6$ and $\beta = (4-e)/(e-2)$.

Chapter 8

1. By using Gauss's multiplication formula and $\Gamma(1) = 1$, we have

$$\frac{1}{n} \sum_{i=1}^{n} \ln \Gamma\left(\frac{i}{n}\right) = \frac{1}{n} \ln \left(\prod_{i=1}^{n-1} \Gamma\left(\frac{i}{n}\right)\right)$$

$$= \frac{1}{n} \ln \left((2\pi)^{(n-1)/2} n^{-1/2}\right)$$

$$= \frac{1}{n} \left(\frac{n-1}{2} \ln(2\pi) - \frac{1}{2} \ln n\right).$$

Letting $n \to \infty$ yields the desired answer.

Remark. If $\int_0^{\pi/2} \ln \sin x \, dx = -\pi \ln 2/2$ has been proved, a short proof can be derived by applying the reflection formula.

2. Using $\Gamma(x+1) = x\Gamma(x)$ and Leibniz's rule, we have

$$\frac{d}{dx} \int_x^{x+1} \ln \Gamma(t) dt = \ln \Gamma(x+1) - \ln \Gamma(x) = \ln x.$$

It is clear that $(x \ln x - x)' = \ln x$. Thus

$$\int_x^{x+1} \ln \Gamma(t) dt = x \ln x - x + C,$$

where C is an arbitrary constant. Now setting $x = 0$ yields $C = \frac{1}{2} \ln(2\pi)$.

3. Notice that
$$\frac{\pi/2}{\text{agm}(1, 1/\sqrt{2})} = \int_0^{\pi/2} \frac{d\theta}{\sqrt{1 - \sin^2\theta/2}}.$$

Substitution of $x = \sin^2\theta/(2 - \sin^2\theta)$ yields
$$\frac{\pi/2}{\text{agm}(1, 1/\sqrt{2})} = \sqrt{2} \int_0^1 \frac{dx}{\sqrt{1-x^4}}.$$

Next, substitution of $u = x^4$ gives
$$\frac{\pi/2}{\text{agm}(1, 1/\sqrt{2})} = \frac{\sqrt{2}}{4} \int_0^1 u^{1/4-1}(1-u)^{1/2-1}\, du.$$

Appealing to the beta function, we have
$$\frac{\pi/2}{\text{agm}(1, 1/\sqrt{2})} = \frac{\sqrt{2}}{4} B(1/4, 1/2).$$

Finally, since $\Gamma(1/2) = \sqrt{\pi}$ and
$$B(1/4, 1/2) = \frac{\Gamma(1/4)\Gamma(1/2)}{\Gamma(3/4)} = \frac{\Gamma^2(1/4)\Gamma(1/2)}{\pi/\sin(\pi/4)} = \frac{\Gamma^2(1/4)\sqrt{2}}{2\sqrt{\pi}},$$

we obtain the proposed answer.

6. By the reflection formula, we have
$$\frac{\pi}{\sin \pi x} = \int_0^\infty \frac{t^{x-1}}{1+t}\, dt = \int_0^1 \frac{t^{x-1}}{1+t}\, dt + \int_1^\infty \frac{t^{x-1}}{1+t}\, dt$$
$$= \int_0^1 \frac{t^{x-1} + t^{-x}}{1+t}\, dt = \int_0^1 (t^{x-1} + t^{-x}) \left[\sum_{k=0}^n (-1)^k t^k + \frac{(-1)^{n+1} t^{n+1}}{1+t} \right] dt$$
$$= \sum_{k=-n}^n \frac{(-1)^k}{x-k} + R_n,$$

where
$$|R_n| \leq \left| \int_0^1 (t^{n+x} + t^{n-x+1})\, dt \right| \leq \frac{1}{n+x+1} + \frac{1}{n-x+2}.$$

11. Take the logarithmic derivative of the Weierstrass product formula.

Chapter 9

2. Let the required sum be S. Appealing to Gauss's pairing trick, we have
$$2S = \sum_{k=0}^n \binom{n}{k} \cos kx \sin(n-k)x + \sum_{k=0}^n \binom{n}{n-k} \cos(n-k)x \sin kx$$
$$= \sum_{k=0}^n \binom{n}{k} (\cos kx \sin(n-k)x + \cos(n-k)x \sin kx)$$
$$= \sum_{k=0}^n \binom{n}{k} \sin nx = 2^n \sin nx.$$

Therefore $S = 2^{n-1} \sin nx$.

Solutions to Selected Exercises

3. Using $\tan\alpha - \tan\beta = \sin(\alpha-\beta)/(\cos\alpha\cos\beta)$.

4. Let $a_n = 4n\left(\frac{1\cdot 3\cdots(2n-1)}{2\cdot 4\cdots 2n}\right)^2$. Then
$$a_{n+1} - a_n = \frac{1}{n+1}\left(\frac{1\cdot 3\cdots(2n-1)}{2\cdot 4\cdots 2n}\right)^2.$$
Appealing to Wallis's formula, telescoping yields that $\lim_{n\to\infty} a_n = 4/\pi$.

7. Inverting the order of summation, we find that
$$\sum_{k=2}^{\infty}(\zeta(k)-1) = \sum_{k=2}^{\infty}\left(\sum_{n=2}^{\infty}\frac{1}{n^k}\right) = \sum_{n=2}^{\infty}\left(\sum_{k=2}^{\infty}\frac{1}{n^k}\right)$$
$$= \sum_{n=2}^{\infty}\frac{1/n^2}{1-1/n} \quad \text{(the inner sum is a geometric series)}$$
$$= \sum_{n=2}^{\infty}\frac{1}{n(n-1)} = \sum_{n=2}^{\infty}\left(\frac{1}{n-1}-\frac{1}{n}\right) = 1.$$

9. (b). Let $f(x) = ax(x-1)$. The identity follows from (9.22) directly.

(c). Notice that
$$\sum_{k=1}^{\infty}(-1)^{k+1}\arctan\left(\frac{2}{k^2}\right) = \sum_{k=0}^{\infty}\arctan\left(\frac{2}{(2k+1)^2}\right) - \sum_{k=1}^{\infty}\arctan\left(\frac{2}{(2k)^2}\right).$$
We have
$$\sum_{k=0}^{\infty}\arctan\left(\frac{2}{(2k+1)^2}\right) = \sum_{k=0}^{\infty}(\arctan(2k+2)-\arctan(2k)) = \frac{\pi}{2},$$
$$\sum_{k=1}^{\infty}\arctan\left(\frac{2}{(2k)^2}\right) = \sum_{k=1}^{\infty}(\arctan(2k+1)-\arctan(2k-1)) = \frac{\pi}{4},$$
from which the required identity follows.

15. If the leading coefficient of $Q(x)$ is 1, by using the identity
$$\arctan x = \frac{1}{2i}\ln\frac{1+ix}{1-ix},$$
we have
$$\sum_{k=1}^{\infty}\arctan\frac{P(k)}{Q(k)} = \frac{1}{2i}\sum_{k=1}^{\infty}\ln\frac{Q(k)+iP(k)}{Q(k)-iP(k)} = \frac{1}{2i}\ln\prod_{k=1}^{\infty}\frac{Q(k)+iP(k)}{Q(k)-iP(k)}.$$
Let a_j ($1 \le j \le n$) be the roots of $Q(x)-iP(x)$. Since $P(x)$ and $Q(x)$ have real coefficients, the roots of $Q(x)+iP(x)$ are complex conjugates of the a_j. Appealing to Exercise 3 of Chapter 8, we obtain
$$\sum_{k=1}^{\infty}\arctan\frac{P(k)}{Q(k)} = \frac{1}{2i}\ln\prod_{k=1}^{\infty}\frac{(n-\bar{a}_1)(n-\bar{a}_2)\cdots(n-\bar{a}_n)}{(n-a_1)(n-a_2)\cdots(n-a_n)}$$
$$= \frac{1}{2i}\ln\prod_{k=1}^{\infty}\frac{\Gamma(1-a_1)\Gamma(1-a_2)\cdots\Gamma(1-a_n)}{\Gamma(1-\bar{a}_1)\Gamma(1-\bar{a}_2)\cdots\Gamma(1-\bar{a}_n)}$$
$$= \sum_{j=1}^{n}\arg(\Gamma(1-a_j)) \quad (\text{mod } 2\pi).$$

For example, let $P(x) = 2$ and $Q(x) = x^2$. Thus a_j are the roots of $x^2 - 2i$, which are $1+i$ and $-1-i$. Therefore

$$\sum_{k=1}^{\infty} \arctan \frac{2}{k^2} = \arg(\Gamma(-i)) + \arg(\Gamma(2+i)) \quad \text{(using } \Gamma(1+x) = x\Gamma(x))$$
$$= \arg(\Gamma(-i)) + \arg(1+i) + \arg(i) + \arg(\Gamma(i))$$
$$= \arg(1+i) + \arg(i) = \frac{3\pi}{4}.$$

The sum of the first few terms shows that no modification by a multiple of 2π is needed.

Chapter 10

3. Integrating the binomial theorem

$$\sum_{k=0}^{n} \binom{n}{k} x^k = (1+x)^n$$

and then manipulating it yields

$$\sum_{k=0}^{n} \binom{n}{k} \frac{x^k}{k+1} = \frac{(x+1)^{n+1} - 1}{(n+1)x}.$$

By the multisection formula we have

$$\sum_{k=0}^{\lfloor n/2 \rfloor} \binom{n}{2k} \frac{x^{2k}}{2k+1} = \frac{(1+x)^{n+1} - (1-x)^{n+1}}{2(n+1)x}.$$

4. By the multisection formula we have

$$\sum_{j=0}^{n} \binom{2n}{2j} x^{2j} = \frac{1}{2}\left[(1+x)^{2n} + (1-x)^{2n}\right].$$

Taking $x = \sqrt{ai}$ yields

$$\sum_{j=0}^{n} \binom{2n}{2j} (-a)^j = \frac{1}{2}\left[(1+\sqrt{ai})^{2n} + (1-\sqrt{ai})^{2n}\right].$$

Let $\theta = \arctan(\sqrt{a})$. Rewrite

$$1 \pm \sqrt{ai} = \sqrt{1+a}\, e^{\pm i\theta}.$$

We have

$$\sum_{j=0}^{n} \binom{2n}{2j} (-a)^j = \frac{1}{2}(1+a)^n (e^{i2n\theta} + e^{-i2n\theta}) = (1+a)^n \cos(2n\theta).$$

5. By the multisection formula with $k=2$ and $m=0$,

$$\sum_{n=1}^{\infty} F_{2n} x^{2n} = \frac{1}{2}\left(\frac{x}{1-x-x^2} - \frac{x}{1+x-x^2}\right) = \frac{x^2}{1 - 3x^2 + x^4}.$$

Solutions to Selected Exercises

Replacing x^2 by x yields

$$\sum_{n=1}^{\infty} F_{2n} x^n = \frac{x}{1 - 3x + x^2},$$

which is the generating function of $\{F_{2n}\}$. Similarly, with $k = 2$ and $m = 1$, the multisection formula gives

$$\sum_{n=1}^{\infty} F_{2n+1} x^{2n+1} = \frac{1}{2}\left(\frac{x}{1-x-x^2} + \frac{x}{1+x-x^2}\right) = \frac{x(1-x^2)}{1-3x^2+x^4}.$$

Dividing by x and then replacing x^2 by x yields the generating function of $\{F_{2n+1}\}$ as

$$\sum_{n=1}^{\infty} F_{2n+1} x^n = \frac{1-x}{1-3x+x^2}.$$

7. For $|x| < 1$, start with

$$\sum_{n=1}^{\infty} \frac{x^n}{n} = -\ln(1-x).$$

Multiplying both sides by x and then integrating gives

$$\sum_{n=1}^{\infty} \frac{x^{n+2}}{n(n+2)} = \frac{1}{4}(2x + x^2 + 2\ln(1-x) - 2x^2 \ln(1-x))$$

or

$$\sum_{n=1}^{\infty} \frac{x^n}{n(n+2)} = \frac{2x + x^2 + 2\ln(1-x) - 2x^2 \ln(1-x)}{4x^2}.$$

Now, using the multisection formula yields

$$\sum_{n=1}^{\infty} \frac{x^{4n+1}}{(4n+1)(4n+3)} = \frac{1}{8}\left[\left(\frac{1}{x^2} - 1\right) \ln \frac{1-x}{1+x} + 2\left(1 + \frac{1}{x^2}\right) \arctan x\right].$$

Letting $x = 1$, we get a byproduct

$$\sum_{n=1}^{\infty} \frac{1}{(4n+1)(4n+3)} = \frac{\pi}{8}.$$

Chapter 11

2. Let $G(x) = \sum_{n=0}^{\infty} u_n x^n$. Then

$$G(x) = a + bx + \sum_{n=2}^{\infty} u_n x^n$$

$$= a + bx + p \sum_{n=2}^{\infty} u_{n-1} x^n - q \sum_{n=2}^{\infty} u_{n-2} x^n$$

$$= a + bx + px \sum_{n=1}^{\infty} u_n x^n - qx^2 \sum_{n=0}^{\infty} u_n x^n$$

$$= a + bx + px(G(x) - a) - qx^2 G(x).$$

Solving for $G(x)$ yields

$$G(x) = \frac{a + (b - ap)x}{1 - px + qx^2}.$$

7. We begin to derive a closed form for $b_n := na_n$. By the assumptions, we have

$$b_1 = 1, \; b_2 = 2, \text{ and } b_n = n - \sum_{k=1}^{n-1}(n-k-1)b_k \quad \text{for } n > 2.$$

Let $G(x) = \sum_{n=1}^{\infty} b_n x^n$. Then

$$G(x) = \sum_{n=1}^{\infty} \left(n - \sum_{k=1}^{n-1}(n-k-1)b_k \right) x^n$$

$$= \sum_{n=1}^{\infty} nx^n - \sum_{n=1}^{\infty} \left(\sum_{k=1}^{n-1}(n-k-1)b_k \right) x^n$$

$$= \frac{x}{(1-x)^2} - \frac{x^2}{(1-x)^2} G(x),$$

where the Cauchy product has been used in the second sum. Therefore,

$$G(x) = \frac{x}{x^2 + (1-x)^2}.$$

To determine the coefficient of x^n in $G(x)$, we factor

$$x^2 + (1-x)^2 = 2(x-\alpha)(x-\beta),$$

where $\alpha = (1+i)/2, \beta = (1-i)/2$. Appealing to partial fractions and geometric series expansion, we have

$$G(x) = \frac{1}{2i}\left(\frac{\alpha}{x-\alpha} - \frac{\beta}{x-\beta}\right) = \frac{1}{2i}\sum_{n=1}^{\infty}[(1+i)^n - (1-i)^n]x^n.$$

Thus $b_n = [(1+i)^n - (1-i)^n]/2i$. Now, by using $1 \pm i = \sqrt{2}e^{\pm \pi i/4}$, we obtain

$$a_n = \frac{b_n}{n} = \frac{2^{n/2}}{n} \cdot \frac{e^{n\pi i/4} - e^{-n\pi i/4}}{2i} = \frac{2^{n/2}\sin(n\pi/4)}{n}.$$

Remark. It is interesting to see that

$$\sum_{n=1}^{\infty} \frac{a_n}{2^n} = \sum_{n=1}^{\infty} \frac{\sin(n\pi/4)}{2^{n/2}\,n} = \frac{\pi}{4}.$$

8. Let

$$S_1 = \sum_{k=1}^{n} F_k \quad \text{and} \quad S_2 = \sum_{k=1}^{n} F_k^2.$$

Appealing to

$$\sum_{k=1}^{n} F_k = F_{n+2} - 1 \quad \text{and} \quad \sum_{k=1}^{n} F_k^2 = F_n F_{n+1},$$

we have

$$\sum_{k=1}^{n} \frac{F_n F_{n+1} - F_k^2}{F_{n+2} - F_k - 1} = \sum_{k=1}^{n} \frac{S_2 - F_k^2}{S_1 - F_k}.$$

Using the power mean inequality

$$a_1^2 + a_2^2 + \cdots + a_m^2 \geq \frac{(a_1 + a_2 + \cdots + a_m)^2}{m}$$

Solutions to Selected Exercises

yields
$$S_2 - F_k^2 \geq \frac{(S_1 - F_k)^2}{n-1}.$$
Therefore
$$\sum_{k=1}^{n} \frac{S_2 - F_k^2}{S_1 - F_k} \geq \sum_{k=1}^{n} \frac{S_1 - F_k}{n-1} = S_1 = F_{n+2} - 1.$$

Chapter 12

4. Recall that Fibonacci polynomials are defined by
$$f_n(x) = xf_{n-1}(x) + f_{n-2}(x) \quad (n \geq 2)$$
with $f_1(x) = 1$ and $f_2(x) = x$. Similar to Binet's formula, we have
$$f_n(x) = \frac{\alpha^n(x) - \beta^n(x)}{\alpha(x) - \beta(x)}$$
where $\alpha(x), \beta(x) = (x \pm \sqrt{x^2+4})/2$. Let $x = 2i\cos z$. Then $\alpha(x) = ie^{-iz}, \beta = ie^{iz}$. Therefore,
$$f_n(x) = i^{n-1} \frac{e^{inz} - e^{-inz}}{e^{iz} - e^{-iz}} = i^{n-1} \frac{\sin nz}{\sin z}. \qquad (*)$$
Appealing to (5.15), we obtain
$$f_n(x) = (2i)^{n-1} \prod_{k=1}^{n-1} \left(\cos z - \cos\frac{k\pi}{n}\right).$$
Since $\cos z = x/(2i)$, we get
$$f_n(x) = \prod_{k=1}^{n-1} \left(x - 2i\cos\frac{k\pi}{n}\right),$$
and so
$$F_n = f_n(1) = \prod_{k=1}^{n-1} \left(1 - 2i\cos\frac{k\pi}{n}\right).$$
Regrouping the factors above yields the proposed factorization without the complex number i. The third equality follows from $*$ by setting $x = 1$.

9. Let
$$\alpha, \beta = \frac{a \pm \sqrt{a^2+4b}}{2}.$$
First, by using induction on k, we prove that
$$(u_k x + bu_{k-1})^i (u_{k+1}x + bu_k)^{n-i}$$
$$= \sum_{i_1,i_2,\ldots,i_k} \binom{n-i}{i_1}\binom{n-i_1}{i_2}\cdots\binom{n-i_{k-1}}{i_k} a^{kn-i-2i_1-\cdots-2i_{k-1}-i_k} b^{i_1+\cdots+i_k} x^{n-i_k}.$$
Notice that
$$x^i(ax+b)^{n-i} = \sum_{0 \leq j \leq n-i} \binom{n-i}{j} a^{n-i-j} b^j x^{n-j},$$
which implies the proposed result is true for $k=1$. Now suppose the equation is true for some positive integer k. Substituting $a + bx^{-1}$ for x and multiplying by x^n, we see the left side of the equation becomes
$$(au_k + bu_{k-1}x + bu_k)^i (au_{k+1} + bu_k x + bu_{k+1})^{n-i},$$

which is equivalent to
$$(u_{k+1}x + bu_k)^i (u_{k+2}x + bu_{k+1})^{n-i}.$$

Expanding the right side of the equation and simplifying, we obtain

$$\sum_{i_1,i_2,\ldots,i_{k+1}} \binom{n-i}{i_1}\binom{n-i_1}{i_2}\cdots\binom{n-i_k}{i_{k+1}} a^{(k+1)n-i-2i_1-\cdots-2i_k-i_{k+1}} b^{i_1+\cdots+i_{k+1}} x^{n-i_{k+1}}$$

as desired. Next, multiplying both sides of the equation by x^i and summing over i, we have

$$\sum_{i=0}^n (u_k x + bu_{k-1})^i (u_{k+1}x + bu_k)^{n-i} x^i$$

$$= \sum_{i,i_1,i_2,\ldots,i_k} \binom{n-i}{i_1}\binom{n-i_1}{i_2}\cdots\binom{n-i_{k-1}}{i_k} a^{kn-i-2i_1-\cdots-2i_{k-1}-i_k} b^{i_1+\cdots+i_k} x^{n+i-i_k}.$$

The coefficient of x^n on the right side of the equation is $\operatorname{tr}(A_{n+1}^k)$, while the coefficient of x^n on the left side of the equation is

$$\sum_{i+s+t=n} \binom{i}{s}\binom{n-i}{t} u_k^s (bu_{k-1})^{i-s} u_{k+1}^t (bu_k)^{n-i-t}$$

$$= \sum_{i+s\leq n} \binom{i}{s}\binom{n-i}{s} (bu_{k-1})^{i-s} u_k^s u_{k+1}^{n-i-s} (bu_k)^s.$$

Let a_n be the right side above. Then

$$\sum_{n=0}^\infty a_n x^n = \sum_{i,s=0}^\infty \binom{i}{s} b^i u_{k-1}^{i-s} u_k^{2s} x^{i+s} \sum_{i+s=n}^\infty \binom{n-i}{s} (u_{k+1}x)^{n-i-s}$$

$$= \sum_{i,s=0}^\infty \binom{i}{s} b^i u_{k-1}^{i-s} u_k^{2s} x^{i+s} (1-u_{k+1}x)^{-s-1}$$

$$= \sum_{s=0}^\infty b^s u_k^{2s} x^{2s} (1-u_{k+1}x)^{-s-1} \sum_{i\geq s} \binom{i}{s} (bu_{k-1}x)^{i-s}$$

$$= \sum_{s=0}^\infty b^s u_k^{2s} x^{2s} (1-u_{k+1}x)^{-s-1} (1-bu_{k-1}x)^{-s-1}$$

$$= \frac{1}{1-(u_{k+1}+bu_{k-1})x + (bu_{k+1}u_{k-1}-bu_k^2)x^2}.$$

Appealing to

$$u_{k+1} + bu_{k-1} = \alpha^k + \beta^k, \quad bu_{k+1}u_{k-1} - bu_k^2 = \alpha^k \beta^k,$$

we find that

$$\frac{1}{1-(u_{k+1}+bu_{k-1})x + (bu_{k+1}u_{k-1}-bu_k^2)x^2} = \frac{1}{\alpha^k-\beta^k}\left(\frac{\alpha^k}{1-\alpha^k x} - \frac{\beta^k}{1-\beta^k x}\right).$$

Thus,
$$\operatorname{tr}(A_{k+1}^k) = a_n = u_{kn+k}/u_k.$$

Finally, let the eigenvalues of A_{n+1} be $\lambda_0, \lambda_1, \ldots, \lambda_n$ and let the characteristic polynomial be

$$P(x) = \det(xI - A_{n+1}).$$

Then

$$\frac{P'(x)}{P(x)} = \sum_{j=0}^{n} \frac{1}{x - \lambda_j} = \sum_{k=0}^{\infty} x^{-k-1} \sum_{j=0}^{n} \lambda_j^k$$

$$= \sum_{k=0}^{\infty} x^{-k-1} \operatorname{tr}(A_{n+1}^k) = \sum_{k=0}^{\infty} x^{-k-1} \frac{\alpha^{nk+k} - \beta^{nk+k}}{\alpha^k - \beta^k}$$

$$= \sum_{k=0}^{\infty} x^{-k-1} \sum_{j=0}^{n} \alpha^{jk} \beta^{(n-j)k} = \sum_{j=0}^{n} \frac{1}{x - \alpha^j \beta^{n-j}}.$$

Therefore

$$P(x) = \prod_{j=0}^{n} (x - \alpha^j \beta^{n-j}),$$

which implies that the eigenvalues of A_{n+1} are

$$\alpha^n, \alpha^{n-1}\beta, \ldots, \alpha\beta^{n-1}, \beta^n.$$

Remark. This solution contains kernels of sophisticated ideas used to study the sequences defined by the second-order recursive relations.

Chapter 13

7. Let

$$D(x_1, x_2, \ldots, x_n) = \det_{1 \le i,j \le n} (x_i^{j-1}).$$

If $x_{i_1} = x_{i_2}$ with $i_1 \ne i_2$, then $D = 0$. Thus,

$$(x_{i_1} - x_{i_2}) \mid D.$$

Repeating this process yields

$$\prod_{1 \le i < j \le n} (x_j - x_i) \mid D.$$

On the other hand, let $(\sigma_1, \sigma_2, \ldots, \sigma_n)$ be a permutation of $(0, 1, \ldots, n-1)$. By the definition of the determinant,

$$D = \sum (-1)^{\sigma_1 + \cdots + \sigma_n} x_1^{\sigma_1} x_2^{\sigma_2} \cdots x_n^{\sigma_n}$$

and $\sigma_1 + \sigma_2 + \cdots + \sigma_n = 1 + 2 + \cdots + (n-1) = n(n-1)/2$. Then D is a polynomial in the x_i's of degree $n(n-1)/2$ and so

$$D = c \prod_{1 \le i < j \le n} (x_j - x_i)$$

for some constant c. To determine the constant, comparing coefficients of $x_1^0 x_2^1 \cdots x_n^{n-1}$ on both sides of the proposed identity yields $c = 1$ as desired.

Remark. The essence of the proof above consists of

- identifying the factors
- determining the degree
- computing the multiplicative constant

Along the same lines, we have

$$\det_{1\leq i,j\leq n} (p_j(x_i)) = a_1 a_2 \cdots a_n \prod_{1\leq i<j\leq n} (x_j - x_i).$$

Remark. Krattenthaler in 1990 published the following extension of the Vandermonde formula: given arbitrary values for $x_1, \ldots, x_n, a_2, \ldots, a_n$, and $b_2, \ldots, b_n$, let $a_{ij} = (x_i + a_n) \cdots (x_i + a_{j+1})(x_i + b_j) \cdots (x_i + b_2)$. Then

$$\det_{1\leq i,j\leq n} (a_{ij}) = \prod_{1\leq i<j\leq n} (x_i - x_j) \prod_{2\leq i\leq j\leq n} (b_i - a_j).$$

11. (a). If $f(x)g(x) = 1$, then the Cauchy product yields

$$\sum_{k=0}^{n} a_k b_{n-k} = \begin{cases} 1, & \text{if } n = 0, \\ 0, & \text{if } n > 0. \end{cases}$$

This can be viewed as a system of linear equations in $n+1$ unknowns $b_0, b_1, \ldots, b_n$. By Cramer's Rule, we have $b_n = (-1)^n D(a_0, a_1, \ldots, a_n)/a_0^{n+1}$.

(b). Notice that

$$\sum_{n=0}^{\infty} \frac{(-1)^n x^{2n}}{(2n+1)!} = \frac{1}{x} \sum_{n=0}^{\infty} \frac{(-1)^n x^{2n+1}}{(2n+1)!} = \frac{\sin x}{x}.$$

Appealing to the result of (a), we have

$$D(x^2) = \sum_{n=0}^{\infty} (-1)^n D_n(a_0, a_1, \ldots, a_n)(-x^2)^n = \frac{x}{\sin x}.$$

Moreover, since

$$\sum_{n=0}^{\infty} \frac{x^{2n}}{(2n+1)!} = \frac{\sinh x}{x} = \frac{e^x - e^{-x}}{2x},$$

we get

$$D(-x^2) = \sum_{n=0}^{\infty} (-1)^n D_n(a_0, a_1, \ldots, a_n) x^{2n} = \frac{2x}{e^x - e^{-x}} = \frac{2xe^x}{e^{2x} - 1}.$$

Chapter 14

1. Setting $z = be^{i\theta}/a$ in (5.16) and then multiplying by a^n on both sides yields

$$a^n - b^n e^{in\theta} = \prod_{k=0}^{n-1} \left(a - be^{i(\theta + 2k\pi/n)}\right).$$

Taking modulus on both sides gives

$$a^{2n} - 2a^n b^n \cos n\theta + b^{2n} = \prod_{k=0}^{n-1} (a^2 - 2ab\cos(\theta + 2k\pi/n) + b^2).$$

Dividing this equation by $a^n b^n$ and regrouping gives

$$\frac{a^{2n} + b^{2n}}{a^n b^n} - 2\cos n\theta = \prod_{k=0}^{n-1} \left(\frac{a^2 + b^2}{ab} - 2\cos(\theta + 2k\pi/n)\right).$$

Solutions to Selected Exercises

Let $a/b = e^{ix}$. Then $b/a = e^{-ix}$,

$$\frac{a^2 + b^2}{ab} = 2\cos x \quad \text{and} \quad \frac{a^{2n} + b^{2n}}{a^n b^n} = 2\cos nx.$$

Therefore,

$$\cos nx - \cos n\theta = 2^{n-1} \prod_{k=0}^{n-1} (\cos x - \cos(\theta + 2k\pi/n)).$$

Now, the proposed identity follows by taking the logarithmic derivatives with respect to x.

5. Let $\omega = e^{2\pi i/n}$. Then

$$\sum_{k=0}^{n-1} \sin^{2a}\left(\frac{bk\pi}{n} + x\right) = \frac{1}{(2i)^{2a}} \sum_{k=0}^{n-1} (\omega^{bk/2} e^{ix} - \omega^{-bk/2} e^{-ix})^{2a}$$

$$= \frac{(-1)^a}{2^{2a}} \sum_{k=0}^{n-1} \sum_{j=0}^{2a} \binom{2a}{j} \omega^{(2a-j)bk/2} e^{(2a-j)ix} (-1)^j \omega^{-jbk/2} e^{-jix}$$

$$= \frac{(-1)^a}{2^{2a}} \sum_{j=0}^{2a} \binom{2a}{j} e^{2ix(a-j)} \sum_{k=0}^{n-1} \omega^{bk(a-j)}.$$

Recall that

$$\sum_{k=0}^{n-1} \omega^{bk(a-j)} = \begin{cases} n, & \text{if } n|b(a-j), \\ 0, & \text{otherwise} \end{cases}$$

Since $0 < ab < n$, we see that $n|b(a-j)$ if only if $j = a$, where the above sum becomes

$$\sum_{k=0}^{n-1} \sin^{2a}\left(\frac{bk\pi}{n} + x\right) = \frac{n}{2^{2a}} \binom{2a}{j}.$$

The cosine identity follows from this by replacing x by $x + \pi/2$.

7. Appealing to (5.15)

$$\frac{\sin n\theta}{\sin \theta} = 2^{n-1} \prod_{k=1}^{n-1} \left(\cos\theta - \cos\left(\frac{k\pi}{n}\right)\right).$$

Then

$$\sin\theta \sin n\theta = 2^{n-1} \sin^2\theta \prod_{k=1}^{n-1}\left(\cos\theta - \cos\left(\frac{k\pi}{n}\right)\right) = -2^{n-1} \prod_{k=0}^{n}\left(\cos\theta - \cos\left(\frac{k\pi}{n}\right)\right).$$

Let $\theta_k = k\pi/n$, $(k = 0, 1, \ldots, n)$. By using the partial fraction, we have

$$\frac{n}{\sin\theta \sin n\theta} = \sum_{k=0}^{n} \frac{a_k}{\cos\theta - \cos\theta_k}, \tag{1}$$

where the constants a_k are determined by

$$a_k = n \lim_{\theta \to \theta_k} \frac{\cos\theta - \cos\theta_k}{\sin\theta \sin n\theta} = \begin{cases} (-1)^{k+1}, & k = 1, 2, \ldots, n-1, \\ (-1)^{k+1}/2, & k = 0, n. \end{cases}$$

If n is even, replacing θ by $\pi - \theta$ in (1) gives

$$\frac{n}{\sin\theta \sin n\theta} = \sum_{k=0}^{n} \frac{a_k}{\cos\theta + \cos x_\theta}. \tag{2}$$

Now, adding (1) and (2) we arrive at the desired answer (a). If n is odd, replacing θ by $\pi - \theta$ in (1) gives

$$\frac{n}{\sin\theta \sin n\theta} = -\sum_{k=0}^{n} \frac{a_k}{\cos\theta + \cos\theta_k}. \tag{3}$$

The desired answer (b) follows from adding (1) and (3).

8. Let

$$S_{2p+1} = \sum_{k=0}^{n-1} (-1)^k \sec^{2p+1}\left(\frac{k\pi}{n}\right)$$

and its generating function be

$$\mathcal{G}(x) = \sum_{p=0}^{\infty} S_{2p+1} x^{2p+1}.$$

Appealing to Exercise 6(b),

$$\frac{n\cos\theta}{\sin\theta \sin n\theta} = \sum_{p=0}^{\infty} \sum_{k=0}^{n-1} (-1)^k \frac{\cos^{1+2p}\theta}{\cos^{1+2p}(k\pi/n)},$$

With $x = \cos\theta$, this shows that

$$\mathcal{G}(x) = \frac{nx \sin(n\pi/2)}{\sqrt{1-x^2}} \sec(n \arcsin x).$$

Remark. If n is even, by using Exercise 6(b), we can find the generating function of $\sum_{k=0, k\neq n/2}^{n-1} (-1)^k \sec^{2p}\left(\frac{k\pi}{n}\right)$ as

$$(-1)^{n/2}\left(1 - \frac{nx}{\sqrt{1-x^2}} \csc(n \arcsin x)\right).$$

9. Recall that the generating function in this case is

$$G(x) = \frac{nx}{\sqrt{1-x^2}} \tan(n \arcsin x).$$

Since

$$\frac{1 - e^{2ni\theta}}{1 + e^{2ni\theta}} = \frac{(1 - e^{2ni\theta})/2}{1 - (1 - e^{2ni\theta})/2}$$

$$= \sum_{m>0} 2^{-m}(1 - e^{2ni\theta})^m$$

$$= \sum_{m>0} 2^{-m} \sum_{k=0}^{m} (-1)^k \binom{m}{k} e^{2kni\theta},$$

we obtain

$$G(x) = \frac{nxi}{\sqrt{1-x^2}} \frac{1 - e^{2ni\theta}}{1 + e^{2ni\theta}}$$

$$= n \sum_{m>0} \sum_{k=0}^{m} 2^{-m} \binom{m}{k} (-1)^k \times \sum_{j>0} (-1)^j \binom{2kn}{2j-1} x^{2j} (1-x^2)^{kn-j}.$$

Solutions to Selected Exercises

Appealing to the binomial theorem, the last summation can be expressed as

$$\sum_{p>0}(-1)^p x^{2p} \sum_{j=0}^{p}\binom{2kn}{2j-1}\binom{kn-j}{p-j}.$$

The desired answer follows from application of the *Vandermonde convolution formula*

$$\sum_{k=0}^{n}\binom{1+2x-r}{r+2k}\binom{x-k-r}{n-k} = 2^{2n+r}\binom{x+n}{2n+r}.$$

12. For $n=2$, clearly the eigenvalue of A is 1. If $n \geq 3$, let the characteristic polynomial of A be

$$P(\lambda) = \det(\lambda I - A) = \lambda^{n-1} + \sum_{k=1}^{n-1}(-1)^k c_k \lambda^{n-k-1},$$

where c_k is the sum of all the principal minors of A of order k. Now, for $k \geq 3$ and $n \geq 4$, we show that $c_k = 0$. To see this, define $B = (b_{ij})$ where

$$b_{ij} = \sin x_i \sin y_j \cos(x_i - y_j), \quad (i, j = 1, 2, 3).$$

It is easy to verify that $\det(B) = 0$ for all x_i and y_j. This implies that any 3-rowed minor of A vanishes, which clearly proves that $c_k = 0$. Thus,

$$P(\lambda) = \lambda^{n-3}\left\{\lambda^2 - \operatorname{tr} A + [(\operatorname{tr} A)^2 - \operatorname{tr} A^2]/2\right\}.$$

By using the trigonometric summation formulas, we have

$$\operatorname{tr} A = \frac{n}{2}, \quad \operatorname{tr} A^2 = \frac{10n^2}{64},$$

and so the eigenvalues of A are 0 ($n-3$ multiple), $n/8$ and $3n/8$.

Remark. A related problem is Putnam Problem 1999-B5: Let $\theta = 2\pi/n$ and $a_{ij} = \cos(i\theta + j\theta)$. Determine the eigenvalues of $A = (a_{ij})$.

Chapter 15

2. Appealing to Wallis's formula for an odd power, we have

$$\int_0^{\pi/2}\cos^{2n+1}x\,dx = \frac{(n!)^2 2^{2n}}{(2n+1)!},$$

and so

$$\sum_{n=0}^{\infty}\frac{(n!)^2 2^{n+1}}{(2n+1)!} = 2\int_0^{\pi/2}\cos x \sum_{n=0}^{\infty}\left(\frac{1}{2}\cos^2 x\right)^n dx$$

$$= \int_0^{\pi/2}\frac{\cos x}{1-\cos^2 x/2}\,dx = \pi.$$

One can also prove this identity via the gamma and beta functions:

$$\sum_{n=0}^{\infty}\frac{(n!)^2 2^{n+1}}{(2n+1)!} = \sum_{n=0}^{\infty} 2^{n+1}\Gamma(n+1)\Gamma(n+1)/\Gamma(2n+2)$$

$$= \sum_{n=0}^{\infty} 2^{n+1} B(n+1, n+1) = \sum_{n=0}^{\infty} 2^{n+1}\int_0^1 x^n(1-x)^n dx$$

$$= 2\int_0^1 \sum_{n=0}^{\infty}(2x)^n(1-x)^n dx = 2\int_0^1 [1-2x(1-x)]^{-1} dx$$

$$= 2\arctan 2(x-1/2)\big|_0^1 = \pi.$$

Remark. The reader may recognize that

$$\frac{(n!)^2 \, 2^{n+1}}{(2n+1)!} = \int_{-1}^{1} x^n P_n(x) \, dx,$$

where $P_n(x)$ is the *Legendre polynomial* of degree n. Appealing to its generating function

$$\sum_{n=0}^{\infty} P_n(x) t^n = (1 - 2xt + t^2)^{-1/2},$$

one obtains

$$\sum_{n=0}^{\infty} \frac{(n!)^2 \, 2^{n+1}}{(2n+1)!} = \int_{-1}^{1} \sum_{n=0}^{\infty} x^n P_n(x) \, dx = \int_{-1}^{1} (1-x^2)^{-1/2} \, dx = \pi.$$

5. The desired result is obtained by setting $a = 1$ in the following generalization. Let $y = e^{a \arcsin x} = \sum_{n=0}^{\infty} a_n x^n$. Then $y' = ay/\sqrt{1-x^2}$, or

$$(1 - x^2)(y')^2 = a^2 y^2.$$

Differentiating both sides with respect to x and then dividing both sides by $2y'$ yields

$$(1 - x^2) y'' - xy' - a^2 y = 0.$$

Substituting the power series into this differential equation and equating coefficients of like powers of x, we have

$$a_{n+2} = \frac{n^2 + a^2}{(n+1)(n+2)} a_n, \quad \text{for } n \geq 0.$$

Appealing to $a_0 = 1$ and $a_1 = a$, for $n \geq 1$, by a simple inductive argument, we obtain

$$a_{2n} = \frac{a^2 (a^2 + 2^2)(a^2 + 4^2) \cdots (a^2 + (2n-2)^2)}{(2n)!},$$

and

$$a_{2n+1} = \frac{a^2 (a^2 + 1^2)(a^2 + 3^2) \cdots (a^2 + (2n-1)^2)}{(2n+1)!}.$$

Setting $a = 1$ yields the proposed power series.

Remark. Consider

$$e^{a \arcsin x} = 1 + (\arcsin x) a + \frac{\arcsin^2 x}{2!} a^2 + \frac{\arcsin^3 x}{3!} a^3 + \cdots.$$

Equating coefficients of a^k on both sides gives the power series for $\arcsin^k x, k \geq 1$. For example, for $k = 4$, we obtain

$$\frac{1}{4!} \arcsin^4 x = \sum_{n=2}^{\infty} \frac{b_{2n}}{(2n)!} x^{2n},$$

where

$$b_{2n} = \sum_{k=1}^{n-1} \frac{2^2 4^2 \cdots (2n-2)^2}{(2k)^2} = 2^{2(n-1)} [(n-1)!]^2 \sum_{k=1}^{n-1} \frac{1}{(2k)^2}.$$

Solutions to Selected Exercises

6. Substituting $x = \sin(t/2)$ yields
$$S = \frac{1}{2} \int_0^{\pi/3} t^2 \cot(t/2)\, dt.$$

Integrating by parts gives
$$S = \int_0^{\pi/3} t^2 d(\ln(2\sin(t/2))\, dt = -2 \int_0^{\pi/3} t \ln(2\sin(t/2))\, dt.$$

Recall that
$$\ln(1-z) = -\sum_{n=1}^{\infty} \frac{z^n}{n}, \quad |z| < 1.$$

Setting $z = e^{it}$ and exporting the real part gives
$$\ln(2\sin(t/2)) = -\sum_{n=1}^{\infty} \frac{\cos nt}{n}, \quad 0 < t < 2\pi,$$

and so
$$S = 2 \sum_{n=1}^{\infty} \frac{1}{n} \int_0^{\pi/2} t \cos nt\, dt.$$

Integrating by parts twice leads to
$$S = \frac{2\pi}{3} \sum_{n=1}^{\infty} \frac{\sin(n\pi/3)}{n^2} + 2 \sum_{n=1}^{\infty} \frac{\cos(n\pi/3)}{n^3} - 2\zeta(3).$$

Appealing to
$$\sum_{n=1}^{\infty} \frac{\cos(n\pi/3)}{n^3} = \frac{1}{3}\zeta(3),$$

we obtain
$$\zeta(3) = \frac{\pi}{2} \sum_{n=1}^{\infty} \frac{\sin(n\pi/3)}{n^2} - \frac{3}{4} S.$$

7. Rewrite the identity as
$$e^{x^2} \int_0^x e^{-t^2}\, dt = \sum_{n=1}^{\infty} \frac{n!}{(2n)!} (2x)^{2n-1}.$$

The function on the left is the unique solution of the initial value problem
$$y' - 2xy = 1 \quad y(0) = 0.$$

Set $y(x) = \sum_{n=0}^{\infty} a_n x^n$. Substituting the series into the differential equation and equating coefficients of like powers of x yields
$$a_{n+1} = \frac{2}{n+1} a_{n-1} \quad \text{for all } n \geq 1.$$

Appealing to $a_0 = 0$ and $a_1 = 1$, we have $a_{2n} = 0$ and
$$a_{2n-1} = \frac{n!}{(2n)!} 2^{2n-1}.$$

Remark. You may also show that each side of the proposed identity is a solution of the initial value problem
$$2y'' - xy' - 2y = 0, \quad y(0) = 1, \quad y'(0) = 0.$$

Since the coefficients of the differential equation are continuous on $\mathbb{R}$, the proposed identity follows from the existence-uniqueness theorem of ode.

10. Dividing (15.15) by θ gives

$$\sum_{n=1}^{\infty} \frac{\zeta(2n)\theta^{2n-1}}{\pi^{2n}} = \frac{1}{2}\left(\frac{1}{\theta} - \cot\theta\right).$$

Integrating this identity from 0 to $\pi/2$ yields

$$\sum_{n=1}^{\infty} \frac{\zeta(2n)}{(2n)2^{2n}} = \frac{1}{2}\int_0^{\pi/2}\left(\frac{1}{\theta} - \cot\theta\right)d\theta = \frac{1}{2}\ln\left(\frac{\theta}{\sin\theta}\right)\Big|_0^{\pi/2} = \frac{1}{2}\ln(\pi/2).$$

Next, integrating (15.15) from 0 to $\pi/2$ yields

$$\sum_{n=1}^{\infty} \frac{\zeta(2n)\pi}{(2n+1)2^{2n+1}} = \frac{1}{2}\left(\frac{\pi}{2} - \int_0^{\pi/2}\theta\cot\theta\,d\theta\right).$$

Further, integrating by parts, we have

$$\int_0^{\pi/2}\theta\cot\theta\,d\theta = \theta\ln\sin\theta\Big|_0^{\pi/2} - \int_0^{\pi/2}\ln\sin\theta\,d\theta = \frac{1}{2}\pi\ln 2,$$

and so

$$\sum_{n=1}^{\infty} \frac{\zeta(2n)}{(2n+1)2^{2n}} = \frac{1}{2}(1 - \ln 2).$$

Remark. In general, one can sum the series $\sum_{n=1}^{\infty} \frac{\zeta(2n)}{(2n+p)2^{2n}}$ for any odd p via

$$\int_0^{\pi/2} x^{p+1}\csc^2 x\,dx = \frac{(p+1)\pi^p}{2^p}\left(\frac{1}{p} - 2\sum_{n=1}^{\infty}\frac{\zeta(2n)}{(2n+p)2^{2n}}\right).$$

A more challenging problem is to evaluate the series involving $\zeta(2n+1)$ such as

$$\sum_{n=1}^{\infty} \frac{\zeta(2n+1)}{(2n+1)2^{2n}}$$

in closed form. A systematic treatment of these series can be found in Srivastava and Choi's "Series Associated with the Zeta and Related Functions."

Chapter 16

1. Setting $x = 1/2$ in the power series of $\ln(1-x)$ yields $\ln 2 = \sum_{n=1}^{\infty} 1/(2^n n)$. Appealing to the Cauchy product, we have

$$\ln^2 2 = \left(\sum_{n=1}^{\infty}\frac{1}{2^n n}\right)\cdot\left(\sum_{m=1}^{\infty}\frac{1}{2^m m}\right)$$

$$= \sum_{n=1}^{\infty}\left(\sum_{i=1}^{n-1}\frac{1}{(n-i)i}\right)\frac{1}{2^n} \quad \text{(using the partial fraction)}$$

$$= \sum_{n=1}^{\infty}\frac{1}{2^{n-1}n}\sum_{i=1}^{n-1}\frac{1}{i} = \sum_{n=1}^{\infty}\frac{H_{n-1}}{2^{n-1}n},$$

Solutions to Selected Exercises

where H_n are the harmonic numbers. Therefore,

$$\ln^2 2 + 2\sum_{n=1}^{\infty} \frac{1}{2^n n^2} = \sum_{n=1}^{\infty} \frac{H_{n-1}}{2^{n-1} n} + \sum_{n=1}^{\infty} \frac{1}{2^{n-1} n^2}$$

$$= \sum_{n=1}^{\infty} \frac{1}{2^{n-1} n}\left(H_{n-1} + \frac{1}{n}\right)$$

$$= \sum_{n=1}^{\infty} \frac{H_n}{2^{n-1} n}.$$

Now we show that this last series is indeed $\zeta(2)$. In Exercise 6.8, set $a_n = 1/n2^{n-1}$, $f(x) = -2\ln(1 - x/2)$. Then

$$\sum_{n=1}^{\infty} \frac{1}{n 2^{n-1}} H_n = \int_0^1 \frac{2\ln 2 + 2\ln(1 - t/2)}{1 - t}\, dt = 2\int_0^1 \frac{\ln(1+t)}{t}\, dt.$$

But

$$\int_0^1 \frac{\ln(1+t)}{t}\, dt = \int_0^1 \sum_{n=1}^{\infty} \frac{(-1)^{n+1}}{n} t^{n-1}\, dt = \sum_{n=1}^{\infty} \frac{(-1)^{n+1}}{n^2} = \frac{1}{2}\zeta(2).$$

Similarly, the second identity follows from

$$\ln^3 2 = 6\sum_{n=1}^{\infty} \left(\frac{1}{(n+2)2^{n+2}} \sum_{k=1}^{n} \frac{H_k}{k+1}\right).$$

6. Set $t = 1/2$ in Euler's series transformation or show that

$$\sum_{k=1}^{n} \binom{n}{k} \frac{(-1)^{k+1}}{k} = \sum_{k=1}^{n} \binom{n}{k} \int_0^1 (-x)^{k-1}\, dx$$

$$= \int_0^1 \frac{(1-x)^n - 1}{-x}\, dx = \int_0^1 \frac{t^n - 1}{t - 1} = \sum_{k=1}^{n} \frac{1}{k} = H_n.$$

9. Substituting x by $1 - x$ yields

$$\zeta(3) = \int_0^1 \frac{\ln(1-x)\ln x}{x}\, dx = \int_0^1 \frac{\ln(1-x)\ln x}{1-x}\, dx.$$

Averaging these two integrals gives

$$\zeta(3) = \frac{1}{2}\int_0^1 \frac{\ln(1-x)}{x} \cdot \frac{\ln x}{1-x}\, dx.$$

By expanding the two factors in the integrand into the power series respectively, we have

$$\zeta(3) = \frac{1}{2}\int_0^1 \left(-\sum_{i=1}^{\infty} \frac{x^{i-1}}{i}\right)\cdot\left(-\sum_{j=1}^{\infty} \frac{(-x)^{j-1}}{j}\right) dx$$

$$= \frac{1}{2}\sum_{i=1}^{\infty}\sum_{j=1}^{\infty} \frac{1}{ij}\int_0^1 x^{i-1}(1-x)^{j-1}\, dx.$$

Appealing to the beta function, we find that

$$\int_0^1 x^{i-1}(1-x)^{j-1}\,dx = B(i,j) = \frac{(i-1)!(j-1)!}{(i+j-1)!},$$

and so

$$\zeta(3) = \frac{1}{2}\sum_{i=1}^{\infty}\sum_{j=1}^{\infty}\frac{1}{ij^2\binom{i+j-1}{j}}.$$

10. First, we rewrite A_n as an integral:

$$A_n = \sum_{i_1=1}^{\infty}\cdots\sum_{i_n=1}^{\infty}\frac{1}{i_1 i_2\cdots i_n}\int_0^1 x^{i_1+i_2+\cdots+i_n-1}\,dx.$$

Since all summands are positive, the monotone convergence theorem permits us to interchange the order of summation and integration. Appealing to $\sum_{i=1}^{\infty} x^i/i = -\ln(1-x)$, we obtain

$$A_n = \int_0^1 \frac{1}{x}\sum_{i_1=1}^{\infty}\frac{x^{i_1}}{i_1}\cdots\sum_{i_n=1}^{\infty}\frac{x^{i_n}}{i_n}\,dx = (-1)^n\int_0^1 \frac{\ln^n(1-x)}{x}\,dx.$$

Next, in view of the well-known fact that

$$\int_0^1 t^k \ln^n t\,dt \stackrel{x=-\ln t}{=} (-1)^n\int_0^{\infty} x^n e^{-(k+1)x}\,dx = (-1)^n\frac{n!}{(k+1)^{n+1}},$$

we have

$$\int_0^1 \frac{\ln^n t}{1-t}\,dt = \sum_{k=0}^{\infty}\int_0^1 t^k \ln^n t\,dt = (-1)^n n!\zeta(n+1).$$

Therefore,

$$A_n = (-1)^n\int_0^1 \frac{\ln^n x}{1-x}\,dx = n!\zeta(n+1).$$

Chapter 17

2. By using the partial fraction decomposition

$$\frac{1}{i^2-j^2} = -\frac{1}{2i}\left(\frac{1}{j-i} - \frac{1}{j+i}\right),$$

for positive integers i and N with $N > i$, we have

$$S(i,N) = \sum_{j=1, j\neq i}-\frac{1}{2i}\left(\frac{1}{j-i} - \frac{1}{j+i}\right)$$

$$= -\frac{1}{2i}\left(\frac{1}{1-i} + \cdots + \frac{1}{(i-1)-i} + \frac{1}{(i+1)-i} + \cdots + \frac{1}{N-i}\right)$$

$$+ \frac{1}{2i}\left(\frac{1}{1+i} + \cdots + \frac{1}{(i-1)+i} + \frac{1}{(i+1)+i} + \cdots + \frac{1}{N+i}\right)$$

$$= \frac{1}{2i}(H_{N+i} - H_{N-i}) - \frac{3}{4i^2},$$

where H_k are the harmonic numbers. Appealing to

$$H_n = \ln n + \gamma + O(1/n),$$

Solutions to Selected Exercises

we find that
$$\lim_{N\to\infty} (H_{N+i} - H_{N-i}) = \lim_{N\to\infty} \ln \frac{N+i}{N-i} = 0,$$
and so
$$\lim_{N\to\infty} S(i, N) = -\frac{3}{4i^2}.$$
Therefore, the proposed sum is
$$\lim_{K\to\infty} \sum_{i=1}^{K} \left(\lim_{N\to\infty} S(i, N) \right) = -\frac{3}{4} \lim_{K\to\infty} \sum_{i=1}^{K} \frac{1}{i^2} = -\frac{\pi^2}{8}.$$

Remark. Two related problems are to evaluate
$$\sum_{i,j=1,(i,j)=1}^{\infty} \frac{1}{i^2 j^2} \quad \text{and} \quad \sum_{i,j=1,(i,j)=1}^{\infty} \frac{1}{ij(i+j)}.$$

7. (a) Rewriting H_n as an integral of a finite geometric series and then appealing to Leibniz's formula and the power series expansions of $\arctan x$, we have

$$\sum_{n=0}^{\infty} \frac{(-1)^n}{2n+1} H_n = \sum_{n=0}^{\infty} \frac{(-1)^n}{2n+1} \int_0^1 \frac{1-t^n}{1-t} dt$$

$$= 2 \sum_{n=0}^{\infty} \frac{(-1)^n}{2n+1} \int_0^1 \frac{x - x^{2n+1}}{1-x^2} dx \quad (\text{set } t = x^2)$$

$$= \frac{\pi}{2} \int_0^1 \frac{x}{1-x^2} dx - 2 \int_0^1 \frac{\arctan x}{1-x^2} dx$$

$$= -\frac{\pi}{2} \int_0^1 \frac{1-x}{1-x^2} dx + 2 \int_0^1 \left(\frac{\pi}{4} - \arctan x \right) \frac{dx}{1-x^2}$$

$$= -\frac{\pi}{2} \ln 2 + G,$$

where we have used the fact that

$$2 \int_0^1 \left(\frac{\pi}{4} - \arctan x \right) \frac{dx}{1-x^2} = 2 \int_0^1 \arctan \left(\frac{1-x}{1+x} \right) \frac{dx}{1-x^2}$$

$$= \int_0^1 \frac{\arctan u\, du}{u} = \int_0^1 \sum_{n=0}^{\infty} \frac{(-1)^n}{2n+1} u^{2n} du$$

$$= \sum_{n=0}^{\infty} \frac{(-1)^n}{(2n+1)^2} = G.$$

(b). Replacing x by ix in (17.18) yields
$$\sum_{n=1}^{\infty} (-1)^k h_k x^{2k} = -\frac{x}{1+x^2} \arctan x.$$

Integrating this identity from 0 to 1 leads to

$$\sum_{n=1}^{\infty} \frac{(-1)^k}{2k+1} h_k = -\int_0^1 \frac{x}{1+x^2} \arctan x\, dx$$

$$= -\frac{\pi}{8} \ln 2 + \frac{1}{2} \int_0^1 \frac{\ln(1+x^2)}{1+x^2} dx \quad (\text{using integration by parts}).$$

Now, the proposed answer follows from

$$\int_0^1 \frac{\ln(1+x^2)}{1+x^2}\,dx = \frac{\pi}{2}\ln 2 - G.$$

Remark. Bradley assembles a fairly exhaustive list of formulas involving Catalan's constant, see `citeseerx.ist.psu.edu/viewdoc/summary?doi=10.1.1.26.1879`

8. Setting $z = e^{ix}$ in

$$\sum_{n=1}^{\infty} \frac{z^n}{n} = -\ln(1-z) = -\ln|2\sin(x/2)| + \frac{\pi-x}{2}i$$

and then exporting the real and imaginary parts yields

$$\sum_{n=1}^{\infty} \frac{\sin(nx)}{n} = \frac{\pi-x}{2} \quad \text{and} \quad \sum_{n=1}^{\infty} \frac{\cos(nx)}{n} = -\ln|2\sin(x/2)|.$$

Next, multiplying these series together and manipulating gives

$$-\frac{\pi-x}{2}\ln|2\sin(x/2)| = \sum_{m,n=1}^{\infty} \frac{\sin(mx)\cos(nx)}{mn} = \frac{1}{2}\sum_{m,n=1}^{\infty} \frac{\sin(m+n)x}{mn}$$

$$= \frac{1}{2}\sum_{n>k>0} \frac{\sin(nx)}{k(n-k)} = \frac{1}{2}\sum_{n=1}^{\infty} \frac{\sin(nx)}{n}\sum_{k=1}^{n-1}\left(\frac{1}{k}+\frac{1}{n-k}\right)$$

$$= \sum_{n=1}^{\infty} \frac{\sin(nx)}{n}\sum_{k=1}^{n-1}\frac{1}{k},$$

from which the proposed identity follows by applying Parseval's identity.
Remark. If one uses arctanh instead of ln, one is led to

$$\int_0^{\pi} \ln^2(\tan(x/4))\,dx = \frac{\pi^3}{4},$$

and hence to another proof of (17.25):

$$\sum_{k=1}^{\infty} \frac{h_k^2}{k^2} = \frac{\pi^4}{32} = \frac{45}{16}\zeta(4).$$

11. Appealing to (17.18), we have

$$f'(x) = \sum_{k=1}^{\infty} h_k x^{2k-2} = \frac{1}{2x(1-x^2)}\ln\left(\frac{1+x}{1-x}\right).$$

Hence,

$$\frac{2}{(2-x)^2}f'\left(\frac{x}{2-x}\right) = \frac{x-2}{4x(1-x)}\ln(1-x) = -\frac{1}{2x}\ln(1-x) - \frac{\ln(1-x)}{4(1-x)}.$$

Integrating the equation above from 0 to x yields the desired identity.

Chapter 18

2. Rewrite

$$\sum_{n=1}^{\infty} \frac{1}{n^2 \binom{2mn}{mn}} = \frac{m}{2} \sum_{n=1}^{\infty} \frac{1}{n(2mn-1)\binom{2mn-2}{mn-1}}$$

$$= \frac{m}{2} \sum_{n=0}^{\infty} \frac{1}{(n+1)(2mn+2m-1)\binom{2mn+2m-2}{mn-1}}.$$

By using the beta function, we have

$$\sum_{n=1}^{\infty} \frac{1}{n^2 \binom{2mn}{mn}} = \frac{m}{2} \sum_{n=0}^{\infty} \frac{1}{n+1} \int_0^1 t^{mn+m-1} (1-t)^{mn+m-1} \, dt.$$

Now, the proposed identity follows from

$$\sum_{n=0}^{\infty} \frac{t^{mn+m}(1-t)^{mn+m}}{n+1} = -\ln[1 - t^m(1-t)^m].$$

4(c). Recall that $G = \int_0^1 (\arctan x / x) \, dx$. Set $x = \tan(t/2)$. Appealing to the double angle formula of sine, we have

$$G = \frac{1}{4} \int_0^{\pi/2} \frac{t \, dt}{\tan(t/2) \cos^2(t/2)} = \frac{1}{2} \int_0^{\pi/2} \frac{t}{\sin t} \, dt.$$

By using (18.1) we obtain

$$2G = \int_0^{\pi/2} \frac{t}{\sin t} \, dt = \int_0^1 \frac{2u \arcsin u}{\sqrt{1-u^2}} \cdot \frac{du}{2u^2} \qquad (u = \sin t)$$

$$= \int_0^1 \sum_{n=1}^{\infty} \frac{(2u)^{2n}}{n \binom{2n}{n}} \cdot \frac{du}{2u^2} = \sum_{n=1}^{\infty} \frac{2^{2n}}{2n(2n-1)\binom{2n}{n}}$$

$$= \sum_{n=0}^{\infty} \frac{2^{2n}}{(2n+1)^2 \binom{2n}{n}}.$$

Remark. If

$$G = \frac{1}{4} \int_0^{\pi/2} \ln\left(\frac{1+\sin x}{1-\sin x}\right) dx$$

has been proved, applying Wallis's formula of odd power yields the following alternative proof:

$$2G = \int_0^{\pi/2} \sum_{n=0}^{\infty} \frac{1}{2n+1} (\sin x)^{2n+1} \, dx$$

$$= \sum_{n=0}^{\infty} \frac{1}{2n+1} \cdot \frac{2 \cdot 4 \cdot 6 \cdots (2n)}{3 \cdot 5 \cdot 7 \cdots (2n+1)}$$

$$= \sum_{n=0}^{\infty} \frac{2^{2n}}{(2n+1)^2 \binom{2n}{n}}.$$

8 We derive the first identity by using Mathematica. First we set

$$\ln 2 = \sum_{n=1}^{\infty} \frac{an^2 + bn + c}{2^n \binom{3n}{n}}.$$

Feeding the general sum to Mathematica and integrating gives

Sum[(a n^2 + b n + c) (3 n + 1) x^n, {n, 1, Infinity}]

(4 a x + 4 b x + 4 c x + 12 a x^2 - 2 b x^2 - 9 c x^2 + 2 a x^3 - 2 b x^3 + 6 c x^3 - c x^4)/(-1 + x)^4

% /. x -> t^2 (1 - t)/2

1/(-1 + 1/2 (1 - t) t^2)^4 (2 a (1 - t) t^2 + 2 b (1 - t) t^2 + 2 c (1 - t) t^2 + 3 a (1 - t)^2 t^4 - 1/2 b (1 - t)^2 t^4 - 9/4 c (1 - t)^2 t^4 + 1/4 a (1 - t)^3 t^6 - 1/4 b (1 - t)^3 t^6 + 3/4 c (1 - t)^3 t^6 - 1/16 c (1 - t)^4 t^8)

Simplify[%]

-1/(2 - t^2 + t^3)^4 (-1 + t) t^2 (4 a (8 + 12 t^2 - 12 t^3 + t^4 - 2 t^5 + t^6) + (2 - t^2 + t^3) (-4 b (-4 - t^2 + t^3) + c (16 - 10 t^2 + 10 t^3 + t^4 - 2 t^5 + t^6)))

Integrate[%, {t, 0, 1}]

1/15625 (a (2805 - (673 + 14 I) ArcTan[1/3] - (673 - 14 I) ArcTan[1/2] + (673 + 14 I) ArcTan[2] + (673 - 14 I) ArcTan[3] + 42 Log[2]) + 5 (b (405 - (158 - 6 I) ArcTan[1/3] - (158 + 6 I) ArcTan[1/2] + (158 - 6 I) ArcTan[2] + (158 + 6 I) ArcTan[3] - 3 Log[64]) + 25 c (10 - (11 - 2 I) ArcTan[1/3] - (11 + 2 I) ArcTan[1/2] + (11 - 2 I) ArcTan[2] + (11 + 2 I) ArcTan[3] - Log[64])))

Next we make some simplifications based on $\arctan(1/x) = \pi/2 - \arctan x$

FullSimplify[% /. ArcTan[1/3] -> (Pi/2 - ArcTan[3]) /. ArcTan[1/2] -> (Pi/2 - ArcTan[2])]

1/31250 (a(5610 + 673\[Pi] + 84Log[2]) + 5(2b (405 + 79\[Pi] - 3Log[64]) + 25c(20 + 11\[Pi] - 2Log[64])))

Expand[% /. Log[64] -> 6 Log[2]]

(561a)/3125 + (81b)/625 + (2c)/25 + (673a\[Pi])/31250 + (79b\[Pi])/3125 + (11c\[Pi])/250 + (42aLog[2])/15625 - (18bLog[2])/3125 - 6/125cLog[2]

Collect[%, {Log[2], Pi}]

(561a)/3125 + (81b)/625 + (2c)/25 + ((673a)/31250 + (79b)/3125 + (11c)/250)\[Pi] + ((42a)/15625 - (18b)/3125 - (6c)/125)Log[2]

Finally, we solve the system to establish the desired identity.

Solve[{(561a)/3125 + (81b)/625 + (2c)/25 == 0, (673a)/31250 + (79b)/3125 + (11c)/250 == 0, (42a)/15625 - (18b)/3125 - (6c)/125 == 1}, {a, b, c}]

{{a -> -(575/6), b -> 965/6, c -> -(91/2)}}

Solutions to Selected Exercises

Chapter 19

5(b). Let I denote the required integral. Differentiating I with respect to α yields

$$I'(\alpha) = \int_0^\infty \frac{\sin \beta x \cos \alpha x}{x} e^{-kx}\, dx.$$

By using the product to sum formula we have

$$I'(\alpha) = \frac{1}{2}\left(\int_0^\infty \frac{\sin(\alpha+\beta)x}{x} e^{-kx}\, dx - \int_0^\infty \frac{\sin(\alpha-\beta)x}{x} e^{-kx}\, dx\right).$$

Appealing to Example 1, we find that

$$I'(\alpha) = \frac{1}{2}\left(\arctan\frac{\alpha+\beta}{k} - \arctan\frac{\alpha+\beta}{k}\right),$$

and so

$$I = \frac{\alpha+\beta}{2}\arctan\frac{\alpha+\beta}{k} - \frac{\alpha-\beta}{2}\arctan\frac{\alpha-\beta}{k} + \frac{k}{4}\ln\frac{k^2+(\alpha-\beta)^2}{k^2+(\alpha+\beta)^2}.$$

Remark. Letting $k \to 0$, we get

$$\int_0^\infty \frac{\sin \alpha x}{x} \cdot \frac{\sin \beta x}{x}\, dx = \begin{cases} \frac{\pi}{2}\beta, & \text{if } \alpha \geq \beta, \\ \frac{\pi}{2}\alpha, & \text{if } \alpha \leq \beta \end{cases}.$$

In general, if $\alpha > \sum_{i=1}^n \alpha_i$ and all $\alpha_i > 0$, then

$$\int_0^\infty \frac{\sin \alpha x}{x} \cdot \frac{\sin \alpha_1 x}{x} \cdots \frac{\sin \alpha_n x}{x}\, dx = \frac{\pi}{2}\alpha_1\alpha_2\cdots\alpha_n.$$

7. Appealing to

$$\frac{\cos ax - \cos bx}{x} = \int_a^b \sin yx\, dy$$

and

$$\int_0^\infty \frac{\sin yx}{x}\, dx = \frac{\pi}{2},$$

we have

$$\int_0^\infty \frac{\cos ax - \cos bx}{x^2}\, dx = \int_0^\infty \int_a^b \frac{\sin yx}{x}\, dy\, dx$$

$$= \int_a^b \int_0^\infty \frac{\sin yx}{x}\, dx\, dy = \frac{\pi}{2}(b-a).$$

9. This is a Frullani integral with $f(x) = e^x/(1+e^x)$ and so the answer is $(1/2)\ln(p/q)$.

11. Let $f(a)$ denote the right side of the equation. Substituting $x = at$ gives

$$f(a) = \frac{1}{\sqrt{\pi}} \int_0^\infty \frac{e^{-a^2 t^2}}{1+t^2}\, dt.$$

By Leibniz's rule, we have

$$f'(a) = -\frac{2a}{\sqrt{\pi}}\left(\int_0^\infty e^{-a^2 t^2}\, dt - \int_0^\infty \frac{e^{-a^2 t^2}}{1+t^2}\, dt\right).$$

Appealing to
$$\int_0^\infty e^{-a^2 t^2}\, dt = \frac{\sqrt{\pi}}{2a},$$
we find that $f(a)$ satisfies an initial value problem
$$f'(a) - 2af(a) = -1, \qquad f(0) = \frac{\sqrt{\pi}}{2}$$
and so
$$f(a) = e^{a^2} \int_a^\infty e^{-t^2}\, dt,$$
which is identical with the left side of the equation after substituting $t^2 = x^2 + a^2$.

14. Recall that
$$J_0(x) = \sum_{n=0}^\infty \binom{-1/2}{n} \frac{x^{2n}}{(2n)!}.$$
Appealing to the binomial theorem, we obtain
$$\int_0^\infty e^{-sx} J_0(x)\, dx = \sum_{n=0}^\infty \binom{-1/2}{n} \frac{1}{(2n)!} \int_0^\infty e^{-sx} x^{2n}\, dx$$
$$= \sum_{n=0}^\infty \binom{-1/2}{n} \frac{1}{(2n)!\, s^{2n+1}} \int_0^\infty e^{-t} t^{2n}\, dt$$
$$= \sum_{n=0}^\infty \binom{-1/2}{n} \frac{\Gamma(2n+1)}{(2n)!\, s^{2n+1}}$$
$$= \frac{1}{s} \sum_{n=0}^\infty \binom{-1/2}{n} \left(\frac{1}{s^2}\right)^2 = \frac{1}{s} \frac{1}{\sqrt{1+(1/s)^2}} = \frac{1}{\sqrt{1+s^2}}.$$

Remark. This result indicates that the *Laplace transform* of $J_0(x)$ is $1/\sqrt{1+s^2}$.

Chapter 20

2. Since all summands are positive, the monotone convergence theorem permits us to interchange the order of summation and integration. Substituting $x = e^{-t}$ yields
$$\int_0^1 x^{-x}\, dx = \int_0^\infty e^{te^{-t}} e^{-t}\, dt$$
$$= \int_0^\infty \left(\sum_{n=0}^\infty \frac{1}{n!} t^n e^{-nt}\right) e^{-t}\, dt$$
$$= \sum_{n=0}^\infty \frac{1}{n!} \int_0^\infty t^n e^{-(n+1)t}\, dt$$
$$= \sum_{n=1}^\infty \frac{1}{(n-1)!} \int_0^\infty t^n e^{-nt}\, dt \qquad \text{(shifting the index } n \text{ by one)}$$
$$= \sum_{n=1}^\infty \frac{n^{-n}}{(n-1)!} \int_0^\infty s^{n-1} e^{-s}\, ds \qquad \text{(set } s = nt\text{)}$$
$$= \sum_{n=1}^\infty \frac{n^{-n}}{(n-1)!} \Gamma(n) = \sum_{n=1}^\infty n^{-n}.$$

Solutions to Selected Exercises

11. Define
$$f(x) = \sum_{-\infty}^{\infty} e^{-(n+x)^2 \pi s}.$$

Clearly, $f(x+1) = f(x)$; i.e., f is periodic and has period 1. Expanding f as a Fourier series
$$\sum_{-\infty}^{\infty} a_n e^{2n\pi xi},$$
we have
$$a_n = \int_{-\infty}^{\infty} e^{-\pi s x^2} e^{-2n\pi xi} \, dx$$
$$= 2 \int_0^{\infty} e^{-\pi s x^2} \cos(2n\pi x) \, dx \qquad \text{(see (19.4))}$$
$$= \sqrt{\frac{1}{s}} e^{-n^2 \pi / s},$$

and so
$$\sum_{-\infty}^{\infty} e^{-(n+x)^2 \pi s} = \sqrt{\frac{1}{s}} \sum_{-\infty}^{\infty} e^{-n^2 \pi / s} e^{2n\pi xi}.$$

The required identity follows from this equation by setting $x = 0$.

12. *First Proof.* Recall the well-known Poisson summation formula, for $|r| < 1$,
$$1 + 2 \sum_{k=1}^{\infty} r^k \cos(kx) = \frac{1 - r^2}{1 - 2r \cos x + r^2}.$$

Taking $r = e^{-\alpha}, x = \alpha y$ yields
$$\frac{1}{2} + \sum_{k=1}^{\infty} e^{-\alpha k} \cos(\alpha y k) = \frac{1}{2} \frac{1 - e^{-2\alpha}}{1 - 2e^{-\alpha} \cos(\alpha y) + e^{-2\alpha}}$$
$$= \frac{1}{2} \frac{e^{2\alpha} - 1}{1 - 2e^{\alpha} \cos(\alpha y) + e^{2\alpha}}.$$

On the other hand, the partial fraction gives
$$\frac{1}{1 + (y + 2\beta j)^2} = \frac{1}{2i} \left(\frac{1}{y - i + 2\beta j} - \frac{1}{y + i + 2\beta j} \right),$$
where $i = \sqrt{-1}$. Thus, the Euler formula
$$\pi \cot(\pi x) = \sum_{j=-\infty}^{\infty} \frac{1}{x + j}$$
leads to
$$\sum_{j=-\infty}^{\infty} \frac{1}{1 + (y + 2\beta j)^2} = \frac{1}{2i} \sum_{j=-\infty}^{\infty} \left(\frac{1}{y - i + 2\beta j} - \frac{1}{y + i + 2\beta j} \right)$$
$$= \frac{\pi}{4\beta i} \left(\cot\left(\frac{y - i}{2\beta}\right) \pi - \cot\left(\frac{y + i}{2\beta}\right) \pi \right)$$
$$= \frac{\alpha}{4i} \left(\cot\left(\frac{\alpha(y - i)}{2}\right) - \cot\left(\frac{\alpha(y + i)}{2}\right) \right), \qquad \text{(using } \alpha\beta = \pi\text{)}.$$

In view of the trigonometric identity
$$\cot s - \cot t = \frac{\sin(t-s)}{\sin s \sin t} = \frac{2\sin(t-s)}{\cos(t-s) - \cos(t+s)}$$
and $\sin(ix) = -i\sinh x, \cos(ix) = \cosh x$, we have
$$\cot\left(\frac{\alpha(y-i)}{2}\right) - \cot\left(\frac{\alpha(y+i)}{2}\right) = \frac{2i\sinh(\alpha)}{\cosh(\alpha) - \cos(\alpha y)} = 2i\frac{e^{2\alpha}-1}{1-2e^{\alpha}\cos(\alpha y) + e^{2\alpha}},$$
and so
$$\frac{1}{\alpha}\sum_{j=-\infty}^{\infty}\frac{1}{1+(y+2\beta j)^2} = \frac{1}{2}\frac{e^{2\alpha}-1}{1-2e^{\alpha}\cos(\alpha y)+e^{2\alpha}} = \frac{1}{2} + \sum_{k=1}^{\infty}e^{-\alpha k}\cos(\alpha y k).$$

Second Proof. This proof is based on the *Poisson summation formula*, which asserts that
$$T\sum_{k=-\infty}^{\infty} f(kT) = \sum_{j=-\infty}^{\infty} \hat{f}(j/T), \tag{1}$$
where the Fourier transform $\hat{f}$ of f is defined by
$$\hat{f}(\xi) = \int_{-\infty}^{\infty} f(x) e^{-2x\xi\pi i}\, dx.$$
Let $f(x) = e^{-|x|}\cos(xy)$. Then
$$\hat{f}(\xi) = \int_{-\infty}^{\infty} e^{-|x|}\cos(xy)e^{-2x\xi\pi i}\, dx$$
$$= 2\int_0^{\infty} e^{-x}\cos(xy)\cos(2\xi x\pi)\, dx$$
$$= \int_0^{\infty} e^{-x}[\cos(y+2\xi\pi)x + \cos(y-2\xi\pi)x]\, dx$$
$$= \frac{1}{1+(y+2\xi\pi)^2} + \frac{1}{1+(y-2\xi\pi)^2}.$$
Now, the required identity follows from (1) by choosing $T = \alpha$ and some simplifications.

16. To eliminate the absolute value, we divide the interval $[0, \infty)$ into
$$\left[n\frac{\pi}{2}, (n+1)\frac{\pi}{2}\right], \quad (n=0,1,2,\ldots).$$
When $n = 2k$,
$$\int_{k\pi}^{k\pi+\pi/2} \frac{\ln|\cos x|}{x^2}\, dx = \int_0^{\pi/2} \frac{\ln\cos t}{(t+k\pi)^2}\, dt;$$
and when $n = 2k-1$,
$$\int_{k\pi-\pi/2}^{k\pi} \frac{\ln|\cos x|}{x^2}\, dx = \int_0^{\pi/2} \frac{\ln\cos t}{(t-k\pi)^2}\, dt.$$
Therefore,
$$\int_0^{\infty} \frac{\ln|\cos x|}{x^2}\, dx = \int_0^{\pi/2} \ln\cos t \left\{\frac{1}{t^2} + \sum_{k=1}^{\infty}\left[\frac{1}{(t-k\pi)^2} + \frac{1}{(t+k\pi)^2}\right]\right\}\, dt.$$
Noticing the expression in the bracket is indeed $1/\sin^2 t$ (see Exercise 6.11), integrating by parts, we find that
$$\int_0^{\infty} \frac{\ln|\cos x|}{x^2}\, dx = \int_0^{\pi/2} \frac{\ln\cos t}{\sin^2 t}\, dt = -\frac{\pi}{2}.$$

17. First, since
$$\lim_{t \to 0} B_{2n+1}(t) \cot(\pi t) = \frac{1}{\pi} \binom{2n+1}{2n} B_{2n},$$
the integral converges. Next, appealing to the trigonometric series
$$B_{2n+1}(t) = (-1)^{n+1} \frac{2(2n+1)!}{(2\pi)^{2n+1}} \sum_{k=1}^{\infty} \frac{\sin(2k\pi t)}{k^{2n+1}} \quad (0 \le t \le 1; n = 1, 2, \ldots),$$
we have
$$(-1)^{n+1} \frac{(2\pi)^{2n+1}}{(2n+1)!} \int_0^{1/2} B_{2n+1}(t) \cot(\pi t) \, dt = 2 \sum_{k=1}^{\infty} \frac{1}{k^{2n+1}} \int_0^{1/2} \sin(2k\pi t) \cot(\pi t) \, dt.$$
Recall that
$$\frac{\sin(2k+1)t}{\sin t} = 1 + 2 \sum_{i=1}^{k} \cos(2kt).$$
For any positive integer k, we have
$$\int_0^{\pi/2} \frac{\sin(2k+1)t}{\sin t} \, dt = \frac{\pi}{2}, \tag{*}$$
and so
$$\int_0^{1/2} \sin(2k\pi t) \cot(\pi t) \, dt = \frac{1}{2\pi} \left(\int_0^{\pi/2} \frac{\sin(2k+1)t}{\sin t} \, dt + \int_0^{\pi/2} \frac{\sin(2k-1)t}{\sin t} \, dt \right) = \frac{1}{2}.$$
Thus, we prove the formula as proposed.
Remark. Appealing to
$$\sum_{k=1}^{n} \sin(2k-1)t = \frac{1 - \cos(2nt)}{2 \sin t} = \frac{\sin^2(nt)}{\sin t},$$
repeatedly using (*) yields
$$\int_0^{\pi/2} \left(\frac{\sin(nt)}{\sin t} \right)^2 dt = \frac{n\pi}{2}.$$

Chapter 21

1. Let I denote the required integral. Rewrite I as
$$I = \int_0^1 \frac{dx}{(1+x^2)(1+x^\alpha)} + \int_1^\infty \frac{dx}{(1+x^2)(1+x^\alpha)}.$$
The substitution $t = 1/x$ in the second integral yields
$$I = \int_0^1 \frac{dx}{(1+x^2)(1+x^\alpha)} + \int_0^1 \frac{t^\alpha \, dt}{(1+t^2)(1+t^\alpha)}.$$
Thus
$$I = \int_0^1 \frac{(1+x^\alpha) dx}{(1+x^2)(1+x^\alpha)} = \int_0^1 \frac{dx}{1+x^2} = \frac{\pi}{4},$$
which is independent of α.
Remark. Let $f(x)$ be a rational function and let $F(x) = f(x) + f(1/x)/x^2$. Then
$$\int_0^\infty f(x) \, dx = \int_0^1 F(x) \, dx.$$

4. For any real number x, we have
$$\lfloor 2x \rfloor - 2\lfloor x \rfloor = \begin{cases} 0, & \text{if } \lfloor 2x \rfloor \text{ is even,} \\ 1, & \text{if } \lfloor 2x \rfloor \text{ is odd.} \end{cases}$$

Thus,
$$S_n := \sum_{k=1}^{n} \left(\left\lfloor \frac{2\sqrt{n}}{\sqrt{k}} \right\rfloor - 2 \left\lfloor \frac{\sqrt{n}}{\sqrt{k}} \right\rfloor \right)$$
$$= \sum_{k=1}^{n} 1 \quad \text{(where } \lfloor 2\sqrt{n/k} \rfloor \text{ is odd)}$$
$$= \sum_{j=1}^{f(n)} \sum_{k=1}^{n} 1 \quad \text{(where } \lfloor 2\sqrt{n/k} \rfloor = 2j+1\text{),}$$

where $f(n) = \lfloor (\lfloor 2\sqrt{n} \rfloor - 1)/2 \rfloor$. Next, we see that $\lfloor \frac{2\sqrt{n}}{\sqrt{k}} \rfloor = 2j+1$ if and only if
$$\frac{4n}{(2j+2)^2} < k \leq \frac{4n}{(2j+1)^2},$$

and so the inner sum above equals
$$\left\lfloor \frac{4n}{(2j+1)^2} \right\rfloor - \left\lfloor \frac{4n}{(2j+2)^2} \right\rfloor = \frac{4n}{(2j+1)^2} - \frac{4n}{(2j+2)^2} + r_1,$$

where $|r_1| < 1$. Therefore,
$$\frac{1}{n} S_n = 4 \sum_{j=1}^{f(n)} \left(\frac{1}{(2j+1)^2} - \frac{1}{(2j+2)^2} \right) + r_2,$$

where
$$|r_2| \leq \frac{f(n)}{n} |r_1| < \frac{f(n)}{n} \leq \frac{1}{\sqrt{n}}.$$

Finally, we find that
$$\lim_{n \to \infty} \frac{1}{n} S_n = 4 \sum_{j=1}^{\infty} \left(\frac{1}{(2j+1)^2} - \frac{1}{(2j+2)^2} \right)$$
$$= 4 \left[\left(\frac{\pi^2}{8} - 1 \right) - \left(\frac{\pi^2}{24} - \frac{1}{4} \right) \right] = \frac{\pi^2}{3} - 3.$$

7. Let the proposed integral be I. We use the geometric series, an elementary integral
$$-\frac{(xy)^n}{\ln(xy)} = \int_n^{\infty} (xy)^t \, dt, \quad (0 < x, y < 1),$$

the monotone convergence theorem, and the definition of γ:
$$I = \int_0^1 \int_0^1 -\sum_{n=0}^{\infty} (1-x) \frac{(xy)^n}{\ln(xy)} \, dxdy = \sum_{n=0}^{\infty} \int_n^{\infty} \left(\int_0^1 \int_0^1 (1-x)(xy)^t \, dxdy \right) dt$$
$$= \sum_{n=0}^{\infty} \int_n^{\infty} \left(\frac{1}{(t+1)^2} - \frac{1}{(t+1)(t+2)} \right) dt = \sum_{n=0}^{\infty} \left(\frac{1}{n+1} - \ln \frac{n+2}{n+1} \right)$$
$$= \lim_{N \to \infty} (H_N - \ln(N+1)) = \gamma.$$

Solutions to Selected Exercises

13. First, integrating by parts yields

$$I_{n+1} = \frac{1}{3n}\left(\frac{1}{3^{n-1}} - 1\right) + \frac{2(2n-1)}{3n} I_n.$$

Appealing to $I_1 = \sqrt{3}\pi/9$, we see that

$$I_n = a_n + b_n \sqrt{3}\pi,$$

where a_n and b_n are rational numbers. By $b_{n+1} = (2(2n-1)/3n)b_n$, we have

$$b_{n+1} = \frac{1}{3^{n+2}} \binom{2n}{n}.$$

Next, a_n satisfies a recursive relation

$$a_{n+1} = \frac{1}{3n}\left(\frac{1}{3^{n-1}} - 1\right) + \frac{2(2n-1)}{3n} a_n.$$

A telescoping process gives

$$a_{n+1} = \frac{1}{3n}\left(\frac{1}{3^{n-1}} - 1\right) + \sum_{k=2}^{n-2} 2^{k-1}(3^{-n} - 3^{-k}) \frac{(2n-1)(2n-3)\cdots(2n-2k+3)}{n(n-1)\cdots(n-k+1)}.$$

15. One possible formulation: by repeated use of the product to sum identity, we rewrite the integral as

$$2^{-n} \sum \int_0^{2\pi} \cos(a_1 n_1 + a_2 n_2 + \cdots + a_k n_k) x \, dx$$

where $a_i = \pm 1$ for $1 \leq i \leq n$ and the summation is taken over all the 2^n combinations of the values of the a_i. Appealing to the fact that

$$\int_0^{2\pi} \cos(nx) \, dx = \begin{cases} 2\pi, & \text{if } n = 0, \\ 0, & \text{if } n \text{ is any nonzero integer,} \end{cases}$$

the integral is a maximum if the number of solutions $(a_1, a_2, \ldots, a_k)$ to the equation

$$a_1 n_1 + a_2 n_2 + \cdots + a_k n_k = 0$$

is a maximum.

Index

Abel's continuity theorem, 115
Abel's summation formula, 258, 264
absolute integrability, 221
AM-GM inequality, 1, 2, 4–8, 11, 12, 17, 255, 256
    Alzer's proof, 6
    Cauchy's leap-forward fall-back induction, 2
    Hardy's proof via smoothing transformation, 5
    Pólya's proof, 7
    proof by normalization and orderings, 4
    proof by regular induction, 1
    proof via infimums, 17
Apery's constant, 69, 108
Apery-like formula, 102, 183
asymptotic formula, 60, 67, 70, 78, 80, 244
    for a sum of cosecants, 78
    for harmonic numbers, 60, 70, 244
    Stirling approximation, 67

Bernoulli, 55, 61–65, 67, 71, 143, 146, 149, 152, 160, 238
    inequality, 36
    numbers, 55, 61–65, 67, 143, 146, 149, 160
    numbers via determinants, 143
    polynomials, 63–65, 71, 238, 253
    polynomials via determinants, 152
    Raabe's multiplication identity, 71
beta function, 114, 202, 247, 284
big O notation, 75
Binet's formula, 57, 121, 123, 127, 135, 192
binomial coefficients, 62, 106, 160, 163, 201, 202, 208
binomial series, 113, 168
binomial theorem, 112, 125, 135, 270, 290

bounds for normal distribution, 241

Carleman's inequality, 15
Cassini's identity, 139
Catalan's constant, 185, 196, 198, 213, 286
Catalan's identity, 60
Cauchy product, 62, 145, 169, 272, 276, 282
Cauchy-Schwarz inequality, 1, 8–12, 17–19, 256, 258
    proof by AM-GM inequality, 8
    proof by Lagrange's identity, 10
    proof by normalization, 9
    proof via infimums, 18
    Schwarz's proof via quadratic functions, 10
central binomial coefficients, 113, 114, 119, 165, 173
characteristic equation, 121
complete elliptic integral of the first kind, 84
Cramer's rule, 143–148, 276

De Moivre's formula, 40, 43, 45, 74
digamma function, 114, 117, 248
Dini test, 221

elliptic integral singular values, 84, 91
Epstein's zeta function, 91
Euler, 39, 45, 47, 52, 53, 60, 65, 68, 70, 71, 76, 77, 117, 170, 177, 193, 237, 247, 252, 253, 264
    constant, 60, 70, 90, 117, 243, 247, 252
    dilogarithm, 193
    formula for γ, 60, 77, 244
    formula for $\zeta(2k)$, 68
    formula for $\zeta(3)$, 170
    formula for $e^{i\theta}$, 39

functional equation of the zeta function, 177
infinite product for $\cosh x$, 52
infinite product for $\sinh x$, 52
infinite product for cosine, 47, 264
infinite product for sine, 45, 177, 237
partial fraction for cotangent, 68, 76
partial fraction for secant, 71
partial fraction for tangent, 71
Euler formula for $\zeta(2)$, 169, 175–177, 181–183, 185, 186
    proof by complex variables, 183
    proof by double integrals, 177, 185
    proof by Euler, 176
    proof by Fourier series, 183
    proof by power series, 182
    proof by trigonometric identities, 181
    Wilf's family of series, 186
Euler sums, 61, 189, 190, 194, 195, 197, 199
Euler's series transformation, 186, 283
Euler-Maclaurin summation formula, 65, 66, 80, 266
Eves's means via a trapezoid, 26

Fibonacci numbers, 55–59, 69, 101, 107, 118, 121, 131, 148, 192
Fibonomial coefficients, 131, 133, 137, 140
finite product identities for cosine, 42
finite product identities for sine, 41, 42
finite summation identities involving cosine, 44
Fourier series, 183, 291
Fresnel integrals, 225
Frullani integrals, 226, 289
Fubini's theorem, 219–222, 228

Gallian's REU model, 73
gamma function, 83–86, 90, 91, 267
    Chowla-Selberg formula, 84
    Gauss multiplication formula, 83, 90, 267
    Legendre's duplication formula, 85
    reflection formula, 84, 86, 90, 267, 268
gamma product, 83, 85, 86, 88, 89
Gauss, 40, 93, 97, 114, 117, 218, 261, 268
    formula of ψ function, 117
    formula of digamma function, 114
    fundamental theorem of algebra, 40
    pairing trick, 93, 97, 268
    probability integral, 218
    quadrature formula, 261
Gauss arithmetic-geometric mean, *see* means
generating functions, 55–57, 59, 61, 63, 64, 72, 118, 121–126, 128, 129, 133, 134, 137, 154, 156–158, 163, 193, 271, 278
    exponential, 56, 57
    for Bernoulli numbers, 61
    for Bernoulli polynomials, 63
    for Fibonacci cubes, 122
    for Fibonacci numbers, 57
    for Fibonacci squares, 122
    for harmonic numbers, 59
    for powers of Fibonacci numbers, 125
    ordinary, 55
golden ratio, 58, 128, 203
Gosper's algorithm, 105, 106
Gosper's identity, 209, 210

Hölder's inequality, 20
Hadamard product, 253
Hadamard theorem, 147
harmonic numbers, 55, 58, 59, 61, 69, 152, 283, 284
Hermite-Ostrogradski formula, 212
higher transcendent, 201

Identities for Fibonacci powers, 131
Identities for Fibonacci squares and cubes, 131
Inverse Symbolic Calculator, 245
inverse tangent sums, 100

Jensen's inequality, 259
Jordan's inequality, 35, 75

Knuth's inequality, 35
Kober's inequality, 37
Ky Fan's inequality, 21

L'Hôpital monotone rule, 33, 262
L'Hôpital's monotone rule, 34–36
L'Hôpital's rule, 33, 36, 50
Lagrange's identity, 9, 10
Laplace integrals, 224
Laplace transform, 290
leap-forward fall-back induction, 13
Legendre polynomial, 280
Leibniz's rule, 204, 219, 222, 224, 234, 267, 289
Liouville's integral, 90
Lucas numbers, 123, 125, 126, 138

Mathematica, 53, 74, 79, 119, 130, 132, 139, 155, 156, 159, 169, 208, 210, 211, 251, 288
Mathieu's inequality, 81
mean value theorem, 28
means, 25–32

Index

arithmetic mean, 25
centroidal mean, 25
Gauss arithmetic-geometric mean, 12, 31
general definition, 29
geometric mean, 25
harmonic mean, 13, 25, 26
Heronian mean, 25
identric mean, 29, 32
logarithmic mean, 25, 32
power mean, 28
Schwab-Schoenberg mean, 14
slope mean, 30
Stolarsky mean, 29
Minkowski's inequality, 22
Monthly problems, 11, 14, 22, 31, 35, 37, 38, 48, 49, 52, 54, 70, 71, 80, 84, 101, 104, 106, 118, 128, 129, 144, 163, 172, 185, 190, 214, 229, 230, 237, 238, 241, 243–247, 249, 250, 252, 253

Open problems, 31, 32, 79–81, 90, 91, 108, 118, 130, 187, 215, 253

Pappus's construction on means, 25
parametric differentiation, 217, 222–224, 227–229, 231, 234
parametric integration, 167, 217–220, 225, 226, 228
Parseval's identity, 286
partial fraction decomposition of $1 - z^n$, 43
Pell numbers, 126, 127, 138
Pell-Lucas numbers, 127
pentagonal numbers, 116
Poisson integral, 231, 235
Poisson summation formula, 291, 292
polygonal numbers, 116
polylogarithm function, 193
power mean inequality, 13, 272
power series, 67–69, 150, 165, 169, 172, 182, 191
    for $\arcsin x$, 169, 182
    for $\arcsin^2 x$, 165
    for $\arcsin^3 x$, 169
    for $\arcsin^{2n} x$ when $n = 2, 3, 4$, 172
    for $\cot x$, 67
    for $\csc x$, 68
    for $\frac{1}{4} \ln^2 \left(\frac{1+x}{1-x}\right)$, 191
    for $\ln^2(1 - x)$, 69
    for $\ln^3(1 - x)$, 69
    for $\sinh x / \sin x$, 150
    for $\tan x$, 68
    for $e^{\arcsin x}$, 172

power-reduction formula, 124–127
primitive root of unity, 40
principle of the argument, 48
Putnam problems, 14, 30, 51, 52, 90, 107–109, 130, 141, 151, 223, 228, 229, 236, 237, 252, 253, 255, 279

q–binomial coefficient, 141
q–Fibonacci polynomials, 141

Ramanujan q-series, 118
Ramanujan identity for $\zeta(3)$, 173
Ramanujan problem, 213
Ramanujan's constant $G(1)$, 198
random Fibonacci numbers, 58, 72
Riemann hypothesis, 61, 72
Riemann sums, 49, 231, 232
Riemann zeta function, 61, 68, 165, 176, 189, 249, 281
roots of unity, 40, 109
Rubel and Stolarsky formula, 109

sampling theorem, 53, 54
series multisection formula, 110, 270, 271
Simpson's 3/8 rule, 31, 261
Sloane's integer sequences, 72, 133, 140, 142
smoothing transformation, 5
special numbers via determinants, 151
Stirling numbers of the first kind, 70
Stirling numbers of the second kind, 152
Stirling's formula, 67, 244

telescoping sums, 93, 94, 96, 97, 101, 103–106, 264, 269, 295
theta function, 237
trapezoidal rule, 80
Triangle inequality, 19
trigonometric power sums, 45, 153, 154, 156, 157, 159, 160, 163, 164
    alternating sums, 163
    cosecant, 45, 156
    cotangent, 159
    secant, 45, 154
    tangent, 157
    various, 160
trilogarithm function, 251
trilogarithmic identity, 251

uniform convergence, 221

Vandermonde convolution formula, 279
Vandermonde determinant, 150

Wallis's formula, 47, 77, 168, 169, 182, 245, 269, 279, 287
Weierstrass M-test, 221
Weierstrass product formula, 89, 268
weighted arithmetic-geometric mean inequality, 7, 20
Wilker's inequality, 34, 37
WZ method, 106

Young's inequality, 75

zero-order Bessel function, 229

About the Author

Hongwei Chen was born in China, and received his PhD from North Carolina State University in 1991. He is currently a professor of mathematics at Christopher Newport University. He has published more than fifty research articles in classical analysis and partial differential equations.